Informatics
in Proteomics

Informatics
in Proteomics

Edited by

Sudhir Srivastava

CRC Press
Taylor & Francis Group
Boca Raton London New York

CRC Press is an imprint of the
Taylor & Francis Group, an **informa** business

A TAYLOR & FRANCIS BOOK

CRC Press
Taylor & Francis Group
6000 Broken Sound Parkway NW, Suite 300
Boca Raton, FL 33487-2742

First issued in paperback 2019

© 2005 by Taylor & Francis Group, LLC
CRC Press is an imprint of Taylor & Francis Group, an Informa business

No claim to original U.S. Government works

ISBN-13: 978-1-57444-480-3 (hbk)
ISBN-13: 978-0-367-39255-3 (pbk)

Library of Congress Cataloging-in-Publication Data

Catalog record is available from the Library of Congress

Visit the Taylor & Francis Web site at
http://www.taylorandfrancis.com

and the CRC Press Web site at
http://www.crcpress.com

Dedication

Dedicated to my daughters, Aditi and Jigisha, and to my lovely wife, Dr. Rashmi Gopal Srivastava

Foreword

A remarkable development in the post-genome era is the re-emergence of proteomics as a new discipline with roots in old-fashioned chemistry and biochemistry, but with new branches in genomics and informatics. The appeal of proteomics stems from the fact that proteins are the most functional component encoded for in the genome and thus represent a direct path to functionality. Proteomics emphasizes the global profiling of cells, tissues, and biological fluids, but there is a long road from applying various proteomics tools to the discovery, for example, of proteins that have clinical utility as disease markers or as therapeutic targets. Given the complexity of various cell and tissue proteomes and the challenges of identifying proteins of particular interest, informatics is central to all aspects of proteomics. However, protein informatics is still in its early stages, as is the entire field of proteomics.

Although collections of protein sequences have preceded genomic sequence databases by more than two decades, there is a substantial need for protein databases as basic protein information resources. There is a need for implementing algorithms, statistical methods, and computer applications that facilitate pattern recognition and biomarker discovery by integrating data from multiple sources. This book, which is dedicated to protein informatics, is intended to serve as a valuable resource for people interested in protein analysis, particularly in the context of biomedical studies. An expert group of authors has been assembled with proteomics informatics–related expertise that is highly valuable in guiding proteomic studies, particularly since currently the analysis of proteomics data is rather informal and largely dependent on the idiosyncrasies of the analyst.

Several chapters address the need for infrastructures for proteomic research and cover the status of public protein databases and interfaces. The creation of a national virtual knowledge environment and information management systems for proteomic research is timely and clearly addressed. Issues surrounding data standardization and integration are very well presented. They are captured in a chapter that describes ongoing initiatives within the Human Proteome Organization (HUPO). A major strength of the book is in the detailed review and discussion of applications of statistical and bioinformatic tools to data analysis and data mining. Much concern at the present time surrounds the analysis of proteomics data by mass spectrometry for a variety of applications. The book shines in its presentation in several chapters of various approaches and issues surrounding mass spectrometry data analysis.

Although the field of proteomics and related informatics is highly evolving, this book captures not only the current state-of-the-art but also presents a vision for where the field is heading. As a result, the contributions of the book and its component chapters will have long-lasting value.

Sam Hanash, M.D.
Fred Hutchinson Cancer Center
Seattle, Washington

Preface

The biological dictates of an organism are largely governed through the structure and function of the products of its genes, the most functional of which is the proteome. Originally defined as the analysis of the entire protein complement of a cell or tissue, proteomics now encompasses the study of expressed proteins including the identification and elucidation of their structure–function relationships under normal and disease conditions. In combination with genomics, proteomics can provide a holistic understanding of the biology underlying disease processes. Information at the level of the proteome is critical for understanding the function of specific cell types and their roles in health and disease. Bioinformatic tools are needed at all levels of proteomic analysis. The main databases serving as the targets for mass spectrometry data searches are the expressed sequence tag (EST) and the protein sequence databases, which contain protein sequence information translated from DNA sequence data. It is thought that virtually any protein that can be detected on a 2DE gel can be identified through the EST database, which contains over 2 million cDNA sequences. However, ESTs cover only a partial sequence of the protein. This poses a formidable challenge for the proteomic community and necessitates the need for databases with extensive coverage and search algorithms for identifying proteins/peptides with accuracy.

The handling and analysis of data generated by proteomic investigations represent an emerging and challenging field. New techniques and collaborations between computer scientists, biostatisticians, and biologists are called for. There is a need to develop and integrate a variety of different types of databases; to develop tools for translating raw primary data into forms suitable for public dissemination and formal data analysis; to obtain and develop user interfaces to store, retrieve, and visualize data from databases; and to develop efficient and valid methods of data analysis. The sheer volume of data to be collected and processed will challenge the usual approaches. Analyzing data of this dimension is a fairly new endeavor for statisticians, for which there is not an extensive technical statistical literature.

There are several levels of complexity in the investigation of proteomic data, from the day-to-day interpretation of protein patterns generated by individual measurement systems to the query and manipulation of data from multiple experiments or information sources. Interaction with data warehouses represents another level of data interrogation. Users typically retrieve data and formulate queries to test hypotheses and generate conclusions. Formulating queries can be a difficult task requiring extensive syntactic and semantic knowledge. Syntactic knowledge is needed to ensure that a query is well formed and references existing relations and attributes. Semantic knowledge is needed to ensure that a query satisfies user intent. Because a user often has an incomplete understanding of the contents and structure of the data warehouse, it is necessary to provide automated techniques for query formulation that significantly reduce the amount of knowledge required by data warehouse users.

This book intends to provide a comprehensive view of informatic approaches to data storage, curation, retrieval, and mining as well as application-specific bioinformatic tools in disease detection, diagnosis, and treatment.

Rapid technological advances are yielding abundant data in many formats that, because of their vast quantity and complexity, are becoming increasingly difficult to analyze. A strategic objective is to streamline the transfer of knowledge and technology to allow for data from disparate sources to be analyzed, providing new inferences about the complex role of proteomics in disease processes. Data mining, the process of knowledge extraction from data and the exploration of available data for patterns and relationships, is increasingly needed for today's high-throughput technologies. Data architectures that support the integration of biological data files with epidemiologic profiles of human clinical responses need to be developed. The ability to develop and analyze metadata will stimulate new research theories and streamline the transfer of basic knowledge into clinical applications. It is my belief that this book will serve as a unique reference for researchers, biologists, technologists, clinicians, and other health professions as it provides information on the informatics needs of proteomic research on molecular targets relevant to disease detection, diagnosis, and treatment.

The nineteen chapters in this volume are contributed by eminent researchers in the field and critically address various aspects of bioinformatics and proteomic research. The first two chapters are introductory: they discuss the biological rationale for proteomic research and provide a brief overview of technologies that allow for rapid analysis of the proteome. The next five chapters describe the infrastructures that provide the foundations for proteomic research: these include the creation of a national, virtual knowledge environment and information management systems for proteomic research; the availability of public protein databases and interfaces; and the need for collaboration and interaction between academia, industry, and government agencies. Chapter 6 illustrates the power of proteomic knowledge in furthering hypothesis-driven cancer biomarker research through data extraction and curation. Chapter 7 and Chapter 8 provide the conceptual framework for data standardization and integration and give an example of an ongoing collaborative research within the Human Proteome Organization. Chapter 9 identifies genomic and proteomic informatic tools used in deciphering functional pathways. The remaining ten chapters describe applications of statistical and bioinformatic tools in data analysis, data presentation, and data mining. Chapter 10 provides an overview of a variety of proteomic data mining tools, and subsequent chapters provide specific examples of data mining approaches and their applications. Chapter 11 describes methods for quantitative analysis of a large number of proteins in a relatively large number of lung cancer samples using two-dimensional gel electrophoresis. Chapter 12 discusses the analysis of mass spectrometric data by nonparametric inference for high-dimensional comparisons involving two or more groups, based on a few samples and very few replicates from within each group. Chapter 13 discusses bioinformatic tools for the identification of proteins by searching a collection of sequences with mass spectrometric data and describes several critical steps that are necessary for the successful protein identification, which include: (a) the masses of peaks in the mass spectrum corresponding to the monoisotopic peptide masses have to assigned; (b) a collection of sequences have to be searched using a sensitive and selective algorithm; (c) the significance of the results have to be tested; and (d) the function of the identified proteins have to be assigned. In Chapter 14, two types of approaches are described: one based on statistical theories and another on machine learning and computational data mining techniques. In Chapter 15, the author discusses the problems with the currently available disease classifier algorithms and puts forward approaches for scaling the data set, searching for outliers, choosing relevant features, building classification models, and then determining the characteristics of the models. Chapter 16 discusses currently available computer tools that support data collection, analysis, and validation in a high-throughput LC-MS/MS–based proteome research environment and subsequent protein identification and quantification with minimal false-positive error rates. Chapter 17 and Chapter 18 describe experimental designs, statistical methodologies, and computational tools for the analysis of spectral patterns in the diagnosis of ovarian and prostate cancer. Finally, Chapter 19 illustrates how quantitative analysis of fluorescence microscope images augments mainstream proteomics by providing information about the abundance, localization, movement, and interactions of proteins inside cells.

This book has brought together a mix of scientific disciplines and specializations, and I encourage readers to expand their knowledge by reading how the combination of proteomics and bioinformatics is used to uncover interesting biology and discover clinically significant biomarkers. In a field with rapidly changing technologies, it is difficult to ever feel that one has knowledge that is current and definitive. Many chapters in this book are conceptual in nature but have been included because proteomics is an evolving science that offers much hope to researchers and patients alike.

Last, but not least, I would like to acknowledge the authors for their contributions and patience. When I accepted the offer to edit this book, I was not sure we were ready for a book on proteomics as the field is continuously evolving, but the excellent contributions and enthusiasm of my colleagues have allayed my fears. The chapters in the book describe the current state-of-the-art in informatics and reflect the interests, experience, and creativity of the authors. Many chapters are intimately related and therefore there may be some overlap in the material presented in each individual chapter. I would also like to acknowledge Dr. Asad Umar for his help in designing the cover for this book. Finally, I would like to express my sincere gratitude to Dr. Sam Hanash, the past president of HUPO, for his encouragement and support.

Sudhir Srivastava, Ph.D., MPH, MS
Bethesda, Maryland

Contributors

Bao-Ling Adam
Department of Microbiology and
 Molecular Cell Biology
Eastern Virginia Medical School
Norfolk, Virginia, USA

Marcin Adamski
Bioinformatics Program
Department of Human Genetics
School of Medicine
University of Michigan
Ann Arbor, Michigan, USA

Ruedi Aebersold
Institute for Systems Biology
Seattle, Washington, USA

R.C. Beavis
Beavis Informatics
Winnipeg, Manitoba, Canada

David G. Beer
General Thoracic Surgery
University of Michigan
Ann Arbor, Michigan, USA

Guoan Chen
General Thoracic Surgery
University of Michigan
Ann Arbor, Michigan, USA

Chad Creighton
Pathology Department
University of Michigan
Ann Arbor, Michigan, USA

Daniel Crichton
Jet Propulsion Laboratory
California Institute of Technology
Pasadena, California, USA

Cim Edelstein
Division of Public Health Services
Fred Hutchinson Cancer Research Center
Seattle, Washington, USA

Jimmy K. Eng
Division of Public Health Services
Fred Hutchinson Cancer Research Center
Seattle, Washington, USA

J. Eriksson
Department of Chemistry
Swedish University of Agricultural
 Sciences
Uppsala, Sweden

Ziding Feng
Division of Public Health Sciences
Fred Hutchinson Cancer Research
 Center
Seattle, Washington, USA

D. Fenyö
Amersham Biosciences AB
Uppsala, Sweden
The Rockefeller University
New York, New York, USA

R. Gangal
SciNova Informatics
Pune, Maharashtra, India

Gary L. Gilliland
Biotechnology Division
National Institute of Standards and
   Technology
Gaithersburg, Maryland, USA

Samir M. Hanash
Division of Public Health Sciences
Fred Hutchinson Cancer Research
   Center
Seattle, Washington, USA

Ben A. Hitt
Correlogic Systems, Inc.
Bethesda, Maryland, USA

J. Steven Hughes
Jet Propulsion Laboratory
California Institute of Technology
Pasadena, California, USA

Donald Johnsey
National Cancer Institute
National Institutes of Health
Bethesda, Maryland, USA

Andrew Keller
Division of Public Health Sciences
Fred Hutchinson Cancer Research
   Center
Seattle, Washington, USA

Sean Kelly
Jet Propulsion Laboratory
California Institute of Technology
Pasadena, California, USA

Heather Kincaid
Fred Hutchinson Cancer Research
   Center
Seattle, Washington, USA

Jeanne Kowalski
Division of Oncology Biostatistics
Johns Hopkins University
Baltimore, Maryland, USA

Peter A. Lemkin
Laboratory of Experimental and
   Computational Biology
Center for Cancer Research
National Cancer Institute
Frederick, Maryland, USA

Xiao-jun Li
Institute for Systems Biology
Seattle, Washington, USA

Chenwei Lin
Department of Computational Biology
Fred Hutchinson Cancer Research
   Center
Seattle, Washington, USA

Lance Liotta
FDA-NCI Clinical Proteomics Program
Laboratory of Pathology
National Cancer Institute
Bethesda, Maryland, USA

Stephen Lockett
NCI–Frederick/SAIC–Frederick
Frederick, Maryland, USA

Brian T. Luke
SAIC-Frederick
Advanced Biomedical Computing
   Center
NCI Frederick
Frederick, Maryland, USA

Dale McLerran
Division of Public Health Sciences
Fred Hutchinson Cancer Research
   Center
Seattle, Washington, USA

Djamel Medjahed
Laboratory of Molecular Technology
SAIC-Frederick Inc.
Frederick, Maryland, USA

Alexey I. Nesvizhskii
Division of Public Health Sciences
Fred Hutchinson Cancer Research
  Center
Seattle, Washington, USA

Jane Meejung Chang Oh
Wayne State University
Detroit, Michigan, USA

Gilbert S. Omenn
Departments of Internal Medicine
  and Human Genetics
Medical School and School
  of Public Health
University of Michigan
Ann Arbor, Michigan, USA

Emanuel Petricoin
FDA-NCI Clinical Proteomics Program
Office of Cell Therapy
CBER/Food and Drug Administration
Bethesda, Maryland, USA

Veerasamy Ravichandran
Biotechnology Division
National Institute of Standards
  and Technology
Gaithersburg, Maryland, USA

John Semmes
Department of Microbiology and
  Molecular Cell Biology
Eastern Virginia Medical School
Norfolk, Virginia, USA

Ram D. Sriram
Manufacturing Systems Integration
  Division
National Institute of Standards and
  Technology
Gaithersburg, Maryland, USA

Sudhir Srivastava
Cancer Biomarkers Research Group
Division of Cancer Prevention
National Cancer Institute
Bethesda, Maryland, USA

David J. States
Bioinformatics Program
Department of Human Genetics
School of Medicine
University of Michigan
Ann Arbor, Michigan, USA

Mark Thornquist
Division of Public Health Sciences
Fred Hutchinson Cancer Research
  Center
Seattle, Washington, USA

Mukesh Verma
Cancer Biomarkers Research Group
Division of Cancer Prevention
National Cancer Institute
Bethesda, Maryland, USA

Paul D. Wagner
Cancer Biomarkers Research Group
Division of Cancer Prevention
National Cancer Institute
Bethesda, Maryland, USA

Denise B. Warzel
Center for Bioinformatics
National Cancer Institute
Rockville, Maryland, USA

Nicole White
Department of Pathology
Johns Hopkins University
Baltimore, Maryland, USA

Marcy Winget
Department of Population Health and
  Information
Alberta Cancer Board
Edmonton, Alberta, Canada

Yutaka Yasui
Division of Public Health Sciences
Fred Hutchinson Cancer Research Center
Seattle, Washington, USA

Mei-Fen Yeh
Division of Oncology Biostatistics
Johns Hopkins University
Baltimore, Maryland, USA

Zhen Zhang
Center for Biomarker Discovery
Department of Pathology
Johns Hopkins University
Baltimore, Maryland, USA

Contents

1 The Promise of Proteomics: Biology, Applications, and Challenges

Paul D. Wagner and Sudhir Srivastava

CONTENTS

1.1 INTRODUCTION

In the 19th century, the light microscope opened a new frontier in the study of diseases, allowing scientists to look deep into the cell. The science of pathology (the branch of medicine that deals with the essential nature of disease) expanded to include the study of structural and functional changes in cells, and diseases could be attributed to recognizable changes in the cells of the body. At the start of the 21st century, the molecular-based methods of genomics and proteomics are bringing about a new revolution in medicine. Diseases will be described in terms of patterns of abnormal genetic and protein expression in cells and how these cellular alterations affect the molecular composition of the surrounding environment. This new pathology will have a profound impact on the practice of medicine, enabling physicians to determine who is at risk for a specific disease, to recognize diseases before they have invaded tissues, to intervene with agents or treatments that may prevent or

1

delay disease progression, to guide the choice of therapies, and to assess how well a treatment is working.

Cancer is one of the many diseases whose treatment will be affected by these molecular approaches. Currently available methods can only detect cancers that have achieved a certain size threshold, and in many cases, the tumors, however small, have already invaded blood vessels or spread to other parts of the body. Molecular markers have the potential to find tumors in their earliest stages of development, even before the cell's physical appearance has changed. Molecular-based detection methods will also change our definition of cancer. For example, precancerous changes in the uterine cervix are called such because of specific architectural and cytological changes. In the future, we may be able to define the expression patterns of specific cellular proteins induced by human papillomavirus that indicate the cells are beginning to progress to cancer. We may also be able to find molecular changes that affect all the tissues of an organ, putting the organ at risk for cancer.

In addition to improving the physician's ability to detect cancers early, molecular technologies will help doctors determine which neoplastic lesions are most likely to progress and which are not destined to do so — a dilemma that confronts urologists in the treatment of prostate cancer. Accurate discrimination will help eliminate overtreatment of harmless lesions. By revealing the metastatic potential of tumors and their corresponding preneoplastic lesions, molecular-based methods will fill a knowledge gap impossible to close with traditional histopathology. If these advances are made and new screening tests are developed, then one day we may be able to identify and eliminate the invasive forms of most malignant epithelial tumors.

1.2 WHY IS PROTEOMICS USEFUL?

Mammalian systems are much more complex than can be deciphered by their genes alone, and the biological dictates of an organism are largely governed through the function of proteins. In combination with genomics, proteomics can provide a holistic understanding of the biology of cells, organisms, and disease processes. The term "proteome" came into use in the mid 1990s and is defined as the protein complement of the genome. Although proteomics was originally used to describe methods for large-scale, high-throughput protein separation and identification,[1] today proteomics encompasses almost any method used to characterize proteins and determine their functions. Information at the level of the proteome is critical for understanding the function of specific cell types and their roles in health and disease. This is because proteins are often expressed at levels and forms that cannot be predicted from mRNA analysis. Proteomics also provides an avenue to understand the interaction between a cell's functional pathways and its environmental milieu, independent of any changes at the RNA level. It is now generally recognized that expression analysis directly at the protein level is necessary to unravel the critical changes that occur as part of disease pathogenesis.

Currently there is much interest in the use of molecular markers or biomarkers for disease diagnosis and prognosis. Biomarkers are cellular, biochemical, and molecular alterations by which normal, abnormal, or simply biologic processes can be recognized or monitored. These alterations should be able to objectively measure and evaluate normal biological processes, pathogenic processes, or pharmacologic responses to a therapeutic intervention. Proteomics is valuable in the discovery of biomarkers as the proteome reflects both the intrinsic genetic program of the cell and the impact of its immediate environment. Protein expression and function are subject to modulation through transcription as well as through translational and posttranslational events. More than one messenger RNA can result from one gene through differential splicing, and proteins can undergo more than 200 types of posttranslation modifications that can affect function, protein–protein and protein–ligand interactions, stability, targeting, or half-life.[2] During the transformation of a normal cell into a neoplastic cell, distinct changes occur at the protein level that range from altered expression, differential modification, changes in specific activity, and aberrant localization, all of which affect cellular function. Identifying and understanding these changes is the underlying theme in cancer proteomics. The deliverables include identification of biomarkers that have utility both for early detection and for determining therapy.

While proteomics has traditionally dealt with quantitative analysis of protein expression, more recently proteomics has been viewed to encompass structural analyses of proteins.[3] Quantitative proteomics strives to investigate the changes in protein expression in different physiological states such as in healthy and diseased tissue or at different stages of the disease. This enables the identification of state- and stage-specific proteins. Structural proteomics attempts to uncover the structure of proteins and to unravel and map protein–protein interactions. Proteomics provides a window to pathophysiological states of cells and their microenvironments and reflects changes that occur as disease-causing agents interact with the host environment. Some examples of proteomics are described below.

1.3 GENE–ENVIRONMENT INTERACTIONS

Infectious diseases result from interactions between the host and pathogen, and understanding these diseases requires understanding not only alterations in gene and protein expressions within the infected cells but also alterations in the surrounding cells and tissues. Although genome and transcriptome analyses can provide a wealth of information on global alterations in gene expression that occur during infections, proteomic approaches allow the monitoring of changes in protein levels and modifications that play important roles in pathogen–host interactions. During acute stages of infection, pathogen-coded proteins play a significant role, whereas in the chronic infection, host proteins play the dominating role. Viruses, such as hepatitis B (HBV), hepatitis C (HCV), and human papillomavirus (HPV), are suitable for proteomic analysis because they express only eight to ten major genes.[4,5] Analyzing a smaller number of genes is easier than analyzing the proteome of an organism with thousands of genes.[6–8] For example, herpes simplex virus type 1 (HSV-1) infection induces severe alterations of the translational apparatus, including phosphorylation of ribosomal proteins and the association of several nonribosomal proteins with the ribosomes.[9–12] Whether ribosomes themselves could contribute to the HSV-1–induced translational control of host and viral gene expression has been investigated. As a prerequisite to test this hypothesis, the investigators undertook the identification of nonribosomal proteins associated with the ribosomes during the course of HSV-1 infection. Two HSV-1 proteins, VP19C and VP26, that are associated to ribosomes with different kinetics were identified. Another nonribosomal protein identified was the poly(A)-binding protein 1 (PAB1P). Newly synthesized PAB1P continued to associate to ribosomes throughout the course of infection. This finding attests to the need for proteomic information for structural and functional characterization.

Approximately 15% of human cancers (about 1.5 million cases per year, worldwide) are linked to viral, bacterial, or other pathogenic infections.[13] For cancer development, infectious agents interact with host genes and sets of infectious agent-specific or host-specific genes are expressed. Oncogenic infections increase the risk of cancer through expression of their genes in the infected cells. Occasionally, these gene products have paracrine effects, leading to neoplasia in neighboring cells. More typically, it is the infected cells that become neoplastic. These viral, bacterial, and parasitic genes and their products are obvious candidates for pharmacologic interruptions or immunologic mimicry, promising approaches for drugs and vaccines. By understanding the pathways involved in the infectious agent–host interaction leading to cancer, it would be possible to identify targets for intervention.

1.4   ORGANELLE-BASED PROTEOMICS

Eukaryotic cells contain a number of organelles, including nucleoli, mitochondria, smooth and rough endoplasmic reticula, Golgi apparatus, peroxisomes, and lysosomes. The mitochondria are among the largest organelles in the cell. Mitochondrial dysfunction has been frequently reported in cancer, neurodegenerative diseases, diabetes, and aging syndromes.[14–16] The mitochondrion genome (16.5 Kb) codes only for a small fraction (estimated to be 1%) of the proteins housed within this organelle. The other proteins are encoded by the nuclear DNA (nDNA) and transported into the mitochondria. Thus, a proteomic approach is needed to fully understand the nature and extent of mutated and modified proteins found in the mitochondria of diseased cells. According to a recent estimate, there are 1000 to 1500 polypeptides in the human mitochondria.[17–20] This estimate is based on several lines of evidence, including the existence of at least 800 distinct proteins in yeast and *Arabidopsis thaliana* mitochondria[18,19] and the identification of 591 abundant mouse mitochondrial proteins.[20]

Investigators face a number of challenges in organelle proteome characterization and data analysis. A complete characterization of the posttranslational modifications that mitochondrial proteins undergo is an enormous and important task, as all of these modifications cannot be identified by a single approach. Differences in posttranslational modifications are likely to be associated with the onset and progression of various diseases. In addition, the mitochondrial proteome, although relatively simple, is made up of complex proteins located in submitochondrial compartments. Researchers will need to reduce the complexity to subproteomes by fractionation and analysis of various compartments. A number of approaches are focusing on specific components of the mitochondria, such as isolation of membrane proteins, affinity labeling, and isolation of redox proteins,[21] or isolation of large complexes.[22]

Other approaches may combine expression data from other species, such as yeast, to identify and characterize the human mitochondrial proteome.[23,24]

The need to identify mitochondrial proteins associated with or altered during the development and progression of cancer is compelling. For example, mitochondrial dysfunction has been frequently associated with transport of proteins, such as cyto-chrome c. Mitochondrial outer membrane permeabilization by pro-apoptotic proteins, such as Bax or Bak, results in the release of cytochrome c and the induction of apoptosis. An altered ratio of anti-apoptotic proteins (e.g., Bcl-2) to pro-apoptotic proteins (e.g., Bax and Bak) promotes cell survival and confers resistance to therapy.[25]

1.5 CANCER DETECTION

Molecular markers or biomarkers are currently used for cancer detection, diagnosis, and monitoring therapy and are likely to play larger roles in the future. In cancer research, a biomarker refers to a substance or process that is indicative of the presence of cancer in the body. It might be a molecule secreted by the malignancy itself, or it can be a specific response of the body to the presence of cancer. The biological basis for usefulness of biomarkers is that alterations in gene sequence or expression and in protein expression and function are associated with every type of cancer and with its progression through the various stages of development.

Genetic mutations, changes in DNA methylation, alterations in gene expression, and alterations in protein expression or modification can be used to detect cancer, determine prognosis, and monitor disease progression and therapeutic response. Currently, DNA-based, RNA-based, and protein-based biomarkers are used in cancer risk assessment and detection. The type of biomarker used depends both on the application (i.e., risk assessment, early detection, prognosis, or response to therapy) and the availability of appropriate biomarkers. The relative advantages and disad-vantages of genomic and proteomic approaches have been widely discussed, but since a cell's ultimate phenotype depends on the functions of expressed proteins, proteomics has the ability to provide precise information on a cell's phenotype. Tumor protein biomarkers are produced either by the tumor cells themselves or by the surrounding tissues in response to the cancer cells.

More than 80% of human tumors (colon, lung, prostate, oral cavity, esophagus, stomach, uterine, cervix, and bladder) originate from epithelial cells, often at the mucosal surface. Cells in these tumors secrete proteins or spontaneously slough off into blood, sputum, or urine. Secreted proteins include growth factors, angiogenic proteins, and proteases. Free DNA is also released by both normal and tumor cells into the blood and patients with cancer have elevated levels of circulating DNA. Thus, body fluids such as blood and urine are good sources for cancer biomarkers. That these fluids can be obtained using minimally invasive methods is a great advantage if the biomarker is to be used for screening and early detection.

From a practical point of view, assays of protein tumor biomarkers, due to their ease of use and robustness, lend themselves to routine clinical practice, and histor-ically tumor markers have been proteins. Indeed, most serum biomarkers used today are antibody-based tests for epithelial cell proteins. Two of the earliest and most widely used cancer biomarkers are PSA and CA25. Prostate-specific antigen (PSA) is a secreted protein produced by epithelial cells within the prostate. In the early 1980s it was found that sera from prostate cancer patients contain higher levels of PSA than do the sera of healthy individuals. Since the late 1980s, PSA has been used to screen asymptomatic men for prostate cancer and there has been a decrease in mortality rates due to prostate cancer. How much of this decrease is attributable to screening with PSA and how much is due to other factors, such as better therapies, is uncertain. Although PSA is the best available serum biomarker for prostate cancer and the only one approved by the FDA for screening asymptomatic men, it is far from ideal. Not all men with prostate cancer have elevated levels of PSA; 20 to 30% of men with prostate cancer have normal PSA levels and are misdiagnosed. Conversely, because PSA levels are increased in other conditions, such as benign prostatic hypertrophy and prostatitis, a significant fraction of men with elevated levels of PSA do not have cancer and undergo needless biopsies.

The CA125 antigen was first detected over 20 years ago; CA125 is a mucin-like glycoprotein present on the cell surface of ovarian tumor cells that is released into the blood.[26] Serum CA125 levels are elevated in about 80% of women with epithelial ovarian cancer but in less than 1% of healthy women. However, the CA125 test only returns a positive result for about 50% of Stage I ovarian cancer patients and is, therefore, not useful by itself as an early detection test.[27] Also, CA125 is elevated in a number of benign conditions, which diminishes its usefulness in the initial diagnosis of ovarian cancer. Despite these limitations, CA125 is considered to be one of the best available cancer serum markers and is used primarily in the management of ovarian cancer. Falling CA125 following chemotherapy indicates that the cancer is responding to treatment.[28] Other serum protein biomarkers, such as alpha fetoprotein (AFP) for hepatocellular carcinoma and CA15.3 for breast cancer, are also of limited usefulness as they are elevated in some individuals without cancer, and not all cancer patients have elevated levels.

1.6   WHY PROTEOMICS HAS NOT SUCCEEDED IN THE PAST: CANCER AS AN EXAMPLE

The inability of these protein biomarkers to detect all cancers (false negatives) reflects both the progressive nature of cancer and its heterogeneity. Cancer is not a single disease but rather an accumulation of several events, genetic and epigenetic, arising in a single cell over a long period of time. Proteins overexpressed in late stage cancers may not be overexpressed in earlier stages and, therefore, are not useful for early cancer detection. For example, the CA125 antigen is not highly expressed in many Stage I ovarian cancers. Also, because tumors are heterogeneous, the same sets of proteins are not necessarily overexpressed in each individual tumor. For example, while most patients with high-grade prostate cancers have increased levels of PSA, approximately 15% of these patients do not have an elevated PSA level. The reciprocal problem of biomarkers indicating the presence of cancer when none is present (false positives) results because these proteins are not uniquely produced by tumors. For example, PSA is produced by prostatitis (inflammation of the prostate) and benign prostatic hyperplasia (BPH), and elevated CA125 levels are caused by endometriosis and pelvic inflammation.

The performance of any biomarker can be described in terms of its specificity and sensitivity. In the context of cancer biomarkers, sensitivity refers to the proportion of case subjects (individuals with confirmed disease) who test positive for the biomarker, and specificity refers to the proportion of control subjects (individuals without disease) who test negative for the biomarker. An ideal biomarker test would have 100% sensitivity and specificity; i.e., everyone with cancer would have a positive test, and everyone without cancer would have a negative test. None of the currently available protein biomarkers achieve 100% sensitivity and specificity. For example, as described above, PSA tests achieve 70 to 90% sensitivity and only about 25% specificity, which results in many men having biopsies when they do not have detectable prostrate cancer. The serum protein biomarker for breast cancer CA15.3 has only 23% sensitivity and 69% specificity. Other frequently used terms are positive predictive value (PPV), the chance that a person with a positive test has cancer, and negative predictive value (NPV), the chance that a person with a negative test does not have cancer. PPV is affected by the prevalence of disease in the screened population. For a given sensitivity and specificity, the higher the prevalence, the higher the PPV. Even when a biomarker provides high specificity and sensitivity, it may not be useful for screening the general population if the cancer has low prevalence. For example, a biomarker with 100% sensitivity and 95% specificity has a PPV of only 17% for a cancer with 1% prevalence (only 17 out of 100 people with a positive test for the biomarker actually have cancer) and 2% for a cancer with 0.1% prevalence. The prevalence of ovarian cancer in the general population is about 0.04%. Thus, a biomarker used to screen the general population must have significantly higher specificity and sensitivity than a biomarker used to monitor an at-risk population.

1.7 HOW HAVE PROTEOMIC APPROACHES CHANGED OVER THE YEARS?

Currently investigators are pursuing three different approaches to develop biomarkers with increased sensitivity and specificity. The first is to improve on a currently used biomarker. For instance, specificity and sensitivity of PSA may be improved by measurement of its complex with alpha(1)-antichymotrypsin; patients with benign prostate conditions have more free PSA than bound, while patients with cancer have more bound PSA than free.[29] This difference is thought to result from differences in the type of PSA released into the circulation by benign and malignant prostatic cells. Researchers are also trying to improve the specificity and sensitivity of PSA by incorporating age- and race-specific cut points and by adjusting serum PSA concentration by prostatic volume (PSA density). The second approach is to discover and validate new biomarkers that have improved sensitivity and specificity. Many investigators are actively pursuing new biomarkers using a variety of new and old technologies. The third approach is to use a panel of biomarkers, either by combining several individually identified biomarkers or by using mass spectrometry to identify a pattern of protein peaks in sera that can be used to predict the presence of cancer or other diseases. High-throughput proteomic methodologies have the potential to revolutionize protein biomarker discovery and to allow for multiple markers to be assayed simultaneously.

In the past, researchers have mostly used a one-at-time approach to biomarker discovery. They have looked for differences in the levels of individual proteins in tissues or blood from patients with disease and from healthy individuals. The choice of proteins to examine was frequently based on biological knowledge of the cancer and its interaction with surrounding tissues. This approach is laborious and time consuming, and most of the biomarkers discovered thus far do not have sufficient sensitivity and specificity to be useful for early cancer detection. A mainstay of protein biomarker discovery has been two-dimensional gel electrophoresis (2DE). The traditional 2DE method is to separately run extracts from control and diseased tissues or cells and to compare the relative intensities of the various protein spots on the stained gels. Proteins whose intensities are significantly increased or decreased in diseased tissues are identified using mass spectrometry. For example, 2DE was recently used to identify proteins that are specifically overexpressed in colon cancer.[30] The limitations of the 2DE approach are well known: the gels are difficult to run reproducibly, a significant fraction of the proteins either do not enter the gels or are not resolved, low-abundance proteins are not detected, and relatively large amounts of sample are needed. A number of modifications have been made to overcome these limitations, including fractionation of samples prior to 2DE, the use of immobilized pH gradients, and labeling proteins from control and disease cells with different fluorescent dyes and then separating them on the same gel (differential in-gel electrophoresis; DIGE). An additional difficulty is contamination from neighboring stromal cells that can confound the detection of tumor-specific markers. Laser capture microdissection (LCD) can be used to improve the specificity of 2DE, as it allows for the isolation of pure cell populations; however, it further reduces the amount of sample available for analysis. Even with these modifications, 2DE is a relatively low throughput methodology that only samples a subset of the proteome, and its applicability for screening and diagnosis is very limited.

A number of newer methods for large-scale protein analysis are being used or are under development. Several of these rely on mass spectrometry and database interrogation. Mass spectrometers work by imparting an electrical charge to the analytes (e.g., proteins or peptides) and then sending the charged particles though a mass analyzer. A time of flight (TOF) mass spectrometer measures the time it takes a charged particle (protein or peptide) to reach the detector; the higher the mass the longer the flight time. A mixture of proteins or peptides analyzed by TOF generates a spectrum of protein peaks. TOF mass spectrometers are used to analyze peptide peaks generated by protease digestion of proteins resolved on 2DE. A major advance in this methodology is matrix-assisted laser desorption ionization (a form of soft ionization), which allows for the ionization of larger biomolecules such as proteins and peptides. TOF mass spectrometers are also used to identify peptides eluted from HPLC columns.

With tandem mass spectrometers (MS/MS), a mixture of charged peptides is separated in the first MS according to their mass-to-charge ratios, generating a list of peaks. In the second MS, the spectrometer is adjusted so that a single mass-to-charge species is directed to a collision cell to generate fragment ions, which are then separated by their mass-to-charge ratios. These patterns are compared to databases to identify the peptide and its parent protein. Liquid chromatography combined with MS or MS/MS (LC-MS and LC-MS/MS) is currently being used as an alternative to 2DE to analyze complex protein mixtures. In this approach, a mixture of proteins is digested with a protease, and the resulting peptides are then fractionated by liquid chromatography (typically reverse-phase HPLC) and analyzed by MS/MS and database interrogation. A major limitation to this approach is the vast number of peptides generated when the initial samples contain a large number of proteins. Even the most advanced LC-MS/MS systems cannot resolve and analyze these complex peptide mixtures, and currently it is necessary to either prefractionate the proteins prior to proteolysis or to enrich for certain types of peptides (e.g., phosphorylated, glycoslylated, or cysteine containing) prior to liquid chromatography.

Although the use of mass spectrometry has accelerated the pace of protein identification, it is not inherently quantitative and the amounts of peptides ionized vary. Thus, the signal obtained in the mass spectrometer cannot be used to measure the amount of protein in the sample. Several comparative mass spectrometry methods have been developed to determine the relative amounts of a particular peptide or protein in two different samples. These approaches rely on labeling proteins in one sample with a reagent containing one stable isotope and labeling the proteins in the other sample with the same reagent containing a different stable isotope. The samples are then mixed, processed, and analyzed together by mass spectrometry. The mass of a peptide from one sample will be different by a fixed amount from the same peptide from the other sample. One such method (isotope-coded affinity tags; ICAT) modifies cysteine residues with an affinity reagent that contains either eight hydrogen or eight deuterium atoms.[31] Other methods include digestion in ^{16}O and ^{18}O water and culturing cells in ^{12}C- and ^{13}C-labeled amino acids.

Although the techniques described thus far are useful for determining proteins that are differently expressed in control and disease, they are expensive, relatively low throughput, and not suitable for routine clinical use. Surface-enhanced laser description ionization time-of-flight (SELDI-TOF) and protein chips are two proteomic approaches that have the potential to be high throughput and adaptable to clinical use. In the SELDI-TOF mass spectrometry approach, protein fractions or body fluids are spotted onto chromatographic surfaces (ion exchange, reverse phase, or metal affinity) that selectively bind a subset of the proteins (Ciphergen® Protein-Chip Arrays). After washing to remove unbound proteins, the bound proteins are ionized and analyzed by TOF mass spectrometry. This method has been used to identify disease-related biomarkers, including the alpha chain of haptoglobin (Hp-alpha) for ovarian cancer[32] and alpha defensin for bladder cancer. Other investigators are using SELDI-TOF to acquire proteomic patterns from whole sera, urine, or other body fluids. The complex patterns of proteins obtained by the TOF mass spectrometer are analyzed using pattern recognition algorithms to identify a set of protein peaks that can be used to distinguish disease from control. With this approach, protein identification and characterization are not necessary for development of clinical assays, and a SELDI protein profile may be sufficient for screening. For example, this method has been reported to identify patients with Stage I ovarian cancer with 100% sensitivity and 95% specificity.[27] Similar, albeit less dramatic, results have been reported for other types of cancer.[28,33-36] At this time, it is uncertain whether SELDI protein profiling will prove to be as valuable a diagnostic tool as the initial reports have suggested. A major technical issue is the reproducibility of the protein profiles. Variability between SELDI-TOF instruments, in the extent of peptide ionization, in the chips used to immobilize the proteins, and in sample processing, can contribute to the lack of reproducibility. There is concern that the protein peaks identified by SELDI and used for discriminating between cancer and control are not derived from the tumor per se but rather from the body's response to the cancer (epiphenomena) and that they may not be specific for cancer; inflammatory conditions and benign pathologies may elicit the same bodily responses.[37,38] Most known tumor marker proteins in the blood are on the order of ng/ml (PSA above 4 ng/ml and alpha fetoprotein above 20 ng/ml are considered indicators of, respectively, prostate and hepatocellular cancers). The SELDI-TOF peptide peaks typically used to distinguish cancer from control are relatively large peaks representing proteins present in the serum on the order of μg to mg/ml; these protein peaks may result from cancer-induced proteolysis or posttranslational modification of proteins normally present in sera. Although identification of these discriminating proteins may not be necessary for this "black-box" approach to yield a clinically useful diagnostic test, identifying these proteins may help elucidate the underlying pathology and lead to improved diagnostic tests. Potential advantages of the SELDI for clinical assays are that it is high throughput, it is relatively inexpensive, and it uses minimally invasive specimens (blood, urine, sputum).

Interest in protein chips in part reflects the success of DNA microarrays. While these two methodologies have similarities, a number of technical and biological differences exist that make the practical application of protein chips or arrays challenging. Proteins, unlike DNA, must be captured in their native conformation and are easily denatured irreversibly. There is no method to amplify their concentrations, and their interactions with other proteins and ligands are less specific and of variable affinity. Current bottlenecks in creating protein arrays include the production (expression and purification) of the huge diversity of proteins that will form the array elements, methods to immobilize proteins in their native states on the surface, and lack of detection methods with sufficient sensitivity and accuracy. To date, the most widely used application of protein chips are antibody microarrays that have the potential for high-throughput profiling of a fixed number of proteins. A number of purified, well-characterized antibodies are spotted onto a surface and then cell extracts or sera are passed over the surface to allow for the antigen to bind to the specific, immobilized antibodies. The bound proteins are detected either by using secondary antibodies against each antigen or by using lysates that are tagged with fluorescent or radioactive labels. A variation that allows for direct comparison between two different samples is to label each extract with a different fluorescent dye, which is then mixed prior to exposure to the antibody array. A significant problem with antibody arrays is lack of specificity; the immobilized antibodies cross react with proteins other than the intended target. The allure of protein chips is their potential to rapidly analyze multiple protein markers simultaneously at a moderate cost.

As discussed earlier, most currently available cancer biomarkers lack sufficient sensitivity and specificity for use in early detection, especially to screen asymptomatic populations. One approach to improve sensitivity and specificity is to use a panel of biomarkers. It is easy to envision how combining biomarkers can increase sensitivity if they detect different pathological processes or different stages of cancer, and one factor to consider in developing such a panel is whether the markers are complementary. However, simply combining two biomarkers will more than likely decrease specificity and increase the number of false positives. Reducing their cutoff values (the concentration of a biomarker that is used as an indication of the presence of cancer) can be useful to reduce the number of false positives. A useful test for evaluating a single biomarker or panel of biomarkers is the receiver operating characteristic (ROC) curve. An ROC curve is a graphical display of false-positive rates and true-positive rates from multiple classification rules (different cutoff values for the various biomarkers). Each point on the graph corresponds to a different classification rule. In addition to analyzing individually measured markers, ROC curves can be used to analyze SELDI-TOF proteomic profiles.[39]

The measurement and analysis of biomarker panels will be greatly facilitated by high-throughput technologies such as protein arrays, microbeads with multiple antibodies bound to them, and mass spectrometry. It is in these areas that a number of companies are concentrating their efforts, as not only must a biomarker or panel of biomarkers have good specificity and sensitivity, there must be an efficient and cost-effective method to assay them.

1.8 FUTURE OF PROTEOMICS IN DRUG DISCOVERY, SCREENING, EARLY DETECTION, AND PREVENTION

Proteomics has benefited greatly from the development of high-throughput methods to simultaneously study thousands of proteins. The successful application of proteomics to medical diagnostics will require the combined efforts of basic researchers, physicians, pathologists, technology developers, and information scientists (Figure 1.1). However, its application in clinics will require development

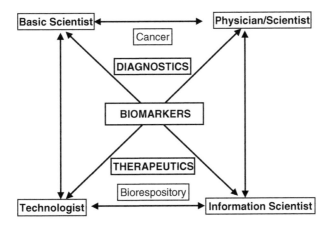

FIGURE 1.1 Application of medical proteomics: Interplay between various disciplines and expertise is the key to developing tools for detection, diagnosis, and treatment of cancer.

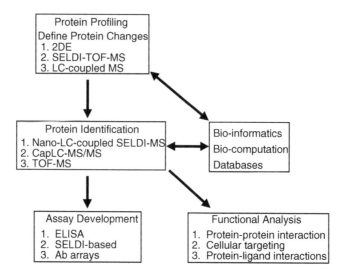

FIGURE 1.2 Strategies in medical proteomics: Steps in identification of detection targets and the development of clinical assays.

of test kits based on pattern analysis, single molecule detection, or multiplexing of several clinical acceptable tests, such as ELISA, for various targets in a systematic way under rigorous quality control regimens (Figure 1.2). Interperson heterogeneity is a major hurdle when attempting to discover a disease-related biomarker within biofluids such as serum. However, the coupling of high-throughput technologies with protein science now enables samples from hundreds of patients to be rapidly compared. Admittedly, proteomic approaches cannot remove the "finding a needle in a haystack" requirement for discovering novel biomarkers; however, we now possess the capability to inventory components within the "haystack" at an unprecedented rate. Indeed, such capabilities have already begun to bear fruits as our knowledge of the different types of proteins within serum is growing exponentially and novel technologies for diagnosing cancers using proteomic technologies are emerging.

Is the development of methods capable of identifying thousands of proteins in a high-throughput manner going to lead to novel biomarkers for the diagnosis of early stage diseases or is the amount of data that is accumulated in such studies going to be overwhelming? The answer to this will depend on our ability to develop and successfully deploy bioinformatic tools. Based on the rate at which interesting leads are being discovered, it is likely that not only will biomarkers with better sensitivity and specificity be identified but individuals will be treated using customized therapies based on their specific protein profile. The promise of proteomics for discovery is its potential to elucidate fundamental information on the biology of cells, signaling pathways, and disease processes; to identify disease biomarkers and new drug targets; and to profile drug leads for efficacy and safety. The promise of proteomics for clinical use is the refinement and development of protein-based assays that are accurate, sensitive, robust, and high throughput. Since many of the proteomic technologies and data management tools are still in their infancy, their validations and refinements are going to be the most important tasks in the future.

REFERENCES

1. Wasinger, V.C., Cordwell, S.J., Cerpa-Poljak, A., et al. Progress with gene-product mapping of the Mollicutes: *Mycoplasma genitalium. Electrophoresis,* 16, 1090–1094, 1995.
2. Banks, R.E., Dunn, M.J., Hochstrasser, D.F., et al. Proteomics: New perspectives, new biomedical opportunities. *Lancet,* 356, 1749–1756, 2000.
3. Anderson, N.L., Matheson, A.D., and Steiner, S. Proteomics: Applications in basic and applied biology. *Curr. Opin. Biotechnol.,* 11, 408–412, 2000.
4. Genther, S.M., Sterling, S., Duensing, S., Munger, K., Sattler, C., and Lambert, P.F. Quantitative role of the human papillomavirus type 16 E5 gene during the productive stage of the viral life cycle. *J. Virol.,* 77, 2832–2842, 2003.
5. Middleton, K., Peh, W., Southern, S., et al. Organization of human papillomavirus productive cycle during neoplastic progression provides a basis for selection of diagnostic markers. *J. Virol.,* 77, 10186–10201, 2003.
6. Verma, M., Lambert, P.F., and Srivastava, S.K. Meeting highlights: National Cancer Institute workshop on molecular signatures of infectious agents. *Dis. Markers,* 17, 191–201, 2001.
7. Verma, M. and Srivastava, S. New cancer biomarkers deriving from NCI early detection research. *Recent Results Canc. Res.,* 163, 72–84; discussion, 264–266, 2003.
8. Verma, M. and Srivastava, S. Epigenetics in cancer: implications for early detection and prevention. *Lancet Oncol.,* 3, 755–763, 2002.
9. Diaz, J.J., Giraud, S., and Greco, A. Alteration of ribosomal protein maps in herpes simplex virus type 1 infection. *J. Chromatogr. B Analyt. Technol. Biomed. Life Sci.,* 771, 237–249, 2002.
10. Greco, A., Bausch, N., Coute, Y., and Diaz, J.J. Characterization by two-dimensional gel electrophoresis of host proteins whose synthesis is sustained or stimulated during the course of herpes simplex virus type 1 infection. *Electrophoresis,* 21, 2522–2530, 2000.
11. Greco, A., Bienvenut, W., Sanchez, J.C., et al. Identification of ribosome-associated viral and cellular basic proteins during the course of infection with herpes simplex virus type 1. *Proteomics,* 1, 545–549, 2001.
12. Laurent, A.M., Madjar, J.J., and Greco, A. Translational control of viral and host protein synthesis during the course of herpes simplex virus type 1 infection: evidence that initiation of translation is the limiting step. *J. Gen. Virol.,* 79, 2765–2775, 1998.
13. Gallo, R.C. Thematic review series. XI: Viruses in the origin of human cancer. Introduction and overview. *Proc. Assoc. Am. Phys,,* 111, 560–562, 1999.
14. Wallace, D.C. Mitochondrial diseases in man and mouse. *Science,* 283, 1482–1488, 1999.
15. Enns, G.M. The contribution of mitochondria to common disorders. *Mol. Genet. Metab.,* 80, 11–26, 2003.
16. Maechler, P. and Wollheim, C.B. Mitochondrial function in normal and diabetic beta-cells. *Nature,* 414, 807–812, 2001.

17. Lopez, M.F. and Melov, S. Applied proteomics: mitochondrial proteins and effect on function. *Circ. Res.,* 90, 380–389, 2002.
18. Kumar, A., Agarwal, S., Heyman, J.A., et al. Subcellular localization of the yeast proteome. *Genes Dev.,* 16, 707–719, 2002.
19. Werhahn, W. and Braun, H.P. Biochemical dissection of the mitochondrial proteome from *Arabidopsis thaliana* by three-dimensional gel electrophoresis. *Electrophoresis,* 23, 640–646, 2002.
20. Mootha, V.K., Bunkenborg, J., Olsen, J.V., et al. Integrated analysis of protein composition, tissue diversity, and gene regulation in mouse mitochondria. *Cell,* 115, 629–640, 2003.
21. Lin, T.K., Hughes, G., Muratovska, A., et al. Specific modification of mitochondrial protein thiols in response to oxidative stress: A proteomics approach. *J. Biol. Chem.,* 277, 17048–17056, 2002.
22. Brookes, P.S., Pinner, A., Ramachandran, A., et al. High throughput two-dimensional blue-native electrophoresis: A tool for functional proteomics of mitochondria and signaling complexes. *Proteomics,* 2, 969–977, 2002.
23. Richly, E., Chinnery, P.F., and Leister, D. Evolutionary diversification of mitochondrial proteomes: Implications for human disease. *Trends Genet.,* 19, 356–362, 2003.
24. Koc, E.C., Burkhart, W., Blackburn, K., Moseley, A., Koc, H., and Spremulli, L.L. A proteomics approach to the identification of mammalian mitochondrial small subunit ribosomal proteins. *J. Biol. Chem.,* 275, 32585–32591, 2000.
25. Newmeyer, D.D. and Ferguson-Miller, S. Mitochondria: Releasing power for life and unleashing the machineries of death. *Cell,* 112, 481–490, 2003.
26. Yin, B.W., Dnistrian, A., and Lloyd, K.O. Ovarian cancer antigen CA125 is encoded by the MUC16 mucin gene. *Int. J. Canc.,* 98, 737–740, 2002.
27. Petricoin, E.F., Ardekani, A.M., Hitt, B.A., et al. Use of proteomic patterns in serum to identify ovarian cancer. *Lancet,* 359, 572–577, 2002.
28. Li, J., Zhang, Z., Rosenzweig, J., Wang, Y.Y., and Chan, D.W. Proteomics and bioinformatics approaches for identification of serum biomarkers to detect breast cancer. *Clin. Chem.,* 48, 1296–1304, 2002.
29. Martinez, M., Espana, F., Royo, M., et al. The proportion of prostate-specific antigen (PSA) complexed to alpha(1)-antichymotrypsin improves the discrimination between prostate cancer and benign prostatic hyperplasia in men with a total PSA of 10 to 30 microg/L. *Clin. Chem.,* 48, 1251–1256, 2002.
30. Brunagel, G., Schoen, R.E., and Getzenberg, R.H. Colon cancer specific nuclear matrix protein alterations in human colonic adenomatous polyps. *J. Cell Biochem.,* 91, 365–374, 2004.
31. Gygi, S.P., Rist, B., Gerber, S.A., Turecek, F., Gelb, M.H., and Aebersold, R. Quantitative analysis of complex protein mixtures using isotope-coded affinity tags. *Nat. Biotechnol.,* 17, 994–999, 1999.
32. Ye, B., Cramer, D.W., Skates, S.J., et al. Haptoglobin-alpha subunit as potential serum biomarker in ovarian cancer: Identification and characterization using proteomic profiling and mass spectrometry. *Clin. Canc. Res.,* 9, 2904–2911, 2003.
33. Adam, B.L., Qu, Y., Davis, J.W., et al. Serum protein fingerprinting coupled with a pattern-matching algorithm distinguishes prostate cancer from benign prostate hyperplasia and healthy men. *Canc. Res.,* 62, 3609–3614, 2002.
34. Poon, T.C., Yip, T.T., Chan, A.T., et al. Comprehensive proteomic profiling identifies serum proteomic signatures for detection of hepatocellular carcinoma and its subtypes. *Clin. Chem.,* 49, 752–760, 2003.

35. Kozak, K.R., Amneus, M.W., Pusey, S.M., et al. Identification of biomarkers for ovarian cancer using strong anion-exchange ProteinChips: Potential use in diagnosis and prognosis. *Proc. Natl. Acad. Sci. USA,* 100, 12343–12348, 2003.
36. Petricoin, E.F., III, Ornstein, D.K., Paweletz, C.P., et al. Serum proteomic patterns for detection of prostate cancer. *J. Natl. Canc. Inst.,* 94, 1576–1578, 2002.
37. Diamandis, E.P. Point: Proteomic patterns in biological fluids: Do they represent the future of cancer diagnostics? *Clin. Chem.,* 49, 1272–1275, 2003.
38. Petricoin, E., III and Liotta, L.A. Counterpoint: The vision for a new diagnostic paradigm. *Clin. Chem.,* 49, 1276–1278, 2003.
39. Baker, S.G. The central role of receiver operating characteristic (ROC) curves in evaluating tests for the early detection of cancer. *J. Natl. Canc. Inst.,* 95, 511–515, 2003.

2 Proteomics Technologies and Bioinformatics

Sudhir Srivastava and Mukesh Verma

CONTENTS

2.1 INTRODUCTION: PROTEOMICS IN CANCER RESEARCH

Proteomics is the study of all expressed proteins. A major goal of proteomics is a complete description of the protein interaction networks underlying cell physiology. Before we discuss protein computational tools and methods, we will give a brief background of current proteomic technologies used in cancer diagnosis. For cancer diagnosis, both surface-enhanced laser desorption ionization (SELDI) and two-dimensional gel electrophoresis (2DE) approaches have been used.[1,2] Recently protein-based microarrays have been developed that show great promise for analyzing the small amount of samples and yielding the maximum data on the cell's microenvironment.[3-5]

2.1.1 Two-Dimensional Gel Electrophoresis (2DE)

The recent upsurge in proteomics research has been facilitated largely by stream-lining of 2DE technology and parallel developments in MS for analysis of peptides and proteins. Two-dimensional gel electrophoresis is used to separate proteins based on charge and mass and can be used to identify posttranslationally modified proteins. A major limitation of this technology in proteomics is that membrane proteins contain a considerable number of hydrophobic amino acids, causing them to precip-itate during the isoelectric focusing of standard 2DE.[6] In addition, information regarding protein– protein interactions is lost during 2DE due to the denaturing conditions used in both gel dimensions. To overcome these limitations, two-dimen-sional blue-native gel electrophoresis has been used to resolve membrane proteins. In this process, membrane protein complexes are solubilized and resolved in the native forms in the first dimension. The separation in the second dimension is performed by sodium dodecyl sulfate polyacrylamide gel electrophoresis (SDS-PAGE), which denatures the complexes and resolves them into their separate subunits. Protein spots are digested with trypsin and analyzed by matrix-assisted laser ionization desorption time-of-flight mass spectrometry (MALDI-TOF MS). The 2DE blue-native gel electrophoresis is suitable for small biological samples and can detect posttranslational modifications (PTMs) in proteins. Common PTMs include phosphorylation, oxidation and nitrosation, fucosylation and galactosylation, reaction with lipid-derived aldehydes, and tyrosine nitration. Improvements are needed to resolve low-molecular-mass proteins, especially those with isoelectric points below pH 3 and above pH 10. This technique has low throughput (at the most 30 samples can be run simultaneously), and most of the steps are manual. Automatic spot-picking also needs improvement.

2.1.2 Mass Spectrometry

Mass spectrometry (MS) is an integral part of the proteomic analysis. MS instruments are made up of three primary components: the source, which produces ions for analysis; the mass analyzer, which separates the ions based on their mass-to-charge ratios (m/z); and the detector, which quantifies the ions resolved by the analyzer. Multiple subtypes of ion sources, analyzers, and detectors have been developed, and different components can be combined to create different instruments, but the principle remains the same— the spectrometers create ion mixtures from a sample and then resolve them into their component ions based on their m/z values. Significant improvements have been made in spectrometric devices during the past two decades, allowing precise analysis of biomolecules too fragile to survive earlier instrumentation. For ionization of peptides and proteins, these ionization sources are usually coupled to time-of-flight (TOF)[2,7,8] spectrometers. Historically, MS has been limited to the analysis of small molecules. Larger biomolecules, such as peptides or proteins, simply do not survive the harsh ionization methods available to create the ions. ESI (electrospray ionization),[9] MALDI, and SELDI techniques permit a gentler ionization of large biomolecules, called soft ionization, without too much fragmentation of the principal ions. ESI and MALDI were both developed during the late 1980s and were the foundation for the emergence of MS as a tool of investigation of biological samples. Although MALDI equipment is expensive, quantitative high throughput can be achieved (about 100 samples per day can be run by a single laboratory).

SELDI, developed in the early 1990s, is a modification of the MALDI approach to ionization. All the ionization techniques described above are sensitive in the picomole-to-femtomole range that is required for application to biological samples, carbohydrates; oligonucleotides; small polar molecules; and peptides, proteins, and posttranslationally modified proteins.

Tandem mass analyzers are instruments used for detailed structural analysis of selected peptides. An example of this kind of analyzer is ABI's QSTAR® (Applied Biosystems, Foster City, CA), a hybrid system that joins two quadrupoles in tandem with a TOF analyzer.[10] Particular tryptic peptide fragments can be sequentially selected and subfragmented in the two quadrupoles, and then the subfragments can be measured in the analyzer. The resulting pattern is somewhat like the sequence-ladder pattern obtained in DNA sequencing. Although the analysis of the protein pattern is more complex than DNA sequencing, software is available that allows the direct determination of the amino acid sequence of peptides. Based on the peptide sequence information, it is possible to identify the parent protein in the database.

2.1.3 Isotope-Coded Affinity Tags (ICAT)

Isotope-coded affinity tags (ICAT)[11] is a technology that facilitates quantitative proteomic analysis. This approach uses isotope tagging of thiol-reactive group to label reduced cysteine residues, and a biotin affinity tag to isolate the labeled peptides. These two functional groups are joined by linkers that contain either eight hydrogen atoms (light reagent) or eight deuterium atoms (heavy reagent). Proteins in a sample (cancer) are labeled with the isotopically light version of the ICAT reagent, while proteins in another sample (control) are labeled by the isotopically heavy version of the ICAT reagent. The two samples are combined, digested to generate peptide fragments, and the cysteine-containing peptides are enriched by avidin affinity chromatography. This results in an approximately tenfold enrichment of the labeled peptides. The peptides may be further purified and analyzed by reverse-phase liquid chromatography, followed by MS. The ratio of the isotopic molecular mass peaks that differ by 8 Da provides a measure of the relative amounts of each protein in the original samples. This technology is good for detection of differentially expressed proteins between two pools. Recently the method has been modified to include ^{16}O and ^{18}O water and culture cells in ^{12}C- and ^{13}C-labeled amino acids. Problems with ICAT include its dependency on radioactive materials, its low throughput (about 30 samples per day), it only detects proteins that contain cysteine, and labeling decreases over time (see also Chapter 16).

2.1.4 Differential 2DE (DIGE)

Differential 2DE (DIGE) allows for a comparison of differentially expressed proteins in up to three samples. In this technology, succinimidyl esters of the cyanine dyes, Cy2, Cy3, and Cy5, are used to fluorescently label proteins in up to three different pools of proteins. After labeling, samples are mixed and run simultaneously on the same 2DE.[12] Images of the gel are obtained using three different excitation/emission filters, and the ratios of different fluorescent signals are used to find protein differences among the samples. The problem with DIGE is that only 2% of the lysine residues in the proteins can be fluorescently modified, so that the solubility of the labeled proteins is maintained during electrophoresis. An additional problem with this technology is that the labeled proteins migrate with slightly higher mass than the bulk of the unlabeled proteins. DIGE technology is more sensitive than silver stain formulations optimized for MS. SYPRO Ruby dye staining detects 40% more protein spots than the Cy dyes.

2.1.5 PROTEIN-BASED MICROARRAYS

DNA microarrays have proven to be a powerful technology for large-scale gene expression analysis. A related objective is the study of selective interactions between proteins and other biomolecules, including other proteins, lipids, antibodies, DNA, and RNA. Therefore, the development of assays that could detect protein-directed interactions in a rapid, inexpensive way using a small number of samples is highly desirable. Protein-based microarrays provide such an opportunity. Proteins are separated using any separation mode, which may consist of ion exchange liquid chromatography (LC), reverse-phase LC, or carrier ampholyte–based separations, such as Rotophor. Each fraction obtained after the first dimensional separation can be further resolved by other methods to yield either purified protein or fractions containing a limited number of proteins that can directly be arrayed or spotted. A robotic arrayer is used for spotting provided the proteins remain in liquid form throughout the separation procedure. These slides are hybridized with primary antibodies against a set of proteins and the resulting immune complex detected. The resulting image shows only these fractions that react with a specific antibody. The use of multidimensional techniques to separate thousands of proteins enhances the utility of protein microarray technology. This approach is sensitive enough to detect specific proteins in individual fractions that have been spotted directly without further concentration of the proteins in individual fraction. However, one of the limitations of the nitrocellulose-based array chip is the lack of control over orientation in the immobilization process and optimization of physical interactions between immobilized macromolecules and their corresponding ligands, which can affect sensitivity of the assay.

Molecular analysis of cells in their native tissue microenvironment can provide the most desirable situation of *in vivo* states of the disease. However, the availability of low numbers of cells of specific populations in the tissue poses a challenge. Laser capture microdissection (LCM) helps alleviate this matter as this technology is capable of procuring specific, pure subpopulations of cells directly from the tissue. Protein profiling of cancer progression within a single patient using selected longitudinal study sets of highly purified normal, premalignant, and carcinoma cells provides the unique opportunity to not only ascertain altered protein profiles but also to determine at what point in the cancer progression these alterations in protein patterns occur. Preliminary results from one such study suggest complex cellular communication between epithelial and stroma cells. A majority of the proteins in this study are signal transduction proteins.[5] Protein-based microarrays were used in this study. Advantages and disadvantages of some proteomic-relevant technologies are listed in Table 2.1.

TABLE 2.1
Comparisons of Various Proteomic Technologies

Characteristics	ELISA	2DE PAGE	IsotopeCoded Affinity Tag (ICAT)™	Multidimensional Protein Identification Technology (MudPIT)™	Proteomic Pattern Diagnostics	Protein Microarrays
	Chemiluminescence or fluorescence-based	2DE serological proteome analysis (SERPA); 2DGE + serum immunoblotting	ICAT/LC-EC-MS/MS; ICAT/LCMS/MS/MALDI	2D LC-MS/MS[a]	MALDI-TOF; SELDI-TOF; SELDIT-OF/QStar™	Antibody arrays: chemiluminescence multi-ELISA platforms; glass fluorescence based (Cy3Cy5); tissue arrays
Sensitivity	Highest	Low, particularly for less abundant proteins; sensitivity limited by detection method; difficult to resolve hydrophobic proteins	High	High	Medium sensitivity with diminishing yield at higher molecular weights; improved with fitting of high-resolution QStar mass spectrometer to SELDI	Medium to highest (depending on detection system)
Direct identification of markers	N/A	Yes	Yes	Yes	No; possible with additional high-resolution MS	Possible when coupled to MS technologies; or probable, if antibodies have been highly defined by epitope mapping and neutralization
Use	Detection of single, well-characterized specific analyte in plasma/serum, tissue; gold standard of clinical assays	Identification and discovery of biomarkers not a direct means for early detection in itself	Quantification of relative abundance of proteins from two different cell states	Detection and ID of potential biomarkers	Diagnostic pattern analysis in body fluids and tissues (LCM); potential biomarker identification	Multiparametric analysis of many analytes simultaneously

(Continued)

TABLE 2.1
Comparisons of Various Proteomic Technologies (Continued)

Characteristics	ELISA	2DE PAGE	IsotopeCoded Affinity Tag (ICAT)™	Multidimensional Protein Identification Technology (MudPIT)™	Proteomic Pattern Diagnostics	Protein Microarrays
Throughput	Moderate	Low	Moderate/low	Very low	High	High
Advantages/drawbacks	Very robust, well-established use in clinical assays; requires well-characterized antibody for detection; requires extensive validation not amenable to direct discovery; calibration (standard) dependent; FDA regulated for clinical diagnostics	Requires a large number of samples; all identifications require validation and testing before clinical use; reproducible and more quantitative combined with fluorescent dyes; not amenable for high throughput or automation; multiple proteins may be positioned at the same location on the gel	Robust, sensitive, and automated; suffers from the demand for continuous on-the-fly selection of precursor ions for sequencing; coupling with MALDI promises to overcome this limitation and increase efficiency of proteomic comparison of biological cell states; still not highly quantitative and difficult to measure subpg/ml concentrations	Significantly higher sensitivity than 2D-PAGE; much larger coverage of the proteome for biomarker discovery; not reliable for low abundance proteins and low-molecular-weight fractions	SELDI protein identification not necessary for biomarker pattern analysis; reproducibility problematic, improved with QStar addition; revolutionary tool; 1-2 μl of material needed; upfront fractionation of protein mixtures and downstream purification methods necessary to obtain absolute protein quantification; MALDI crystallization of protein can lack reproducibility and be matrix dependent; high MW proteins requires MS/MS	Format is flexible; can be used to assay for multiple analytes in a single specimen or a single analyte in a number of specimens; requires prior knowledge of analyte being measured; limited by antibody sensitivity and specificity; requires extensive crossvalidation for antibody crossreactivity; requires use of an amplified tag detection system; requires more sample to measure low abundant proteins; needs to be measured undiluted
Bioinformatic needs	Moderate, standardized	Moderate; mostly home grown, some proprietary	Moderate	Moderate	Moderate to extensive; home grown, not standardized	Extensive, home grown; not standardized

a LCM: Laser Capture Microdissection

TABLE 2.2
Database Search Tools for 2DE and MS

Name of the Software	Web Site
Delta2D[a]	www.decodon.com/Solutions/Delta2D.html
GD Impressionist[a]	www.genedata.com/productsgell/Gellab.html
Investigator HT PC Analyzer[a]	www.genomicsolutions.com/proteomics/2dgelanal.html
Phortix 2D[a]	www.phortix.com/products/2d_products.htm
Z3 2D-Gel Analysis System[a]	www.2dgels.com
Mascot	www.matrixscience.com
MassSearch	www.Cbrg.inf.ethz/Server/MassSearch.html
MS-FIT	www.Prospector.ucsf.edu
PeptIdent	www.expasy.ch/tools/peptident.html

[a] Software for 2DE.

2.2 CURRENT BIOINFORMATICS APPROACHES IN PROTEOMICS

Most biological databases have been generated by the biological community, whereas most computational databases have been generated by the mathematical and computational community. As a result, biological databases are not easily acquiescent to automated data mining methods and are unintelligible to some computers, and computational tools are nonintuitive to biologists. A list of database search tools is presented in Table 2.2, and some frequently used databases to study protein-protein interaction are shown in Table 2.3. A number of bioinformatic approaches have been discussed elsewhere in the book (see Chapters 10 and 14); therefore, we have described only the basic principles of some of these approaches.

An important goal of bioinformatics is to develop robust, sensitive, and specific methodologies and tools for the simultaneous analysis of all the proteins expressed by the human genome, referred to as the human proteome, and to establish "biosignature" profiles that discriminate between disease states. Artifacts can be introduced into spectra from physical, electrical, or chemical sources. Each spectrum in

TABLE 2.3
Database for Protein Interaction

Name of the Database	Web Site
CuraGen	Portal.curagen.com
DIP	Dipdoe-mbi.ucla.edu
Interact	Bioinf.man.ac.uk/interactso.htm
MIPS	www.mips.biochem.mpg.de
ProNet	Pronet.doublewist.com

MALDI or SELDI-TOF could be composed of three components: (1) true peak signal, (2) exponential baseline, and (3) white noise.

Low-level processing is usually used to disentangle these components, remove systematic artifacts, and isolate the true protein signal.

A key for successful biomarker discovery is the bioinformatic approach that enables thorough, yet robust, analysis of a massive database generated by modern biotechnologies, such as microarrays for genetic markers and time-of-flight mass spectrometry for proteomic spectra.

Prior to a statistical analysis of marker discovery, TOF-MS data require a pre-analysis processing: this enables extraction of relevant information from the data. This can be thought of as a way to standardize and summarize the data for a subsequent statistical analysis. For example, based on some eminent properties of the data, pre-analytical processing first identifies all protein signals that are distinguishable from noise, then calibrates mass (per charge) values of proteins for potential measurement errors, and finally aggregates, as a single signal, multiple protein signals that are within the range of measurement errors. The above discussion is specifically relevant to serum-based analysis prone to all types of artifacts and errors. Serum proteomic pattern analysis is an emerging technology that is increasingly employed for the early detection of disease, the measurement of therapeutic toxicity and disease responses, and the discovery of new drug targets for therapy. Various bioinformatics algorithms have been used for protein pattern discovery, but all studies have used the SELDI ionization technique along with low-resolution TOF-MS analysis. Earlier studies demonstrated proof-of-principle of biomarker development for prostate cancer using SELDI-TOF, but some of the studies relied on the isolation of actual malignant cells from pathology specimens.[13–16] Body-fluid-based diagnostics, using lavage, effluent, or effusion material, offers a less invasive approach to biomarker discovery than biopsy or surgical-specimen-dependent approaches.[17] Additionally, serum-based approaches may offer a superior repository of biomarkers because serum is easy and inexpensive to obtain.[18–21]

Several preprocessing and postprocessing steps are needed in the protein chip data analysis. For data analysis we must process the mass spectra in such a way that it is conducive to downstream multidimensional methods (clustering and classification, for example). The binding to protein chip spots used for general profiling is specific only to a class of proteins that share a physical or chemical property that creates an affinity for a given protein chip array surface. As a result, mass spectra can contain hundreds of protein expression levels encoded in their peaks.

Bioinformatics tools have promise in aiding early cancer detection and risk assessment. Some of the useful areas in bioinformatics tools are pattern clustering, classification, array analysis, decision support, and data mining. A brief application of these approaches is described below.

2.2.1 CLUSTERING

Two major approaches to clustering methods are bottom-up and top-down. An example of the bottom-up approach includes hierarchical clustering where each gene has its own profile.[22] The basis of the clustering is that closest pairs are clustered first followed by successive clustering with other clusters and finally combining into one matrix. The representation of clustering is accomplished by a dendogram (hierarchical tree) and this dendogram can be cut at multiple levels for easy interpretation. The top-down clustering, in contrast to the bottom-up approach, starts with a specified number of clusters and cluster centers. Observation assignment is done based on the closest cluster center to help partitioning of the data.

2.2.2 ARTIFICIAL NEURAL NETWORKS

For multifactorial classification and multivariate nonlinear regression, artificial neural networks (ANN) are excellent tools. The method has evolved in 40 years in terms of its mathematical nature and capabilities. Because of the popularity of personal computers and their routine use in data analysis, ANN-based analysis is a useful approach. ANN is utilized by stock market analysis staff, scientists, medical imaging technicians, and mathematicians. However, the problem with the ANN is that several decisions related to the choice of ANN structures and parameters are subjective. Other problems include lack of theoretical recommendations for the size of training data and unavailability of the optimum size. Overtraining is another problem associated with the ANN, which often results in memorization rather than generalization of the data. Bootstrap sampling or random sampling with replacement to produce a large number of individual neural networks provides an alternative to resolve such problems. The structure of the objective function in the space of the ANN parameters is usually very complicated because ANN is a strongly nonlinear approximation.

Despite the problems in the design and training of ANNs, there has been a marked increase in application of ANNs in biomedical areas including cancer research. The consensus among scientists indicates use of a panel of markers instead of single marker for cancer detection and/or risk assessment.[19] ANN is inherently suited in this regard as it is capable of performing simultaneous analysis of large amounts of diverse information. The classification performance of the ANN is often evaluated using the receiver operating characteristics curves (ROC). In the ROC curve the y-axis and x-axis represent sensitivity and specificity, respectively. The area under the ROC curve represents how well the independent variable separating two dichotomous classes performs.

2.2.3 SUPPORT VECTOR MACHINE (SVM)

SVM is a supervised algorithm for classification of data that projects data into higher dimensional space where two classes are linearly separable. SVM selects a hyperplane in the space of the data points that separates two classes of data and maximizes the width of separating band (also called margin) between the hyperplane and the data points.

Finding disease-associated proteins is a natural first step in analyzing expression data. Methods such as Mann–Whitney and the Kruskal–Wallis analysis eliminate any assumption on the distribution of the peak intensity data.[23,24] These tests give an indication of group mean differences between control and cases that may not always be helpful if the distribution of data creates a large spread. The multidimensional analysis involves two categories: unsupervised learning in the form of cluster analysis, and supervised learning in the form of classification methods. These approaches have been applied for expression data analysis, visualization, self-organizing maps, and SVM.[25,26] The software examines each peak cluster present in the spectra and assesses it for classification. Furthermore, the peak intensity is also assigned to indicate a value discriminatory between normal and disease samples. The classification tree so generated is an attractive tool for protein expression studies—it is easier to interpret than comparable black-box classifiers such as nearest-neighbor and neural network classifiers. The openness of a tree-based model is useful in identifying diagnostic and therapeutic targets.

2.3 PROTEIN KNOWLEDGE SYSTEM

BLASTAP algorithm is used to identify distinct proteins identified during the protein analysis.[27] This analysis is based on clustering of all proteins sharing greater than 95% identity over an aligned region. The advantage of this clustering is that it allows the grouping of identical sequences, splice variants, and sequence fragments and possible paralogs. Functional annotation of proteins is achieved using UniProt (www.uniprot.org) and PIR (www.pir.georgetown.edu). Both of these programs provide protein database, data mining tools, and sequence analysis tools.[28,29]

These databases also provide comprehensive, value-added descriptions of proteins and serve as a framework for data integration, especially in a distributed networking environment. All the proteins in these databases consist of links to over 50 databases of protein sequences, structures, families, functions and pathways, protein–protein interactions, structural modifications, gene ontologies, taxonomy, and major posttranslational modifications. At the level of superfamily, domain, and motif, the protein family classification helps in grouping of proteins in different functional categories and removing any redundancy.

2.4 MARKET OPPORTUNITIES IN COMPUTATIONAL PROTEOMICS

Unlike the "fixed" genome, the proteome remains flexible throughout the human life, especially in the disease state; thus it is important proteomic computational tools that truly reflect disease-associated changes and make interpretation of the data easy. This is where "market opportunities" exist; an arbitrary projection of future market in this area is presented in Figure 2.1.[30] As indicated in the figure, the market for computational proteomics is expected to rise sharply. Both fully integrated pharmaceutical companies and small venture-funded startups are exhibiting interest in the technology. The focus of both industrial entities is on "innovation." Many times a great idea developed by a small company with few resources gets fragmented. Making steps in protein analysis automatic and at miniature scale might help in future. The problem that the market for proteomic computation faces is that a small number of expensive software is generated, all of which is targeted to a select few companies. If software were designed for multiple users, then software companies would make a profit.

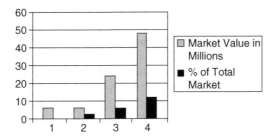

FIGURE 2.1 Bioinformatics-based market forecast. 1: 2005; 2: 2010; 3: 2015; 4: 2020. (From Razvi, E., *Biotechniques,* March suppl., 54–62, 2002. With permission.)

In a recent survey, the modeling and simulation market was about $100 million. Both chemoinformatics and bioinformatics are growing rapidly and are expected to grow to several billion dollars in the next 10 years. The U.S. and Europe cover most of the market. Two sectors where these computational tools are being utilized the most are pharmaceuticals and agriculture bioinformatics.

2.5 CHALLENGES

A large number of small-scale and large-scale experiments have contributed to expanding our understanding of the nature of the interaction of different pathways in cancer development. The most common challenges are biological and bioinformatics related. The following exemplify the biological issues:

1. Sample integrity: material degradation/standardization of serum collection
2. Sample quality in tumor-burdened or diseased patient (icteric, lipemic, etc.)
3. Limited sample volume for repeat analysis
4. Enormous dynamic range of expressed proteins
5. Difficulty in mining low abundance proteins

However, the necessary data integration and/or data mining across experiments have problems of fragmentation, and it is difficult to utilize publicly available protein interaction data. Currently the data exist in different formats in databases, on authors' Web sites, or sometimes only in printed publications. HUPO (Human Proteomic Organization) has taken this challenge and started developing new computational tools and compiling data.[31] Data generated by one group has been reanalyzed by some other groups.[32]

As with any data-rich enterprise, informatics issues become apparent on several proteomics fronts. Sample documentation, implementation of rigorous standards, and proper annotation of gene function are important. It is crucial that software development is linked at an early stage through agreed documentation, XML-based definitions, and controlled vocabularies that allow different tools in order to exchange primary data sets in a high-throughput way. Efforts in interaction databases, systems biology software, and infrastructure are being made for future proteomics initiatives. In the future, the development of statistically sound methods for assignment of protein identities from incomplete mass spectral data will be critical for automated deposition into databases. Lessons learned from analysis of DNA microarray data, including clustering, compendium, and pattern-matching approaches, should be seriously considered for proteomic analysis. It is encouraging that HUPO and the European Bioinformatics Institute have together started an initiative on the exchange of protein–protein interaction and other proteomic data (http://psidev.sourceforge.net/).

2.6 CONCLUSION

Proteomic bioinformatics may dramatically change how a disease is detected, monitored, and managed. Tremendous progress has been made in the past few years in generating large-scale data sets for protein–protein interactions, organelle composition, protein activity patterns, and protein profiles in cancer patients. But further technological improvements, organization of international proteomics projects, and open access to results are needed for proteomics to fulfill its potential. Because most drug targets are proteins, it is unavoidable that proteomics will enable drug discovery, development, and clinical practice. The form(s) in which proteomics will best fulfill this mandate is in a state of flux owing to multiple factors including the varied technological platforms in different stages of implementation. Whatever the future holds, proteomics will yield great returns in biology and medicine.[33]

REFERENCES

1. Vlahou, A., Laronga, C., Wilson, L., et al. A novel approach toward development of a rapid blood test for breast cancer. *Clin. Breast Canc.,* 4, 203–209, 2003.
2. Shin, B.K., Wang, H., and Hanash, S. Proteomics approaches to uncover the repertoire of circulating biomarkers for breast cancer. *J. Mammary Gland Biol. Neoplasia,* 7, 407–413, 2002.
3. Knezevic, V., Leethanakul, C., Bichsel, V.E., et al. Proteomic profiling of the cancer microenvironment by antibody arrays. *Proteomics,* 1, 1271–1278, 2001.
4. Madoz-Gurpide, J. and Hanash, S.M. Molecular analysis of cancer using DNA and protein microarrays. *Adv. Exp. Med. Biol.,* 532, 51–58, 2003.
5. Madoz-Gurpide, J., Wang, H., Misek, D.E., Brichory, F., and Hanash, S.M. Protein based microarrays: A tool for probing the proteome of cancer cells and tissues. *Proteomics,* 1, 1279–1287, 2001.
6. Zhang, W., Zhou, G., Zhao, Y., and White, M.A. Affinity enrichment of plasma membrane for proteomics analysis. *Electrophoresis,* 24, 2855–2863, 2003.

7. Naour, F.L., Brichory, F., Beretta, L., and Hanash, S.M. Identification of tumor-associated antigens using proteomics. *Technol. Canc. Res. Treat.,* 1, 257–262, 2002.
8. Hanash S. Disease proteomics. *Nature,* 422, 226–232, 2003.
9. Malik, G., Ward, M.D., Gupata, S.K., Trosset, M.W., Grizzle, W.E., Adams, B.L., and Semmes, O.J. Serum levels of an isoform of apolipoprotein A-II as a potential marker for prostate cancer. *Clin. Can. Res.,* 11,1073–1085, 2005.
10 Nielsen, M.L., Bennett, K.L., Larsen, B., Moniatte, M., and Mann, M. Peptide end sequencing by orthogonal MALDI tandem mass spectrometry. *J. Proteome Res.,* 1, 63–71, 2002.
11. Yan, W., Lee, H., Deutsch, E.W., Lazaro, C.A., Tang, W., Chen, E., Fausto, N., Katz, M.G., and Abersold, R. A dataset of human liver proteome identified by protein profiling via isotope-coded affinity tag (ICAT) and tandem mass spectrometry. *Mol. Cell. Proteomics,* 3, 1039–1041, 2004.
12. Srinivas, P.R., Verma, M., Zhao, Y., and Srivastava, S. Proteomics for cancer biomarker discovery. *Clin. Chem.,* 48, 1160–1169, 2002.
13. Lehrer, S., Roboz, J., Ding, H., et al. Putative protein markers in the sera of men with prostatic neoplasms. *BJU Int.,* 92, 223–225, 2003.
14. Yasui, Y., Pepe, M., Thompson, M.L., et al. A data-analytic strategy for protein biomarker discovery: profiling of high-dimensional proteomic data for cancer detection. *Biostatistics,* 4, 449–463, 2003.
15. Issaq, H.J., Veenstra, T.D., Conrads, T.P., and Felschow, D. The SELDI-TOF MS approach to proteomics: Protein profiling and biomarker identification. *Biochem. Biophys. Res. Commun.,* 292, 587–592, 2002.
16. Wellmann, A., Wollscheid, V., Lu, H., et al. Analysis of microdissected prostate tissue with ProteinChip arrays — A way to new insights into carcinogenesis and to diagnostic tools. *Int. J. Mol. Med.,* 9, 341–347, 2002.
17. Diamandis, E.P. Mass spectrometry as a diagnostic and a cancer biomarker discovery tool: Opportunities and potential limitations. *Mol. Cell Proteomics,* 3, 367–378, 2004.
18. Verma, M., Dunn, B.K., Ross, S., et al. Early detection and risk assessment: Proceedings and recommendations from the Workshop on Epigenetics in Cancer Prevention. *Ann. NY Acad. Sci.,* 983, 298–319, 2003.
19. Verma, M. and Srivastava, S. New cancer biomarkers deriving from NCI early detection research. *Recent Results Canc. Res.,* 163, 72–84; discussion 264–266, 2003.
20. Negm, R.S., Verma, M., and Srivastava, S. The promise of biomarkers in cancer screening and detection. *Trends Mol. Med.,* 8, 288–293, 2002.
21. Verma, M., Wright, G.L., Jr., Hanash, S.M., Gopal-Srivastava, R., and Srivastava, S. Proteomic approaches within the NCI early detection research network for the discovery and identification of cancer biomarkers. *Ann. NY Acad. Sci.,* 945, 103–115, 2001.
22. Xiao, X., Liu, D., Tang, Y., et al. Development of proteomic patterns for detecting lung cancer. *Dis. Markers,* 19, 33–39, 2003.
23. Fung, E.T. and Enderwick, C. ProteinChip clinical proteomics: Computational challenges and solutions. *Biotechniques,* Suppl. 34–41, 2002.
24. Fung, E., Diamond, D., Simonsesn, A.H., and Weinberger, S.R. The use of SELDI ProteinChip array technology in renal disease research. *Methods Mol. Med.,* 86, 295–312, 2003.
25. Golub, T.R., Slonim, D.K., Tamayo, P., et al. Molecular classification of cancer: Class discovery and class prediction by gene expression monitoring. *Science,* 286, 531–537, 1999.

26. Moler, E.J., Chow, M.L., and Mian, I.S. Analysis of molecular profile data using generative and discriminative methods. *Physiol. Genomics,* 4, 109–126, 2000.

27. Heckel, D., Comtesse, N., Brass, N., Blin, N., Zang, K.D., and Meese, E. Novel immunogenic antigen homologous to hyaluronidase in meningioma. *Hum. Mol. Genet.,* 7, 1859–1872, 1998.

28. Apweiler, R., Bairoch, A., Wu, C.H., et al. UniProt: The Universal Protein knowledgebase. *Nucleic Acids Res.,* 32, D115–D119, 2004.

29. Garavelli, J.S. The RESID database of protein modifications: 2003 developments. *Nucleic Acids Res.,* 31, 499–501, 2003.

30. Razvi, E. Market opportunity in computational proteomics. *Biotechniques,* 54–62, 2002.

31. Hermjakob, H., Montecchi-Palazzi, L., Bader, G., et al. The HUPO PSI's molecular interaction format — A community standard for the representation of protein interaction data. *Nat. Biotechnol.,* 22, 177–183, 2004.

32. Sorace, J.M. and Zhan, M. A data review and re-assessment of ovarian cancer serum proteomic profiling. *BMC Bioinformatics,* 4, 24, 2003.

33. Tyers, M. and Mann, M. From genomics to proteomics. *Nature,* 422, 193–197, 2003.

3 Creating a National Virtual Knowledge Environment for Proteomics and Information Management

*Daniel Crichton, Heather Kincaid, Sean Kelly,
Sudhir Srivastava, J. Steven Hughes,
and Donald Johnsey*

CONTENTS

3.1 INTRODUCTION

Biomedical research generates an enormous amount of data located in geographically distributed data repositories. Modern discovery processes are dependent on having a rich environment for information management that provides the informatics infrastructure necessary to capture, correlate, and distribute data obtained within and across studies. Often, data generated within a particular study or assay is captured and managed without reference to any standard principles of information management. This limits the ability of researchers to correlate data across experiments, potentially reducing opportunities for new discovery. Interoperability and efficient use of biomarkers, genomics, and proteomics data stored in disparate systems is difficult because of differing semantic and technology architectures that manage the data. As a result, it is challenging to create tools for automated understanding and analysis; the heterogeneity of the environment and the different methods for capturing the data increase the expense of research and reduce the returns. Ultimately, discoveries are delayed that could potentially eradicate disease, improve health, and better the quality of life.

The National Cancer Institute created a network of collaborating institutions focused on the discovery and validation of cancer biomarkers called the Early Detection Research Network (EDRN).[1] Informatics plays a key role in this network by creating a virtual knowledge environment that provides scientists with real-time access to distributed data sets located at research institutions across the nation. The distributed and heterogeneous nature of the collaboration makes data sharing across institutions very difficult. EDRN has developed a national informatics infrastructure to enable seamless capture, access, sharing, and discovery of science data resources across participating cancer research centers.

A key requirement of establishing the informatics infrastructure for the EDRN was to allow researchers located at independent facilities to continue to perform their studies using their existing systems and data collection methods. The EDRN informatics team therefore architected a data grid solution[2] that enabled distributed information management based on a well-defined data and technology architecture. The concept behind the knowledge system is that each of the 31 institutions participating in EDRN is a potential "peer" of the knowledge system, conceptually contributing data using a common ontological model* for early cancer detection. This aggregation of data across all institutions creates the knowledge space that encompasses the EDRN data.

The National Cancer Institute, the Fred Hutchinson Cancer Research Center, and the National Aeronautic and Space Administration's (NASA) Jet Propulsion Laboratory (JPL) leveraged JPL's metadata-based data grid initiative, called the Object Oriented Data Technology (OODT), framework toward creating a national virtual specimen-sharing system.[3] The software framework, originally developed to support location, access, and sharing of planetary science data, allows for data resources and products to be widely distributed over the Internet, enabling distributed discovery and access of science data within a multi-institution environment. The

---

* An ontological model defines concepts and their relationships.

team, as part of its efforts to build a comprehensive data grid framework, identified several key project goals[4] that include

1. Location transparency when finding distributed data products
2. The separation of the software and data architectures
3. The encapsulation of data nodes to hide uniqueness
4. The use of metadata for all messages exchanged between distributed services
5. A standard data dictionary for describing data resources
6. The use of ubiquitous software interfaces across multiple data systems that provide interoperability via a common "grid-like" query mechanism
7. The establishment of a standard data model for describing any data resource regardless of its location

A key to establishing a scientific knowledge system is defining the associated scientific processes that support the discovery and validation of cancer biomarkers. The EDRN established working groups across the institutions in order to flush out the associated discovery and validation processes that would be used across the network to perform the data collection and analysis. Basic scientists, clinicians, epidemiologists, biostatisticians, and computer scientists represented these cross-disciplinary working groups. The key processes that were identified enabled the computer scientists to generate "use cases," an important element necessary in constructing an information system.

The EDRN knowledge system has been broken into three critical pieces: science processes, data architecture, and technology. Each of these has been critical to developing an informatics knowledge infrastructure supporting the scientific discovery process. The technology architecture is based on the application of open standards technology to a layered component model that enables the software objects to be "plugged" together with well-defined interfaces. The EDRN data architecture implements a common data dictionary and model that helps to not only enable search and retrieval of legacy data repositories, but provides a standard for the development of new informatics tools.

3.2 A DATA ARCHITECTURE FOR DISTRIBUTED INFORMATION MANAGEMENT

The data architecture provides the ontological model necessary to construct the knowledge system. A data architecture is critical to effectively search heterogeneous distributed data systems and enable correlative science. It defines the common data elements and their relationships within the knowledge space and enables interoperability between distributed institutions by providing a common semantic language for communication.

Information management is often described as the management of information objects. These objects are described by both a data object and a representational object. A data object in itself is often rendered useless unless there is some representational information that can be applied in order to interpret its meaning, structure, and behavior. For example, if the character "3" is described as a value of temperature, the knowledge communicated is more complete and useful. Furthermore, if the units Celsius are applied, then it can be correctly interpreted.

In most domains, whether medical research, space science, or engineering, a description of a value is still insufficient in communicating knowledge; a description of the description itself is critical. Again, what is temperature? This implies that descriptors of the meta data—also known as data about metadata, or alternatively meta-metadata—are needed in order to correctly capture the appropriate attributes of a temperature measurement. For example, temperature could be described using a definition, the data type of its values, and the units in which the values were measured. This would allow the value of, say, 37.2, to be more precisely described as the real value of a radiative surface in physical contact with a mercury-based oral thermometer for human use that reacts to average surrounding molecular vibration using a centigrade scale with zero as the freezing point of distilled water. This level of precision communicates even more knowledge. Data modeling is important for describing relationships between information objects. For example, by relating our value of temperature to observed patients, it could be placed in context as the body temperature of a specific patient currently in the intensive care unit at a specific hospital in a specific room. There is ongoing research in several domains related to the construction of standard information models that can be used to describe data. The semantic web community,[5] for example, describes the use of relationships as an ontological model that is necessary for giving meaning to data. The EDRN developed a data architecture for its knowledge system along with an overarching ontological model for describing the cancer data that was captured and shared. This model served as an import link between distributed databases across the country.[12]

Several standards have been developed that support the definition of a data architecture. ISO/IEC 11179[6] provides a standard definition for describing data elements. This enables consistency when developing data dictionaries. The ISO/IEC standard recommends that a data element consists of attributes for four key categories: identification, definitional, representational, and administrative. EDRN uses ISO/IEC 11179 in conjunction with an Internet standard called Dublin Core.[7] Dublin Core provides a minimal set of data elements that should be part of every data dictionary. Within EDRN, a model for describing specimen repositories and associated epidemiological data was created that described the heterogeneous data that was captured in databases located at various collaborating research laboratories.

While the technical components of the data architecture are critical to enabling interoperability at the system level, it is important to note that creation of a common model is often the result of a highly cross-disciplinary team. EDRN developed working groups that focused on defining the common data elements (CDEs) for each of the objects of the common data model along with their associated relationships.[8] Specific objects, for example, include organ specimen models such as lung, breast, and prostate. The ISO/IEC 11179 standards provide guidance on how to structure and define the data elements.

3.2.1 Descriptors for Describing Metadata

The most basic component of an architecture for data is the set of descriptors needed to describe the data descriptors. These are the attributes that are used to describe the data elements that were discussed earlier. They are also known as meta attributes or meta-metadata.

ISO/IEC 11179 (11179), as mentioned earlier, is a framework for the specification and standardization of data elements. It provides a base set of descriptors (or attributes) needed to describe data elements. As an international standard, it provides a common basis for data element definition and classification across many areas of interest. Table 3.1 shows the basic attributes and the four categories under which they exist.

The *identifying* category identifies a data element. That is, the attribute identifier uniquely identifies a data element within an area of interest. The *definitional* category describes the semantic aspects of a data element and consists of a textual description that communicates knowledge about the data element that typically is not captured by any of the basic attributes. The *relational* category captures associations among

TABLE 3.1
ISO/IEC Basic Attributes with EDRN CDE Anatomical Site Example

Attribute Category	Name of Data Element Attribute	EDRN Value	Obligation
Identifying	Name	SPECIMEN_ TISSUE_ ANATOMIC-SITE_CODE	Mandatory
	Identifier	N/A	Conditional
	Version	1.0	Conditional
	Registration authority	N/A	Conditional
	Synonymous name	N/A	Optional
	Context	EDRN	Conditional
Definitional	Definition	Anatomical site	Mandatory
Representational	Datatype of data element values	Integer	Mandatory
	Maximum size of data element values	N/A	Mandatory
	Minimum size of data element values	N/A	Mandatory
	Permissible data element values	1 Bladder 2 Bladder 3 Bowel 4 Corpus 5 Cervix … 25 Lymph node 97 Other, specify	Mandatory
Administrative	Comments	N/A	Optional

data elements and between data elements and classification schemes, data element concepts, objects, or entities. The *representational* category captures representational aspects of data elements such as the list of permissible data values and their type. Finally, the *administrative* category provides management and control information.

The "Obligation" column in the table designates whether an attribute is mandatory (always required), conditional (required under certain conditions), or optional (simply allowed).

The application of a general specification such as ISO/IEC 11179 to a specific domain requires specialization. For example, the attribute "datatype of data element values" is defined as "A set of distinct values for representing the data element value." Best practices would suggest that it be constrained by adopting a specific standard that provides an enumeration of data types.

3.2.2 Creating Standard Data Elements

The Dublin Core (DC) initiative developed a set of common data elements for the description of any electronic resource on the web. Data dictionaries across domains should, at some level, have some uniformity in terms of their data elements. This may occur at several levels. This could be at the domain level (i.e., a standard set of data elements for describing cancer biospecimens) or it may be at a more general level such as the Dublin Core initiative. Nevertheless, interoperability within a geographically distributed environment hinges on the ability for distributed resources to relate through the use of common semantics.

The DC initiative specifically addresses commonality in describing data resources and recommends the following list of 15 data elements as presented in Table 3.2.

TABLE 3.2
Dublin Core Data Elements

Dublin Core Element	Definition
Title	A name given to the resource
Creator	An entity primarily responsible for making the content of the resource
Subject and keywords	The topic of the content of the resource
Description	An account of the content of the resource
Publisher	An entity responsible for making the resource available
Contributor	An entity responsible for making contributions to the content of the resource
Date	A date associated with an event in the life cycle of the resource
Resource type	The nature or genre of the content of the resource
Format	The physical or digital manifestation of the resource
Resource identifier	An unambiguous reference to the resource within a given context
Source	A reference to a resource from which the present resource is derived
Language	A language of the intellectual content of the resource
Relation	A reference to a related resource
Coverage	The extent or scope of the content of the resource
Rights management	Information about rights held in and over the resource

It is also important to point out that the Dublin Core data elements can be described using the ISO/IEC 11179 standard. In fact, Table 3.2 uses two key attributes of the ISO/IEC 11179, "Name" and "Definition," to describe the Dublin Core data elements. Several other attributes of ISO/IEC 11179 have not been included, but the Dublin Core Metadata Initiative has completed the full mapping.

3.2.3 METADATA REGISTRIES

A *metadata registry* captures domain vocabularies, the data model, and ontological relationships within a discipline. EDRN developed a web-based metadata registry system to capture and harmonize disparate metadata models used by existing informatics systems running at participating research institutions. An EDRN mapping tool was created to work in conjunction with the metadata registry system; this utility enables both capture of existing definitions and mapping of local definitions to a common EDRN meta model. Each of the participating institutions within the knowledge system defines the translation from their local data models to the general knowledge system model in order to provide semantic consistency across the system. Attributes of the data element including permissible values, units, format, and data type are captured and mapped to one another in order to provide the mapping at the informatics level. This enables the distributed software infrastructure to run a mediation function as part of the process of querying and retrieving data from the distributed EDRN institutions in order to convert local data elements and values to EDRN data elements and values for consistency. In addition, having an online data dictionary as part of a metadata registry enables one to validate the data elements. Validation of data is particularly important when running and capturing results from various studies. Validating the metadata increases the reliability of the results by ensuring compliance in terms of permissible values.

3.2.4 DATA ARCHITECTURE APPROACHES FOR SUPPORTING RESOURCE DISCOVERY

The Dublin Core descriptors are limited in number and are by definition quite general. Science domains, as mentioned earlier, have been developing common data elements for describing data products. While Dublin Core provides a good general foundation, it is often not useful for searching and locating large data repositories given its general descriptors. The use of domain data dictionaries and models is extremely important to sufficiently describe data resources. A data resource may be an image of a genome, a description of a specimen, or a data system itself.

The Object Oriented Data Technology (OODT) software framework defines a common schema for describing data resources, called a profile. A schema for the profile, described using the Extensible Markup Language (XML),[9] is provided in Figure 3.1, and has three groups of descriptors, profile descriptors, resource descriptors, and descriptors from the domain-controlled vocabulary, called the profile elements. The first section, the profile descriptors, simply describes the profile itself and contains system level attributes such as profile identifier, type, and status that

```
<!ELEMENT profiles (profile*)>
<!ELEMENT profile (profAttributes, resAttributes, profElement*)>
  <!ELEMENT profAttributes (profId, profVersion?, profType,
    profStatusId, profSecurityType?, profParentId?, profChildId*,
    profRegAuthority?, profRevisionNote*, profDataDictId?)>
<!ELEMENT resAttributes (Identifier, Title?, Format*,
    Description?, Creator*, Subject*, Publisher*, Contributor*,
    Date*, Type*, Source*, Language*, Relation*, Coverage*,
    Rights*, resContext+, resAggregation?, resClass,
    resLocation*)>
<!ELEMENT profElement (elemId?, elemName, elemDesc?, elemType?,
    elemUnit?, elemEnumFlag,
    (elemValue* | (elemMinValue, elemMaxValue)),
    elemSynonym*, elemObligation?, elemMaxOccurrence?,
    elemComment?)>
```

FIGURE 3.1 XML profile structure.

are useful for the management of a profile. The second section, the resource descriptors, generically describes the resource using the Dublin Core element set along with some additional descriptors to describe the way in which the location fits into a science context along with the location and method for accessing the resource. Finally, the profile element section provides domain-specific descriptors for the resource. These descriptors are constructed from the data elements described within a domain data dictionary.

The following fragment defined using XML illustrates an example of a common data element used by the EDRN supported by the ISO/IEC 11179 attributes:

```
<dataElement>

  <name>ANATOMIC_SITE</name>

  <version>1.0</version>


  <registration_authority>NCI.EDRN</registration_autho
  rity>

  <definition>Anatomical site</definition>

  <dataType>Integer</dataType>

  <unit>Integer</unit>

</dataElement>
```

Permissible values for the data element "ANATOMIC_SITE" could include "bladder (1)", "Lymph node (25)", etc. One of the key benefits of establishing a data dictionary is that it provides a mechanism to validate data objects against the data dictionary. In the above example, the value "334" might not be in the valid set of permissible values.

A profile is an instance of a set of data elements that points to a particular data resource. For example, one might create a profile of a specimen bank located at a research institution in California that includes specimens for bladder cancer. One possible data element used within the profile is "ANATOMIC_SITE" as described above. A profile would describe the data system and provide the mechanism for accessing the information by specifying the "RESOURCE_LOCATION" attribute as described below. A query against this profile would match for those investigators that are interested in finding and sending queries to specimen databases that contain bladder cancer.

```
<profile>
  <profAttributes>
  <profId>1.3.6.1.4.1.1306.2.104.10018791</profId>
  <profVersion>1.0</profVersion>
  <profType>profile</profType>
  </profAttributes>
   <resAttributes>
      <Identifier>Specimen Bank -
A123456</Identifier>
  <Title>California Research Institute Specimen
Bank</Title>
  <Description>null</Description>
  <resContext>NIH.NCI.EARLY-DETECTION</resContext>
  <resClass>dataSystem</resClass>

  <resLocation>urn:oodt:rmi:California.specimen.bank</
resLocation>
  </resAttributes>
  <profElement>
      <elemName>ANATOMIC_SITE</elemName>
  <elemValue>BLADDER</elemValue>
  </profElement>
</profile>
```

The resolution of a query within a distributed environment is a two-phase process. The first phase provides the resource discovery and finds metadata about the resource. Resource discovery occurs within our design as the result of searching distributed profile catalogs that identify the existence of a resource of interest such as a data system, a data product (like an image), or a Web site. The second phase retrieves the resource itself from a remote data system. More details on the software infrastructure that implements this process will be described in a later section.

3.3 A TECHNICAL ARCHITECTURE FOR DISTRIBUTED INFORMATION MANAGEMENT

The Object Oriented Data Technology (OODT) framework that implements the data architecture described in Section 3.2.4 consists of a set of cooperating, distributed peer software components. The major components of the OODT framework implement a metadata (profile) and data (product) model using profile servers and product servers. In addition, a query service directs queries by traversing a network of connected profile and product servers, providing the veneer of a peer-to-peer network. The distributed services provide for the location and description of resources (profile queries) and retrieval of resources (product queries) leveraging the profile metadata model.

3.3.1 DISTRIBUTED FRAMEWORK COMMUNICATION

OODT is a distributed system wherein components may be dispersed geographically across a standard TCP/IP network, such as the Internet. Connectivity between components utilizes a standards-based distributed systems implementation such as Java Remote Method Invocation (RMI) or the Internet Inter-ORB Protocol (IIOP) for CORBA-based communication.

The OODT components support plugins that extend the implementation by performing the work of querying both the metadata catalogs and the data repositories themselves. In this way, the OODT software is a framework.[10] Frameworks are different from traditional libraries in that application programmers extend and implement prescribed software objects and interfaces that directly integrate into the framework. This is in contrast to normal software implementation efforts that may not specify ubiquitous interfaces. More work necessarily falls upon the developers of the framework to support the prescriptive interfaces.

OODT's framework provides three major components:

Profile servers serve scientific metadata and can tell whether a particular resource can provide an answer to a query.
Product servers serve data products in a system-independent format.
Query servers accept profile and product queries and traverse the network of profile and product servers, collecting results. It is possible to access the query service through direct interfaces with the distributed computing interfaces (such as RMI and CORBA invocations), or through an HTTP interface.

The query server is the starting point for all end-user activity with the system. Investigators run a profile query on the query service to determine the profile(s) that describe the resource and its location. For resources that are product servers, researchers can make product queries on the query service to retrieve data. Figure 3.2 shows the typical deployment of the OODT systems in a network.

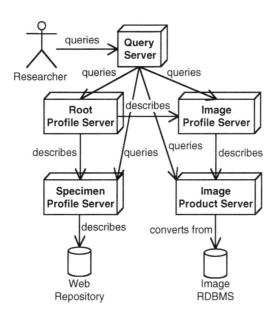

FIGURE 3.2 Typical OODT deployment.

In this example deployment, one system runs a query server and a root profile server. The root profile server contains profiles that describe two other profile servers, the specimen profile server and the image profile server. The image profile server contains profiles that describe resources within a product server that serves images; this product server retrieves resources from a database. The specimen profile server contains profiles that describe specimen databases accessible within the network.

Each profile and product server uses a set of customizable backend implementations in order to process queries. For example, a profile server may retrieve profile metadata from an XML database, a comma-separated values file, a document catalog, a relational database, and so forth. Similarly, a product server could retrieve images stored in a proprietary format as a binary (unstructured) data object in a relational database and return them in standard PNG format. Both profile and product servers specify the interfaces for interchangeable backends; by creating implementations that conform to those interfaces, the framework can enable different and user-customizable behavior that integrate into an overall system. These interfaces are specified using Java's interface mechanism; backend classes implement the profile and/or product interfaces. At run time, profile and product servers consult the system properties to determine which backend classes to load, instantiate, and install as the backend implementations.

In order to handle queries, both profile and product servers manipulate an identical query structure called the XMLQuery. (It's called such only because it can be expressed in XML.) The XMLQuery structure was defined as a neutral approach to define a general query of distributed resources for data grid environments. Represented as both an XML document and as an object of a Java class, the XMLQuery

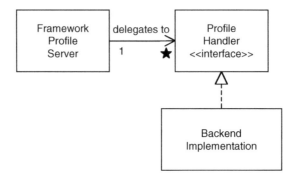

FIGURE 3.3 Delegation architecture of a profile server.

encapsulates the user's query in a query-language–independent fashion along with any results retrieved so far.

3.3.2 PROFILE SERVERS

As described in Section 3.3.1, profiles are metadata descriptions of resources; that is, they "profile" a resource by describing its inception and composition using the common data elements of the data architecture. Profile servers enable discovery of resources by providing the ability to search resource collections. Profile servers answer the question "Where can I go to find out about *X*?"

All profile operations that arrive at a profile server are handled by the server's configured backend implementation. Figure 3.3 demonstrates the implementation of a profile server by identifying a standard profile handler interface along with its implementation.

The profile server is a Java interface; therefore, backend implementations are Java classes that implement the interface. Zero or more backends may be present in a profile server (although a server with zero backends does not make sense since there is no implementation class for the profile server). In general, two kinds of backends can be developed:

Static backends serve profiles that exist statically. These are profiles that typically describe long-lived resources that do not change; thus, the profiles themselves rarely need change, except for occasional corrections to errors. Such a backend generates the profiles using a static set of information or refers to complete profiles in long-term storage, such as profiles serialized as Java objects or XML documents.

Dynamic profile servers create profiles on-the-fly in response to profile queries. Such profiles may describe ephemeral resources or long-lived resources for which having static profiles would be onerous. For example, having profiles for millions of different datasets that vary in only small ways would require too much disk space or memory to maintain; a dynamic profile server can synthesize profiles for such resources based on one copy of the nonchanging information.

A profile server's primary responsibility is to provide a way to run a query against the server's set of profiles. Although users may access a profile server directly via its remote interface, it is far more common for queries to enter the system through the query server, which directs them transparently to and along graphs of appropriate profile servers (when a profile describes another profile server).

Upon receiving a query, the profile server's backend interprets the XMLQuery passed in a way appropriate to the implementation. For example, a backend that stores information in a relational database may convert parts of the XMLQuery into an SQL query. For each matching profile, the backend constructs a list of matching profiles and returns them.

Consider an example of a profile server managing a set of profiles for a specimen collection. Each specimen includes information about its kind, how it is stored, and the kind of person from whom it came. A query for resources that included a certain specimen type could be handled by this server, which would analyze the query in order to return profiles that contained the desired specimen. More complex queries are possible; for example, a query for a specific kind of specimen, stored in a specific way, and from a specific range of demographic characteristics of individuals may result in a smaller set of matching profiles.

3.3.3 Product Servers

Product servers exist to provide a way to retrieve specific data products. Product servers accept the same XMLQuery structure as profile servers, but instead of returning a list of matching profiles, they add matching products to the XMLQuery object and return it. Data products in this sense can be individual data granules, datasets, or collections of datasets, depending on the backend implementation in the product server and the way it handles queries and results.

As with profile servers, product servers can be configured to handle requests by zero or more specific backends that respond to the actual query. Figure 3.4 shows the class architecture.

The backend interface is called a QueryHandler and is deployed as a Java interface, making backend implementations concrete classes that implement the

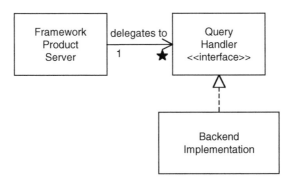

FIGURE 3.4 Delegation architecture of product server.

interface. The QueryHandler interface provides a query method like the ProfileHandler interface; however, instead of returning a list of matching profiles, the method returns the XMLQuery passed in. It is the job of the query method to resolve the queries with any matching results. Query handlers may also perform some processing on the data prior to returning it. This enables custom product servers to be configured and installed for different science domains.

When constructing a query, the user may indicate preferred MIME[11] types.* For example, a user wanting PNG images may list image/png as the only acceptable MIME type. A user preferring PNG images but willing to have JPEG images would list image/png, image/jpeg in that order. A user preferring PNG images but willing to accept any image type would list image/png, image/*. If the user doesn't specify a MIME type when creating the XMLQuery object, the software generates a default list of acceptable MIME types, namely */*, meaning that any type is acceptable. Sophisticated product servers can convert between data types. One mechanism for handling interoperability of legacy data systems is to deploy product servers that convert between file formats that are native to the local data system and the common data formats supported by the larger data grid system.

The XMLQuery's result section indicates the MIME type of each result. For example, a product server that generates image maps from contour data may return products in the image/jpeg format. A product server that retrieves tabular data from a database may return products in the text/tab-separated-values format.

3.3.4 QUERY SERVERS

Query servers manage queries across distributed resources and are the point of entry into an OODT framework installation. Query servers contain the algorithms necessary to traverse the logical P2P model, executing queries at appropriate servers and gathering results. Query servers also simplify the interaction with the user, who is freed from the knowledge of accessing the remote interfaces of profile servers and product servers. Users instead call upon a query server for all profile and product interaction.

The OODT implementation supports several different interfaces to the query service to ensure that it supports both cross-platform and cross-language interoperability. This includes not only interfaces for programming languages such as Java, but interfaces using the web standard http. The http interfaces include

Generic Query Interface: This interface requires that the caller be able to construct and parse XML documents. It provides full access to a Query Server for the retrieval of metadata (profiles) and data (products). To use this interface, the caller constructs an XMLQuery and sends it to the server as an XML string, along with the search type (profile or product) and optionally the name of a specific server to receive the query. The return value is an XML document (MIME type text/XML) with any results.

Product Query Interface: This simpler interface enables access to products, only profiles cannot be retrieved. To use this interface, the caller sends in

---

* A MIME type indicates the data format of a data object.

a keyword query string and an optional list of desired MIME types. The interface constructs a corresponding XMLQuery object and executes the query. It returns the first matching product in its MIME type as a result. Using this interface it is trivial to create web pages where complex data products are retrieved through distributed product services running across the country.

End-users usually will not have foreknowledge of the profile and product servers that exist and the network names/addresses by which they are known and accessed. The OODT framework is typically bootstrapped in such a way that there is a root profile server that contains metadata descriptions of other profile servers. These other profile servers may in turn describe yet more profile servers and/or product servers and other resources.

Using this network architecture, a user just sends in a profile query to the default profile server. The query server will comb the network to gather results and return a list of known product servers that can provide the sought product. The user can then submit a second query (a product query) targeted at a specific product server to retrieve the sought data. Figure 3.5 demonstrates this process.

3.3.5 SOFTWARE DEPLOYMENT

One of the most difficult challenges encountered in managing a large grid of cooperating applications is deployment. Deploying the framework requires a network of computers capable of running the Java virtual machine on which the software is installed. To make deployment more manageable, the software, in addition to running the service, includes self-monitoring components that enable the various servers to be restarted in the event of failure and/or to notify a system administrator. It also

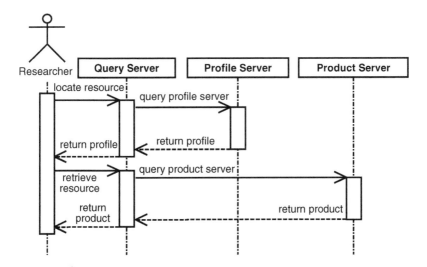

FIGURE 3.5 Typical interaction with the OODT framework.

enables patching and updating of software components and provides a limited degree of remote debugging of servers.

The component that manages all this is called the Server Manager. When installing the OODT framework, the Server Manager is included. A Windows installer automatically sets up the Server Manager at an OODT site. A slightly more complex set of steps is necessary to install the software on Unix and MacOSX.

The Server Manager's primary job is to manage the processes that comprise the OODT framework. In general, however, it can manage any kind of server process and is extensible to support additional kinds. The secondary responsibility of the Server Manager is to allow remote manipulation of server processes: defining, creating, stopping, starting, and diagnosing of processes. Additionally, it allows access (with only permissions that it itself has) to the file system of the system on which it runs so that one can patch and upgrade software installations.

The Server Manager arranges to capture the output of every program it runs and makes that output available on demand. This feature enables debugging of remote processes. Using the graphic management application, a developer can retrieve buffered output from a process for review. Typically, a developer may then patch the program, upload it using the Server Manager, restart the process, and retrieve the output again, repeating until the bug is resolved.

Using the ServerManager, experts from both JPL and the Fred Hutchinson Cancer Research Center remotely support and configure the knowledge system across the nation.

3.4 KNOWLEDGE SYSTEM APPLICATIONS

Two knowledge system applications have been developed to support the EDRN. The first is a virtual specimen repository, called ERNE,[3] and the second is an infrastructure for capturing and operating validation studies, called VSIMS.

3.4.1 EDRN RESOURCE NETWORK EXCHANGE

ERNE, the EDRN Resource Network Exchange, was developed to enable investigators to easily identify the availability of biospecimens and associated epidemiological information needed for their research. Given the heterogeneous and distributed nature of the EDRN, it was important to develop ERNE on top of the EDRN informatics infrastructure. This enabled distributed search and retrieval of specimens managed in databases across the country.[12] ERNE's specific goal is to provide transparent access to existing specimen repositories providing EDRN a virtual knowledge environment despite the distributed nature of the collaboration. Figure 3.6 shows the ERNE deployment as of July 2003.[13]

The EDRN developed and funded deployment of ERNE to specific sites within EDRN. The deployment required that each site install a product server (as described in Section 3.3.3). The product server was configured such that the "backend" was connected to the existing specimen repositories and data systems running at the EDRN research institutions. The product servers performed a mediation function that translated the EDRN CDE-based query into a query that was supported by the local

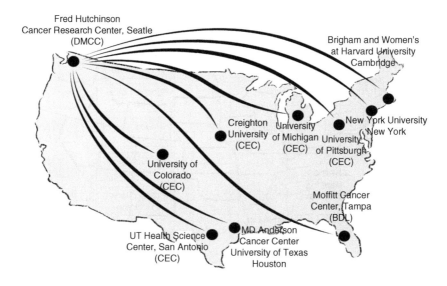

FIGURE 3.6 ERNE distributed deployment.

institution's database. This allowed EDRN to plug together existing specimen databases that describe the contents of participating institutional specimen banks. In addition to deploying the technology, each site received approval from their Institutional Review Boards (IRB) allowing for the sharing of data external to the organization.

ERNE provides scientists with a common portal that is used to initiate queries into the distributed infrastructure. The portal sends queries into the EDRN informatics infrastructure first by determining the databases that can handle the query and second by sending the query to only those databases that manage specimens of interest. This leverages the informatics infrastructure described in Section 3.2 and Section 3.3. The EDRN portal was developed using Java Server Pages (JSP), which connects the informatics middleware to the distributed repositories (see Figure 3.7). This portal allows scientists to specify search criteria using the project's specimen and epidemiological CDEs described in the EDRN data architecture. The portal is dynamic and will limit choices that appear based on specified user selections. When a user logs in they will see all sites and their operational status. If a site's server is down the user cannot select this site as part of the query. The portal also narrows choices based upon sites selected to show only the specimens that are available at those sites. The software summarizes results, showing the numbers of queried specimens available at each institution. Users can obtain more details regarding the specimen and characteristics of the donor by clicking on the "details" link adjacent to the site of interest. Data that identify the specimen donor are not available in any part of the results screens.

ERNE is a success in showing first the importance of establishing a data architecture, and second the importance of establishing predictable interfaces to distributed repositories via middleware. ERNE can be easily scaled to integrate new sites as the EDRN informatics infrastructure expands to new institutions.

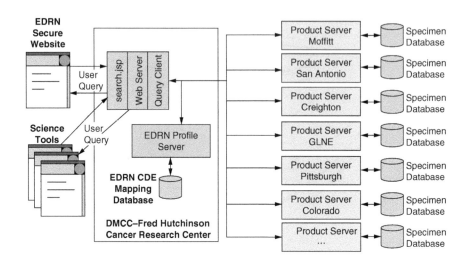

FIGURE 3.7 **(Color insert follows page 204)** ERNE scalable system architecture.

3.4.2 EDRN VALIDATION STUDIES INFORMATION MANAGEMENT SYSTEM

One of the critical components of a knowledge system is its ability to capture data as part of the science data processing and analysis infrastructure. Within EDRN, this occurs as part of the process to identify and validate cancer biomarkers. With this in mind EDRN has designed a secure, Web-based system that includes the main components needed for capturing and preserving the necessary metadata and data objects that integrate into the overall knowledge system architecture. The major components of the system include protocol management tools, communication tools, a data collection and processing system, and a specimen tracking system. All are based on having a robust data architecture as described in Section 3.2. Information maintained in the system is secure and stored separately for each multisite study, allowing multiple protocols to be coordinated centrally through the same data management system. The system, named the Validation Studies Information Management System (VSIMS), is therefore a major component of the EDRN knowledge system.

VSIMS is data driven, enabling adaptability to various validation study requirements. Each multisite study or protocol has its own procedures defined in the system that allow components of VSIMS that need to be modified to meet that protocol's specifications. This allows for protocols to be implemented very quickly and modified easily; a central data management and coordinating center can handle several protocols conducted simultaneously or consecutively.

There are multiple levels of security that make VSIMS a secure system. VSIMS uses 128-bit encryption* for all data transfers and requires that all users of VSIMS

---

* 128-bit encryption is the de facto standard for data encryption over the public Internet.

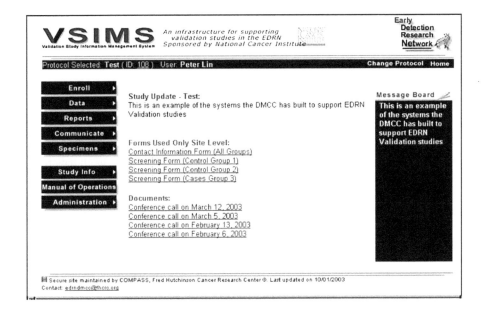

FIGURE 3.8 VSIMS Web portal.

be authenticated prior to entry into the unified portal. Each VSIMS user is required to complete an access application in order to be provided a user account. In addition, user accounts are assigned protocol-specific access that identifies the protocol and associated permissions. Additional security measures include auditing, connection time-outs, and deactivation of inactive accounts.

The VSIMS Web portal, shown in Figure 3.8, dynamically configures itself based on the user's permissions, displaying only those protocols and capabilities provided to the user. Each protocol has its own unique data-driven home page. The protocol home page consists of a vertical list of buttons representing the major system components on the left-hand side of the screen. The study update section is a place to post important study information; the documents section has links to various study documents, minutes from conference calls, or other documents deemed important and allows users to have easy access to study communications. A message board provides space for additional messages to be posted.

EDRN developed a metadata-driven forms entry system that takes advantage of the EDRN Common Data Element (CDE) metadata repository discussed in Section 3.2. The metadata repository includes attributes such as data element name, wording of question, definition, data type, permissible value list, form instruction, etc. The metadata repository enables uniformity in the collection of data (including common data elements and valid values) across multiple studies. Study-specific forms are created and the appropriate CDEs are linked to the forms. The system automatically inserts the name of the data entry person, time and date when data are entered or modified and has a required field to describe the reason for any data changes, thus providing an automated data audit trail. The system also links the following study-based criteria to a specific VSIMS defined protocol: skip patterns; order of questions; choice of the display of the valid values of a question as check boxes or pull-down lists; deletion, addition, and modification of questions and their valid values; range and logic checks; and optional double data entry verification. Use of data-driven forms enables flexibility and adaptability in collecting data for a variety of validation studies. It also allows online data entry forms to be created very quickly.

In addition, EDRN is building a secure data transfer and processing infrastructure that allows data collected at remote locations to be ingested into VSIMS using a secure communications infrastructure. The infrastructure will catalog all data transfers using the EDRN CDEs. Use of the CDEs by EDRN sites helps assure that a consistent catalog can be created that can be later integrated into the overall knowledge environment enabling scientists to locate and access data captured during the validation study. Assay results are also ingested into VSIMS using the secure data transfer and processing infrastructure.

Communication tracking tools have been developed using an open-source product called Scarab. Scarab, developed by Tigris.org, is a software tool geared toward tracking of software development–related issues, such as defect, enhancement, requirement, etc. EDRN modified Scarab to track questions to and from the coordinating center along with data clarifications. For example, a specific question about a procedure in the manual of operations could be sent electronically through this system from a study site and routed to the appropriate person at the coordinating center. This question would then be answered and tracked appropriately, potentially being routed to other sites to communicate consistently about procedures to all sites. Data clarification tracking allows the coordinating center to track data clarifications within the database and document this information. One example could be to clarify a data collection question. The coordinating center would send a question to a study site that would be tracked in the system and would then be cross-referenced in the data clarification tracking module.

Finally, VSIMS includes a specimen tracking system to track the shipment and receipt of specimens between sites for validation studies. It integrates with the EDRN Resource Network Exchange application, as discussed in Section 3.4.1 by providing distributed interfaces to its online specimen catalog. The system utilizes barcoding technology. Each specimen is labeled at the site that collects the specimen with a barcode label with a unique specimen identification number. Shipping and receiving sites then just scan the specimen containers and the system automatically enters the date, time, specimen ID, participant ID, shipping location, and receiving site into the database, minimizing error due to manual data entry.

3.5 CONCLUSIONS

A key to correlating, locating, and analyzing disparate information is creating the architecture that allows for the capture and organization of the data. The approach presented enables the separation of the data and technology architecture layers. This allows for the evolution of these two architectural approaches to occur independently. By focusing on generic methods for describing information resources, this design is able to provide a software framework for information capture and retrieval across many different science domains.

Our experience has shown that development of the data architecture can be very difficult. It requires pairing both scientists or domain experts and computer scientists together in order to develop the domain models that describe the data architecture. Creating the model allows for interoperability if disparate data sources can provide the mapping between locally implemented data systems and the domain model.

The generic application of both the data architecture and the software framework is evident in that the identical system is deployed to NASA's Planetary Data System (PDS) for distribution of data from NASA's planetary missions.[14] Engineers readily translated PDS's metadata model into OODT's profile format and developed profile and product servers that provided planetary data from institutions across the country through an easy-to-use portal.

Virtual clustering of scientific information, leveraging metadata, and implementing through distributed services makes automatic correlation and unified views like these presented possible. While science domains differ, the technical approach to implementing distributed clustering and information retrieval systems can follow a similar development path. As collaborative science research continues to increase, the demand for well-designed information grids will continue. The EDRN knowledge system has shown that a well-designed data architecture is key to enabling interoperability and the creation of a virtual knowledge system. It has also proven that informatics is a critical component for collaborative research networks. As the discovery process progresses, it is clear that informatics will continue to play an even greater role. The EDRN knowledge system will be a core tool that enables scientists to seamlessly access, share, and correlate data as new approaches and paradigms to the discovery process are defined. It provides an infrastructure not only for EDRN, but major collaborative research projects across the National Cancer Institute, providing data to cancer researchers at all stages.

ACKNOWLEDGMENTS

The authors thank the following partner institutions for their involvement in the development of the EDRN informatics infrastructure: H. Lee Moffit Cancer Research Center, University of Texas at San Antonio, Creighton University, University of Colorado, University of Pittsburgh, Dartmouth Medical School, Brigham and Women's Hospital, University of Texas MD Anderson, and New York University.

A portion of this work was performed at the Jet Propulsion Laboratory, California Institute of Technology, which is under contract to the National Aeronautics and Space Administration.

REFERENCES

1. Srivastava, S., Kagan, J., and Verma, M. Early Detection Research Network, 2nd Annual Report, National Cancer Institute, 2002.
2. Kesselman, C., Foster, I., and Tuecke, S. The anatomy of the grid: Enabling scalable virtual organizations. *Inter. J. Supercomput. Appl.,* 15(3), 200–2002, 2001.

3. Crichton, D., Downing, G., Hughes, J., Kincaid, H., and Srivastava, S. An Interoperable Data Architecture for Data Exchange in a Biomedical Research Network. 14th IEEE Conference on Computer Based Medical Systems. Bethesda, MD, July 2001.

4. Crichton, D., Hughes, J., Hyon, J., and Kelly, S. Science Search and Retrieval Using XML. 2nd National Conference of Scientific and Technical Data. Washington, DC, March 2000.

5. Berners-Lee, T., Hendler, J., and Lassila, O. The semantic web. *Sci. Am.*, May 2001.

6. International Organization for Standardization and International Electrotechnical Commission Joint Technical Committee 1, Subcommittee 32. Information Technology — Metadata Registries — Part 1: Framework. Geneva, CH, January 10, 2002.

7. The Dublin Core Metadata Initiative. The Dublin Core Element Set Version 1.1. Dublin Core Metadata Initiative, Dublin, 1999.

8. Winget, M., Baron, J., Spitz, M., Brenner, D., Warzel, D., Kincaid, H., Thornquist, M., and Feng, Z. Development of common data elements: the experience of and recommendations from the early detection research network. *Int. J. Med. Informat.*, 70, 41–48, 2003.

9. Bray, T., Paoli, J., Sperberg-McQueen, C.M., Maler, E., and Vergeau, F. Extensible Markup Language (XML) 1.0 (Second Edition). World Wide Web Consortium, Cambridge, 2000.

10. Gamma, E., Helm, R., Johnson, R., and Vlissides, J. *Design Patterns: Elements of Reusable Object-Oriented Software*. Addison-Wesley, Reading, MA, 1995.

11. Freed, N. and Borenstein, N. *Multipurpose Internet Mail Extensions (MIME) Part One: Format of Internet Message Bodies (RFC2045)*. The Internet Society, Reston, VA, 1996.

12. Tenenbaum, D. Serving up specimens: NASA-NCI project links databases across the country. *J. Natl. Canc. Inst.*, 95(3), 186–187, 2003.

13. Kincaid, H., Crichton, D., Winget, M., Kelly, S., Johnsey, D., Srivastava, S., and Thornquist, M. A National Virtual Specimen Database for Early Cancer Detection. 16th IEEE Conference on Computer-Based Medical Systems, New York, NY, July 2003.

14. Kelly, S., Crichton, D., and Hughes, J. Deploying Object Oriented Data Technology to the Planetary Data System. 34rd Lunar and Planetary Science Conference, 1607, League City, TX, 2003.

4 Public Protein Databases and Interfaces

Jane Meejung Chang Oh

CONTENTS

4.1 INTRODUCTION

4.1.1 THE EVERGROWING WEB OF PROTEOMICS DATA

Over the past several years, a lot of large-scale public databases have been success-fully established—notably in the field of genomics and proteomics — that laid the foundation for the evergrowing web of proteomics data. Today, the amount of proteomics data that is becoming available is growing exponentially and the role of the database has become more significant than ever. Multiple forces drive the explosive growth of proteomics data—not only by advances in proteomics technology, but by innovative computational tools and methods as well.

The advancement of high-throughput proteomics technologies provides a wealth of information about each of the thousands of proteins encoded by a genome. This includes detailed experimental results from

1. Sequencing techniques for experimentally deriving protein/peptide sequences
2. Global (yeast) two-hybrid and protein microarray techniques for determining their localization, modifications, interactions, and activities, as well as their functions
3. Two-dimensional gel electrophoresis (2-D PAGE) and mass spectrometry (MS) approaches for separating and identifying novel proteins and their characterizations
4. X-ray crystallography and nuclear magnetic resonance (NMR) spectroscopy techniques for experimentally deriving high-resolution, three-dimensional (3-D) structures of proteins

Innovative computational tools and methods are utilized to process, analyze, and interpret prodigious amounts of data and to determine protein functions as new proteins are revealed. This includes computational results derived from the automated translation of coding DNA sequences (CDS), the use of sequence alignment tools, and the prediction of homology-based protein domains, structures, and functions.

Thus, information technologies, including Internet-based database management systems and intelligent computational software, are playing a vital and expanding role in proteomics. The blending of proteomics technologies and information technologies is impacting all life science communities and is essential for the ongoing evolution of biomedical research.

4.1.2 PUBLIC PROTEIN DATABASES

Today there is a considerable need to develop comprehensive, systematic mechanisms to analyze the vast amount of proteins that orchestrate various cellular functions and to identify proteins that are associated with disease or affected by pharmacological agents. Genomics and proteomics research has produced a large amount of data pertaining to gene expression at the RNA and protein levels.

Thus, many protein databases have been developed to serve as electronic data repositories for automatically processing and managing proteomics data from different studies, with the support behind the concept of a centralized repository for protein data. Consequently, these protein databases help scientists make intelligent decisions regarding a specific protein, using the information that is publicly and freely available.

4.1.3 WWW AND FTP PROTEOMICS SERVERS

The most compelling informatics driver is the Internet because it makes so much proteomics information available in the public domain. The explosive growth of

Internet users and the number of public protein databases is revolutionizing the way proteomics research is conducted and researchers work.

WWW servers are logical groups of interrelated or interacting elements on the Internet forming a unified whole. By using WWW proteomics servers (e.g., http:// www.expasy.org), users can dynamically obtain the most up-to-date data with minimal knowledge of the structure of databases. This way users do not have to cope with the usual database problems, such as (1) storing large amounts of data, (2) daily updates, and (3) software upgrades. However, net access can be very slow at peak hours due to the limited Internet speed and number of concurrent users that can access the database.

There are many FTP servers (e.g., ftp://ftp.ncbi.nih.gov/repository/) available in the public domain that enable users to anonymously download the databases as well as the computational tools (e.g., protein sequence similarity/homology search, pattern matching programs, etc.) and to search the databases that run on a user's local server. A disadvantage of this is that scientists may get outdated data if the local database is not updated as frequently as the public databases.

WWW and FTP servers on the Internet make public proteomics databases more valuable and useful to researchers by providing a means to quickly extract only the information needed to answer a specific biological question and to easily navigate multiple forms, reports, and files through a simple mouse click.

4.2 CATALOG OF PUBLIC PROTEIN DATABASES

4.2.1 ONLINE CATALOG OF DATABASES

The exponentially growing amount of proteomics information is publicly available online, one just needs to know where to find it. Thus, there have been many attempts to create a comprehensive directory of software and databases in the field of molecular biology, genomics, and proteomics. On many Web pages, an inexhaustive list of databases is provided, along with descriptions and hyperlinks to public molecular biology databases and biology-related Web resources. The following is a list of some widely used Web pages, which are kept constantly up-to-date:

1. The first issue of each year of *Nucleic Acids Research* (http://nar.oupjournals.org/archive/index.dtl) has been a database issue since 1996
2. The Amos' WWW links page (http://www.expasy.org/alinks.html) for a directory of software and databases in the field of proteomics
3. The 123genomics Web page for a directory of genomics and proteomics databases (http://123genomics.homestead.com/files/databases.html)
4. The online publication of "Introduction to Molecular Biology Database" (http://www.ebi.ac.uk/swissprot/Publications/mbd1.html)

In addition to these Web pages, catalog databases were established so that users can access the catalog not only with hyperlinking capability, but search capability as well. Two such databases are (1) the Biocatalog at EBI (http://www.ebi.ac.uk/biocat/)

and (2) the DBCAT at Infobiogen (the French EMBnet node, http://www.infobiogen.fr/services/dbcat/).

The Biocatalog is a directory of general interest in the field of molecular biology and genetics. This is not only a directory of software for protein sequence analysis, protein structure analysis, structure prediction, and pattern identification, but also a catalog of databases and database searching tools. The Biocatalog is arranged in "domains" (e.g., DNA, proteins, genomes, genetic, databases, servers, etc.) and "subdomains." Users can access the catalog by clicking the hyperlinked text in the Web page or by searching, using the EBI SRS server (http://srs.ebi.ac.uk). Additionally, the catalog of databases and the full database comprised of ASCII (text) files can be downloaded from the EBI FTP server (ftp://ftp.ebi.ac.uk/pub/databases/).

The DBCAT is a public database for cataloging databases of DNA, RNA, protein, genomic, protein structure, etc. The release, as of April 9, 2003, of DBCAT has a total of 511 entries, including 94 protein-related databases and 18 protein structure–related databases. Users can download the catalog of a specific domain (e.g., "dbcat_PROT.txt" for a list of protein related databases) from the Infobiogen FTP server (ftp://ftp.infobiogen.fr/pub/db/dbcat/). In addition, users can search DBCAT online with a database name, author, or domain using the Infobiogen SRS (sequence retrieval system) server (http://www.infobiogen.fr/srs).

The next section provides an inexhaustive list of databases for protein sequences, domains, structures, expression, and characterization. The lists and their contents are extracted from the catalogs described here and also from the Web site of the database itself.

4.2.2 LIST OF PUBLIC PROTEIN DATABASES

Public protein databases have been generated by large-scale proteomics experiments, which address research topics that cannot be resolved by DNA analysis. This includes studies of the relative abundance of protein products, posttranslational modifications, subcellular localizations, molecular turnover, protein interactions, and protein functions. The creation of a comprehensive database of genes and gene products laid the foundation for studies of the expression levels and properties of thousands of proteins to go with the thousands of genes identified on genetic maps, therefore offering a global approach to the study of gene expression.

4.2.2.1 Protein Sequence Databases

Protein sequence databases contain the amino acid translations extracted from nucleotide sequence database records that are annotated with one or more coding DNA sequence (CDS) features and the experimental results reported in published literatures. This section provides an inexhaustive list of protein sequence databases including PIR, PRF, RefSeq, Swiss-Prot, and TrEMBL.

NAME: PIR, Protein Information Resource[1,2]
DESC: The PIR is a worldwide protein information resource that is composed of a number of databases and computational tools designed for the identification and analysis of protein sequences. The PIR protein sequence

database (PIR-PSD) contains information concerning all naturally occurring, wild-type proteins whose sequence is known. The PIR nonredundant protein sequence database (PIR-NREF) provides comprehensive, nonredundant data uniquely organized by homology and taxonomy. The WWW server provides keyword searching as well as sequence similarity searching against PIR-PSD and PIR-NREF.

GROUP: National Biomedical Research Foundation, Washington, DC, U.S.

EMAIL: pirmail@georgetown.edu

WWW: http://www-nbrf.georgetown.edu/

QUERY: http://www-nbrf.georgetown.edu/pirwww/search/textpsd.shtml and http://www-nbrf.georgetown.edu/pirwww/search/pirnref.shtml

FTP: ftp://ftp.pir.georgetown.edu/pir_databases/, ftp://ftp.infobiogen.fr/pub/db/pir, ftp://ftp.ebi.ac.uk/databases/pir/, and ftp://ftp.ncbi.nih.gov/repository/PIR/ to download the database.

NAME: PRF/SEQDB, Protein Research Foundation/SEQuence DataBase

DESC: The PRF protein sequence database contains amino acid sequences of peptides and proteins, and also sequences predicted from genes as well as manual annotations with regard to amino acids, peptides, and proteins.

GROUP: Protein Research Foundation, Osaka, Japan

EMAIL: isoyama@prf.or.jp

WWW: http://www.prf.or.jp/

QUERY: http://www.prf.or.jp/en/os.html for Amino Acid Sequence Database search by using short segments (limited to <20 amino acid sequences) as probe.

FTP: ftp://ftp.genome.ad.jp/pub/db/genomenet/ to download database

NAME: RefSeq, Reference Sequence Database[3]

DESC: The RefSeq contains nonredundant sets of sequences, including genomic DNA, transcript RNA, and protein products. The RefSeq NPs is a reference set of protein sequences and the RefSeq XPs is a reference set of *Homo sapiens* model proteins provided by the human genome annotation process.

GROUP: NCBI, National Center for Biotechnology Information, U.S.

EMAIL: info@ncbi.nlm.nih.gov?subject=RefSeq

WWW: http://www.ncbi.nih.gov/RefSeq/

QUERY: http://www.ncbi.nih.gov/Entrez/ for the Entrez-based database retrieval and http://www.ncbi.nih.gov/BLAST to search the database using BLAST algorithms.

FTP: ftp://ftp.ncbi.nih.gov/refseq/ to download the database

NAME: Swiss-Prot[4,5]

DESC: The Swiss-Prot is a curated protein sequence database that strives to provide a high level of annotation (such as the description of the function of a protein, its domains structure, posttranslational modifications, variants, etc.), a minimal level of redundancy, and a high level of integration with other databases.

GROUP: Swiss Institute of Bioinformatics (SIB), Geneva, Switzerland
EMAIL: swiss-prot@expasy.org
WWW: http://www.expasy.ch/sprot/sprot-top.html
QUERY: The WWW server (http://www.expasy.ch/cgi-bin/sprot-search-ful/,
 http://www.infobiogen.fr/srs/, and http://www.ebi.ac.uk/swissprot/) is available for keyword searching and sequence similarity searching. The
 Swiss-Shop (http://www.expasy.org/swiss-shop/) is available for an automatic
 database retrieval service against the noncumulative weekly additions of new
 protein sequences to the Swiss-Prot.
FTP: ftp://ftp.expasy.ch/databases/swiss-prot/, ftp://ftp.infobiogen.fr/pub/db/
 swissprot/, and ftp://ftp.ebi.ac.uk/pub/databases/swissprot/ to download the
 database.

NAME: TrEMBL, Translation from EMBL[6]
DESC: The TrEMBL is a protein sequence database from the EMBL nucleotide
 sequence translations. TrEMBL is split in two main sections: SP-TrEMBL
 and REM-TrEMBL. SP-TrEMBL (Swiss-Prot TrEMBL) contains the entries
 that should be incorporated into Swiss-Prot. REM-TrEMBL (REMaining
 TrEMBL) contains the entries that EMBL does not want to include in
 Swiss-Prot for a variety of reasons.
GROUP: EMBL Outstation - European Bioinformatics Institute (EBI), U.K.
EMAIL: swiss-prot@expasy.org
WWW: http://www.ebi.ac.uk/trembl/
QUERY: http://www.ebi.ac.uk/trembl/access.html and http://srs.embl-heidelberg.
 de:80000/
FTP: ftp://ftp.ebi.ac.uk/pub/databases/trembl/

4.2.2.2 Protein Family Databases

The rapid expansion of the nucleotide sequence databases has caused a massive
influx of data into the protein sequence databases and this has led to the same influx
of data into the protein family databases. This section provides an inexhaustive list
of protein domain, family, motif, and fingerprint databases, which were delineated
by the assessment of computational results derived from automatic classification of
protein sequences using sequence similarity/homology programs.

NAME: InterPro[7]
DESC: The InterPro is a database of protein families, domains, and functional
 sites in which identifiable features found in known proteins can be applied
 to unknown protein sequences. The InterPro contains high-quality annotations and cross references to other protein family databases including Pfam,
 PRINTS, ProDom, SMART, TIGRFAMs, and PROSITE as well as
 Swiss-Prot and TrEMBL. The InterPro database has almost 3,000 families
 classified by expert curators.
GROUP: EMBL Outstation - European Bioinformatics Institute (EBI), U.K.

EMAIL: interhelp@ebi.ac.uk

WWW: http://www.ebi.ac.uk/interpro/ for a service for biological sequence analysis

QUERY: http://www.ebi.ac.uk/interproscan/ for InterProScan sequence similarity search against InterPro.[8]

FTP: ftp://ftp.ebi.ac.uk/pub/databases/interpro/

NAME: iProClass[9,10]

DESC: The iProClass database is a nonredundant protein database organized according to family relationships as defined collectively by PROSITE patterns and PIR superfamilies. PROSITE patterns are defined as sequences (from Swiss-Prot) with the common function (http://pir.georgetown. edu/pirwww/search/pattern_help.html). PIR superfamilies are defined as sequences (from PIR protein sequence database) with the same function in various organisms (http://pir.georgetown.edu/iproclass/description.html).

GROUP: Georgetown University Hospital, Washington, DC, U.S.

EMAIL: pirmail@georgetown.edu

WWW: http://pir.georgetown.edu/iproclass

QUERY: http://pir.georgetown.edu/pirwww/search/searchseq.html

FTP: ftp://ftp.pir.georgetown.edu/pir_databases/iproclass/

NAME: Pfam, Protein Families[11]

DESC: The Pfam consists of two parts: Pfam-A and Pfam-B. The Pfam-A is a comprehensive collection of annotated protein domain families, including multiple sequence alignments and Hidden Markov Models (HMMs) covering many common protein domains. The Pfam-B is a supplement to the Pfam-A and contains a large number of small families automatically clustered from the ProDom database. The Pfam 8.0, which came out in February 2003, contains over 5,193 protein families.

GROUP: The Sanger Centre, Hinxton, U.K.

EMAIL: esr@sanger.ac.uk

WWW: http://www.sanger.ac.uk/Pfam/

QUERY: http://www.sanger.ac.uk/software/Pfam/

FTP: ftp://ftp.sanger.ac.uk/pub/databases/Pfam/

NAME: PIR-ALN[12]

DESC: The PIR-ALN is a database of protein sequence alignments. Alignments are of sequences in the same family (less than 55% different from each other), of sequences representing various families within a superfamily, or of sequence segments corresponding to the same homology domain in different proteins.

GROUP: National Biomedical Research Foundation, Washington, DC, U.S.

EMAIL: pirmail@nbrf.georgetown.edu

WWW: http://www-nbrf.georgetown.edu/pir/alndb.html

QUERY: http://www-nbrf.georgetown.edu/pirwww/search/searchseq.html

FTP:ftp://ftp.pir.georgetown.edu/pir_databases/other_databases/piraln/

NAME: PRINTS[13]

DESC: The PRINTS is a protein motif fingerprint database. Each protein family is represented by a fingerprint, which is a series of ungapped multiple alignments corresponding to the conserved motifs. The PRINTS obtains protein sequences from Swiss-Prot and TrEMBL databases.

GROUP: UMBER, University of Manchester Bioinformatics Education and Research, U.K.

EMAIL: attwood@bioinf.man.ac.uk

WWW: http://umber.sbs.man.ac.uk/dbbrowser/PRINTS/

QUERY: http://umber.sbs.man.ac.uk/dbbrowser/fingerPRINTScan/ for the PRINTS similarity search to find the closet matching PRINTS fingerprints by a user-specified protein sequence.[14]

FTP: ftp://ftp.bioinf.man.ac.uk/pub/prints/, ftp://ftp.ebi.ac.uk/pub/databases/ prints/, ftp://ftp.embl-heidelberg.de/pub/databases/, and ftp://ftp.ncbi.nih. gov/repository/PRINTS

NAME: ProDom[15]

DESC: The ProDom is a database of homologous domain families automatically generated from Swiss-Prot and TrEMBL. The database provides users capabilities for graphical display that link related families through their shared sequences and for pairwise comparison with every sequence in each family. The database has 365,172 entries, as of February 19, 2003.

GROUP: INRA/CNRS, Laboratoire de Biologie Moleculaire, France

EMAIL: proquest@toulouse.inra.fr

WWW: http://protein.toulouse.inra.fr/prodom.html

QUERY: http://protein.toulouse.inra.fr/prodom/current/html/form.php and http://prodes.toulouse.inra.fr/srs6/ for navigation between ProDom, Swiss-Prot, TrEMBL, PROSITE, PFAMA, InterPro, and PDB.

FTP: ftp://ftp.infobiogen.fr/pub/db/prodom/ and ftp://ftp.ebi.ac.uk/pub/databases/ prodom/

NAME: PROSITE[16]

DESC: The PROSITE is a database of protein families and domains and obtains protein sequences from Swiss-Prot. The PROSITE database contains biologically significant sites, patterns, and profiles that help to reliably identify to which known family of protein (if any) a new sequence belongs and to look for small motifs found in nonhomologous contexts. The PROSITE 17.46, which came out on May 11, 2003, contains 1,187 documented entries that describe 1,625 different patterns, rules, and profiles/matrices. The WWW server provides keyword searching as well as pattern match searching for classification of protein sequences.

GROUP: Swiss Institute of Bioinformatics (SIB) Geneva, Switzerland

EMAIL: prosite@expasy.org

WWW: http://www.expasy.org/prosite/

QUERY: The ScanProsite (http://www.expasy.org/tools/scanprosite/) allows users to scan a sequence against PROSITE or a pattern against Swiss-Prot or PDB and visualize matches on structures.[17]

FTP: ftp://ftp.expasy.ch/databases/prosite/, ftp://ftp.infobiogen.fr/pub/db/prosite/, and ftp://ftp.ebi.ac.uk/pub/databases/prosite/

NAME: SMART, Simple Modular Architecture Research Tool[18]

DESC: The SMART allows the identification and annotation of genetically mobile domains and the analysis of domain architectures. The SMART v3.5, which came out on April 28, 2003, contains 654 HMMs found in signaling, extracellular, and chromatin-associated proteins that are detectable. The focus of SMART is to search for evolutionarily conserved protein domains rather than small sites of posttranslational modification. The WWW server provides keyword searching as well as HMM searching for classification of protein sequences.

GROUP: European Molecular Biology Laboratory (EMBL), Heidelberg, Germany

EMAIL: smart@embl.de

WWW: http://smart.embl-heidelberg.de/

QUERY: http://smart.embl-heidelberg.de/index2.cgi for the SMART advanced search and http://dylan.embl-heidelberg.de/alert/ for the SMART alert service to be automatically informed each time a new protein with a defined domain composition is deposited in databases.

NAME: TIGRFAMs[19]

DESC: The TIGRFAMs is a database of protein families based on Hidden Markov Models.

GROUP: The Institute for Genomic Research (TIGR), Rockville, MD, U.S.

EMAIL: tigrfams@tigr.org

WWW: http://www.tigr.org/TIGRFAMs/index.shtml

QUERY: http://www.tigr.org/tigr-scripts/CMR2/find_hmm.spl?db=CMR for TIGRFAMs text search and http://tigrblast.tigr.org/web-hmm/ for TIGRFAMs sequence similarity search.

FTP: ftp://ftp.tigr.org/pub/data/TIGRFAMs/

4.2.2.3 Protein Structure Databases

Protein structure database is an archive of experimentally or computationally determined three-dimensional (3D) structures of biological macromolecules. The three-dimensional structure of a protein is determined by techniques such as X-ray crystallography and nuclear magnetic resonance (NMR). The 3D structure databases contain atomic coordinates, bibliographic citations, and primary and secondary structure information, as well as crystallographic structure factors and NMR experimental data. Four well-established protein structure databases are listed below, although the PDB is the most comprehensive one and the MMDB is the secondary database of the PDB.

NAME: CSD, Cambridge Structural Database[20]

DESC: The CSD contains crystal structure information for 272,066 organic and metal organic compounds, as of November 2002. Each crystallographic entry in the database consists of three distinct type of data: bibliographic reference, 2D chemical connectivity, and 3D structure.

GROUP: Cambridge Crystallographic Data Centre (CCDC), U.K.

EMAIL: data_request@ccdc.cam.ac.uk

WWW: http://www.ccdc.cam.ac.uk/

QUERY: http://www.ccdc.cam.ac.uk/prods/conquest/updates to download new CSD entries.

FTP: http://www.ccdc.cam.ac.uk/free_services/free_downloads/

NAME: MMDB, Molecular Modeling Database[21]

DESC: The MMDB contains the structure and sequence contents of the Brookhaven Protein Data Bank in ASN.1 format. The MMDB contains 3D structures extracted from the PDB. The Entrez is available for simple text searches (http://www.ncbi.nlm.nih.gov/entrez/query.fcgi?db=Structure) by ID, author, and texts, as well as for structure similarity searches against the MMDB. All the protein structures are compared with each other using the VAST algorithm (http://www.ncbi. nlm.nih.gov/Structure/VAST/ vast.shtml) in the same way as the sequence similarity search is used in conjunction with the BLAST algorithm (http://www.ncbi.nlm.nih.gov/ BLAST/).

GROUP: NCBI, National Center for Biotechnology Information, U.S.

EMAIL: bryant@ncbi.nlm.nih.gov

WWW: http://www.ncbi.nlm.nih.gov/Structure/MMDB/mmdb.shtml

QUERY: http://www.ncbi.nlm.nih.gov/Structure/CN3D/cn3d.shtml to views 3D structures using the Web browser.

FTP: ftp://ftp.ncbi.nih.gov/mmdb/

NAME: NDB, Nucleic Acid DataBase[22]

DESC: The NDB is a repository of 3D structural information about nucleic acids. The database contains 2,095 structures, as of May 13, 2003.

GROUP: Rutgers, The State University of New Jersey, U.S.

EMAIL: ndbadmin@ndbserver.rutgers.edu

WWW: http://ndbserver.rutgers.edu/

QUERY: http://ndbserver.rutgers.edu/NDB/structure-finder/index.html to find X-ray crystallographic structures and to search for nucleic acid containing structures determined by either X-ray crystallography or NMR.

FTP: http://ndbserver.rutgers.edu/ftp/NDB/ to download the coordinate files in PDB and mmCIF formats and the structure factor files in CIF format.

NAME: PDB, Protein Data Bank[23,24]

DESC: The PDB is the single worldwide repository for the processing and distribution of experimentally- or computationally-determined three-dimensional structures of biological macromolecules, serving a global community of researchers. The PDB has 21,126 entries, as of April 6, 2003.

GROUP: Protein Data Bank, BNL (Brookhaven National Laboratory), NY, U.S.
EMAIL: pdb@bnl.gov
WWW: http://www.rcsb.org/pdb
QUERY: http://rutgers.rcsb.org/ and http://pdb.ccdc.cam.ac.uk/pdb/
FTP: ftp://ftp.rcsb.org/pub/pdb/, ftp://pdb.ccdc.cam.ac.uk/rcsb, ftp://ftp.
    infobiogen.fr/pub/db/pdb

4.2.2.4 Protein Structural Classification Databases

Protein structure classification databases are a comprehensive database to provide
hierarchical structural classifications of proteins above the superfamily level. Proteins with similar structures (i.e., folds) are considered as members of the same
family or superfamily. Five such databases are listed below, although the CATH and
SCOP databases are the most widely accepted.

NAME: CATH, Class, Architecture, Topology, and Homologous Superfamily[25,26]
DESC: The CATH is a protein structure classification databases that provides
    a novel hierarchical classification of protein domain structures that clusters
    proteins at four major levels: class (C), architecture (A), topology (T), and
    homologous superfamily (H).
GROUP: Biomolecular Structure and Modelling Group, University College
    London, U.K.
EMAIL: cath@bsm.bioc.ucl.ac.uk
WWW: http://www.biochem.ucl.ac.uk/bsm/cath/index.html
QUERY: http://www.biochem.ucl.ac.uk/bsm/cath/class.html
FTP: ftp://ftp.biochem.ucl.ac.uk/pub/cathdata/

NAME: DSSP, Database of Secondary Structures Assignments for All
    Protein[27]
DESC: The DDSP is the Database of Secondary Structure Assignments for
    All Protein entries in the Protein Data Bank (PDB).
GROUP: Sander Group, EMBL, Heidelberg, Germany
EMAIL: sander@embl-heidelberg.de
WWW: http://swift.embl-heidelberg.de/dssp/
FTP: ftp://ftp.embl-heidelberg.de/pub/databases/dssp/

NAME: FSSP, Database of Families of Structurally Similar Proteins[28]
DESC: The FSSP was established based on exhaustive all-against-all 3D
    structure comparison of protein structures currently in the Protein Data
    Bank (PDB).
GROUP: EMBL Outstation - European Bioinformatics Institute (EBI), U.K.
EMAIL: holm@embl-ebi.ac.uk
WWW: http://www.ebi.ac.uk/dali/fssp/
FTP: ftp://ftp.ebi.ac.uk/pub/databases/fssp/

NAME: HSSP, Homology-Derived Secondary Structure Proteins[29]
DESC: The HSSP contains secondary structure of proteins derived by using
the model built by homology.
GROUP: EMBL, Heidelberg, Germany
EMAIL: sander@embl-heidelberg.de
WWW: http://www.sander.embl-heidelberg.de/hssp
FTP: ftp://ftp.embl-heidelberg.de/pub/databases/hssp and ftp://ftp.ebi.ac.uk/pub/
databases/hssp

NAME: SCOP, Structural Classification Of Proteins Database[30]
DESC: The SCOP is a manual, hierarchical classification of proteins of known
structure.
GROUP: MRC Laboratory of Molecular Biology and Centre for Protein
Engineering, U.K.
EMAIL: scop@mrc-lmb.cam.ac.uk
WWW: http://scop.mrc-lmb.cam.ac.uk/scop/
QUERY: http://scop.mrc-lmb.cam.ac.uk/scop/search.cgi, http://scop.berkeley.
edu/, and http://pdb.weizmann.ac.il/scop/

4.2.2.5 Protein–Protein Interaction Databases

Several databases for protein–protein interactions are available although there is a
need to develop a common data standard to share their data among themselves as
well as to allow users to easily retrieve and compare all relevant information from
different databases. Three well-established protein–protein interaction databases are
described below.

NAME: BIND, Biomolecular Interaction Network Database[31]
DESC: The BIND contains interaction, molecular complex, and pathway
records. The database can be used to study networks of interactions, to map
pathways across taxonomic branches, and to generate information for
kinetic simulations.
GROUP: Samuel Lunenfeld Research Institue (SLRI), Toronto, Canada
EMAIL: info@bind.ca
WWW: http://www.bind.ca/
FTP: ftp://ftp.mshri.on.ca/pub/

NAME: DIP, Database of Interacting Proteins[32]
DESC: The DIP lists protein pairs that are known to interact with each other.
It contains 11,000 experimentally determined interactions between proteins
and high-quality annotations by expert curators. Registration is required to
gain access to most of the DIP features. Registration is free to members of
the academic community.
GROUP: UCLA-DOE Institute for Genomics and Proteomics, UCLA, U.S.
EMAIL: dip@mbi.ucla.edu

WWW: http://dip.doe-mbi.ucla.edu/

QUERY: http://dip.doe-mbi.ucla.edu/dip/Search.cgi?SM=3 for text search, http://dip.doe-mbi.ucla.edu/dip/Search.cgi?SM=2 for sequence similarity search, and http://dip.doe-mbi.ucla.edu/dip/Search.cgi?SM=6 for sequence motif search.

NAME: MINT, Molecular INTeractions Database[33]

DESC: The MINT database is a collection of protein–protein and protein–DNA interactions. The database contains experimentally determined functional interactions between proteins, both direct and indirect. Each entry has high-quality annotations of interacting information extracted from the scientific literature by expert curators.

GROUP: Centro di Bioinformatica Molecolare, Universita di Roma, Italy

EMAIL: giovanni.cesareni@uniroma2.it

WWW: http://cbm.bio.uniroma2.it/mint/

QUERY: http://mint.bio.uniroma2.it/mint/search.php to search the database by Swiss-Prot accession number, by protein name or gene, or by keywords and view the interaction data graphically.

4.2.2.6   Posttranslational Modification Databases

The protein sequence databases are the most useful resource to obtain posttranslational modification related information. The feature lines of each sequence entry describe regions or sites of interest in the sequence and list posttranslational modifications, binding sites, enzyme active sites, local secondary structure and other characteristics. This section describes two databases that are dedicated to capture comprehensive data to a specific posttranslational modification.

NAME: O-GlycBase, O-Glycosylated Proteins[34]

DESC: The O-GlycBase contains 242 glycoprotein entries with at least one experimentally verified O- or C-glycosylation site. The database consists of nonredundant sequences, carbohydrate species, and cross references to external databases such as PIR, Swiss-Prot, PDB, PROSITE, etc.

GROUP: Center for Biological Sequence Analysis, Department of Chemistry, The Technical University of Denmark, Lyngby, Denmark

EMAIL: janhan@cbs.dtu.dk

WWW: http://www.cbs.dtu.dk/databases/OGLYCBASE/

NAME: RESID Database of Protein Modifications[35]

DESC: The RESID is a database of protein posttranslational modifications with descriptive, chemical, structural, and bibliographic information, including amino-terminal, carboxyl-terminal, and peptide chain cross-link, pre-, co-, and posttranslational modifications.

GROUP: National Biomedical Research Foundation, Washington, DC, U.S.

EMAIL: garavelli@nbrf.georgetown.edu

WWW: http://www-nbrf.georgetown.edu/resid/get.html
QUERY: http://www-nbrf.georgetown.edu/cgi-bin/resid/
FTP: ftp://ftp.pir.georgetown.edu/pir_databases/, ftp://ftp.ebi.ac.uk/pub/data-
    bases/RESID/ and ftp://ftp.ncifcrf.gov/pub/users/residues/

4.2.2.7 2-D PAGE Databases

High-resolution, two-dimensional polyacrylamide gel electrophoresis (2D PAGE)
currently provides the most comprehensive analysis system of the whole proteome.
A systematic analysis of the whole proteome by 2D PAGE requires computer-based
image analysis tools, which generate qualitative and quantitative information pertaining to experimental design, sample processing, spot detection, gel matching, and
protein identification.[36] Thus, 2D PAGE databases store data derived from the computer-based image analysis tools and provide a user-friendly interface, including a
Web-based interface that allows researchers to search for, and discover, new data,
facts, and findings. Three most widely used 2D PAGE databases are listed below.

NAME: Human 2D PAGE Database[37]
DESC: The first 2D PAGE Database was established in 1981 by the Danish
    Centre for Human Genome Research at the University of Aarhus. The Human
    2D PAGE database contains databases for the study of global cell regulation
    and skin diseases, databases for the study of bladder cancer, and other 2D PAGE
    databases. These databases contain data on proteins identified on various ref-
    erence maps, as well as extensive links to other databases (MEDLINE, Gen-
    Bank, Swiss-Prot, PIR, PDB, OMIM, UniGene, GeneCards, etc.). The WWW
    server (http://proteomics.cancer.dk/) provides access to the databases through
    the functions such as (1) search by clicking the spot of interest on the image;
    (2) search by protein name, keyword, spot number, Mr and pI, and organelle
    or component; and (3) list all proteins in information category.
GROUP: Danish Centre for Human Genome Research, University of Aarhus,
    Denmark
EMAIL: jec@cancer.dk
WWW: http://proteomics.cancer.dk/
QUERY: http://proteomics.cancer.dk/jecelis/human_data_select.html

NAME: Swiss-2D PAGE Database[38]
DESC: The Swiss-2D PAGE database contains data on proteins identified on
    various 2D PAGE (two-dimensional polyacrylamide gel electrophoresis)
    reference maps. Each Swiss-2D PAGE entry contains textual data on one
    protein, including mapping procedures, physiological and pathological
    information, experimental data (isoelectric point, molecular weight, amino
    acid composition, peptide masses), and bibliographical references as well
    as cross references (recorded in the DR lines) to Medline, other federated
    2D databases (COMPLUYEAST-2D PAGE, ECO2DBASE, HSC-2D
    PAGE, PHCI-2D PAGE, PMMA-2D PAGE, Siena-2D PAGE, YEPD),
    Swiss-Prot, and other molecular databases (EMBL, Genbank, PROSITE,

OMIM, etc.). There are three basic database search programs: (1) search by description, accession number, author, and text; (2) search by clicking on a spot (http://www.expasy.org/cgi-bin/map1); and (3) search by spot serial number (http://www.expasy.org/cgi-bin/ch2d-search-sn).
GROUP: Swiss Institute of Bioinformatics, Geneva, Switzerland
EMAIL: Ron.Appel@isb-sib.ch
WWW: http://www.expasy.ch/ch2d/
QUERY: http://www.expasy.ch/ch2d/
FTP: ftp://www.expasy.ch/databases/swiss-2dpage/

NAME: WORLD-2D PAGE Database[39]
DESC: The WORLD-2D PAGE
GROUP: Swiss Institute of Bioinformatics, Geneva, Switzerland
EMAIL: Christine.Hoogland@isb-sib.ch
WWW: http://www.expasy.org/ch2d/2d-index.html

4.2.2.8 Mass Spectrometry Databases

Any evidence of widely accepted public repositories for mass spectrometry–based protein identifications does not exist, although there is a considerable need to develop comprehensive, systematic mechanisms to analyze and compare the vast number of mass spectrometry queries (e.g., Rank, MOWSE score, %Masses Matched, MW, pI, Species, Accession, Protein Name, Submitted Mass, Matched Mass, Petta PPM, Start, End, Peptide Seq, Modifications, Unmatched Masses).[40] So, in order to fulfill the need of such data analysis today, the Proteomics Standards Initiative (PSI) was founded at the HUPO (Human Proteome Organization, http://www.hupo.org) meeting on April 28-29, 2002. The aim of the PSI is to define worldwide community standards for proteomic data representation to make electronic data exchange, comparison, and integration possible. The PSI's mass spectrometry group emphasizes developing a standard representation of experimental spectra in the context of the experimental setup and the analyzed system. Such a format will allow scientists to support publications with detailed experimental results and to enable collaborations beyond the local research team.

4.2.2.9 Other Protein Databases

There are many protein databases appearing and disappearing in public domain. Scientists are encouraged to regularly browse through online catalogs of protein databases (some are described in Section 4.2.1) that are dedicated to molecular biology with an emphasis on data relevant to proteins and kept constantly up-to-date. This section provides an inexhaustive list of some analysis databases that have been developed in order to extend its interpretation of information in the protein sequence and structure databases and to assist scientists in the field of proteomics research.

NAME: COG, Clusters of Orthologous Groups of Proteins Database[41]
DESC: The COG database contains phylogenetic classification of proteins
encoded in 43 complete genomes. Each COG entry consists of individual

proteins and groups of paralogs from at least three major phylogenetic lineages and thus corresponds to an ancient conserved domain.

GROUP: National Center for Biotechnology Information (NCBI), Bethesda, MD, U.S.

EMAIL: info@ncbi.nlm.nih.gov

WWW: http://www.ncbi.nlm.nih.gov/COG/

QUERY: http://www.ncbi.nlm.nih.gov/COG/new/ for the new database search and http://www.ncbi.nlm.nih.gov/COG/old/phylox.html for the phylogenetic patterns search.

FTP: ftp://ftp.ncbi.nih.gov/pub/COG/

NAME: ENZYME Nomenclature Database[42]

DESC: The ENZYME is a repository of information relative to the nomenclature of enzymes.

GROUP: Medical Biochemistry Department, Centre Medical Universitaire, Geneva, Switzerland

EMAIL: enzyme@expasy.org

WWW: http://www.expasy.org/enzyme/

FTP: ftp://ftp.expasy.ch/databases/enzyme/

NAME: GOA, Gene Ontology Annotation Database[43]

DESC: The Gene Ontology Annotation (GOA) database contains a universal ontology to describe molecular functions, biological processes, and cellular components of genes or gene products. Crucial to this database is the integration of internal and external databases and resources using its standard vocabulary to characterize the activities of proteins in the Swiss-Prot, TrEMBL, and InterPro databases.

GROUP: EMBL Outstation - European Bioinformatics Institute (EBI), U.K.

EMAIL: goa@ebi.ac.uk

WWW: http://www.ebi.ac.uk/GOA/

QUERY: http://www.ebi.ac.uk/ego/index.html to access core GO data and up-to-date electronic and manual EBI GO annotations.

FTP: ftp://ftp.ebi.ac.uk/pub/databases/GO/goa/HUMAN to download GO annotation for the human proteome.

NAME: IPI, International Protein Index Database

DESC: The IPI database provides complete proteome sets of human, mouse, and rat proteins from Swiss-Prot, TrEMBL, RefSeq (NPs and XPs), and Ensembl. The database was established automatically by mapping proteins between the different databases by pairwise similarity searches. Each IPI entry consists of a cluster of related sequences and cross references to the constituent databases.

GROUP: EMBL Outstation - European Bioinformatics Institute (EBI), U.K.

EMAIL: ipi_help@ebi.ac.uk

WWW: http://www.ebi.ac.uk/IPI/

QUERY: http://www.ebi.ac.uk/blast2/ to search the IPI database using BLAST algorithms, http://www.ebi.ac.uk/fasta33/ to search the database using FASTA algorithms, and http://srs.ebi.ac.uk to search the database using SRS.

FTP: ftp://ftp.ebi.ac.uk/pub/databases/IPI/current/

NAME: PMD, Protein Mutant Database[44]

DESC: The PMD contains literature describing protein mutations, including information on what kinds of functional and/or structural influences are brought about by amino acid mutation at a specific position of protein. The release of PMD, on March 10, 2003, contains 28,645 entries and 150,645 mutants.

GROUP: Center for Information Biology and DNA Data Bank of Japan, National Institute of Genetics, Japan

EMAIL: pmd-admin@pmd.ddbj.nig.ac.jp

WWW: http://pmd.ddbj.nig.ac.jp/

QUERY: http://pmd.ddbj.nig.ac.jp/~pmd/pmdkey.html for keyword search and http://pmd.ddbj.nig.ac.jp/~pmd/pmdseqblt.html for BLAST sequence similarity search.

FTP: ftp://spock.genes.nig.ac.jp/pub/pmd

NAME: Proteome Analysis Database[45]

DESC: The Proteome Analysis Database contains comprehensive statistical and comparative analyses of the predicted proteomes of fully sequenced organisms. There are 123 proteome sets available and proteome analysis is available for 119 of these, as of April 23, 2003.

GROUP: EMBL Outstation - European Bioinformatics Institute (EBI), U.K.

EMAIL: proteome_help@ebi.ac.uk

WWW: http://www.ebi.ac.uk/proteome/

QUERY: http://www.ebi.ac.uk/proteome/ to get precomputed proteome analysis for a given organism, http://www.ebi.ac.uk/proteome/comparisons. html to perform proteome comparisons between any combination of organisms in the database, and http://www.ebi.ac.uk/fasta33/genomes.html to run a FASTA similarity search against a complete proteome.

FTP: ftp://ftp.ebi.ac.uk/pub/databases/Spproteomes/ to download a proteome set or a list of InterPro matches for a given organism.

NAME: REBASE, The Restriction Enzyme Database[46]

DESC: The REBASE is a collection of information about restriction enzymes; methylases, the microorganisms from which they have been isolated; recognition sequences; cleavage sites; methylation specificity; the commercial availability of the enzymes; and references.

GROUP: New England BioLabs, Beverly, MA, U.S.

EMAIL: roberts@neb.com or macelis@neb.com

WWW: http://rebase.neb.com/rebase/rebase.html

FTP: ftp://ftp.neb.com/pub/rebase/, ftp://ftp.ebi.ac.uk/pub/databases/rebase/, and ftp://ftp.ncbi.nih.gov/repository/REBASE/

NAME: STRING[47]

DESC: The STRING is a database of predicted functional associations among genes and proteins. The database contains 261,033 genes in 89 species.

GROUP: EMBL, Heidelberg, Germany

EMAIL: bork@embl-heidelberg.de

WWW: http://www.bork.embl-heidelberg.de/STRING/

QUERY: http://www.bork.embl-heidelberg.de/STRING/ for the retrieval of interacting genes and proteins by a given gene or by a given protein sequence.

FTP: http://dag.embl-heidelberg.de/newstring_cgi/show_download_page.pl to download the predicted functional links between orthologous groups of genes/proteins.

4.2.3 NUCLEOTIDE SEQUENCE DATABASES

The exponential growth of the GenBank, EMBL, and DDBJ nucleotide databases has led to a massive influx of data into the protein sequence databases. The GenBank, EMBL, and DDBJ databases are maintained by an international collaboration among NCBI, EIB, and NIG. These three organizations exchange data on a daily basis. GenBank (http://www.ncbi.nlm.nih.gov) was set up in 1979 at the LANL (Los Alamos National Laboratory, http://www.lanl.gov), and it has been maintained since 1992 by NCBI (National Center for Biotechnology Information, http://www.ncbi.nlm.nih.gov); U.S. EMBL (http://www.ebi.ac.uk/embl) was created in 1980 at the European Molecular Biology Laboratory in Heidelberg, Germany, and it has been maintained since 1994 by EBI-Cambridge; U.K. DDBJ (DNA Data Bank of Japan, http://www. ddbj.nig.ac.jp) was started in 1984 and has been maintained by a bioinformatics team from the National Institute of Genetics (NIG) in Mishima, Japan.

4.3 DATABASE RETRIEVAL SYSTEMS

There are several database retrieval systems that are composed of an integrated collection of logically related elements to support proteomic research. The integrated database retrieval systems provide users a one-stop shop for accessing almost all widely accepted public proteomic databases and allow users to query multiple heterogeneous data sources as if they were components of a single large database. Such systems provide access, not only to the databases to enable one to logically create an integrated view of remote or local heterogeneous data sources by using simple and advanced database search tools, but also to analytical computational tools and software for the identification of proteins, the analysis of sequence sequences, and the prediction of a tertiary protein structure. Among many systems, four integrated database retrieval systems are as follows.

NAME: DBGET/LinkDB[48]

DESC: The DBGET/LinkDB at GenomeNet (http://www.genome.jp/) is an integrated database retrieval system established by the Institute for Chemical Research, Kyoto University, Japan. Currently, it supports the following databases and gene catalogs: nucleic acid sequences (GenBank, EMBL, and DDBJ), protein sequences (Swiss-Prot, PIR, PRF, and PDB), 3D structures (PDB), sequence motifs (PROSITE, EPD, and TRANSFAC), enzyme reactions (LIGAND), metabolic pathways (PATHWAY), amino acid mutations (PMD), amino acid indices (AAindex), genetic diseases (OMIM), literature (LITDB, Medline), and gene catalogs (*E. coli, H. influenzae, M. genitalium, M. pneumoniae, M. jannaschii, Synechocystis* sp., and *S. cerevisiae*).

GROUP: GenomeNet, Kyoto University Bioinformatics Center, Japan

WWW: http://www.genome.ad.jp/dbget/

NAME: Entrez[49]

DESC: The Entrez is a database retrieval system for searching several linked databases including PubMed (Medline), the nucleotide sequence database (GenBank), protein sequence database (SwissProt, PIR, PRF, PDB, and translations from annotated coding regions in GenBank and RefSeq), structure (three-dimensional macromolecular structures), genome (complete genome assemblies), PopSet (Population study data sets), taxonomy (organisms in GenBank), OMIM (Online Mendelian Inheritance in Man), and many others.

GROUP: NCBI, National Center for Biotechnology Information, U.S.

EMAIL: info@ncbi.nlm.nih.gov

WWW: http://www.ncbi.nlm.nih.gov/Entrez/

NAME: ExPASy, Expert Protein Analysis System[50]

DESC: The ExPASy is dedicated to molecular biology with an emphasis on data relevant to proteins. Using the ExPASy proteomics server users can browse through a number of databases produced in SIB (e.g., Swiss-Prot, PROSITE, Swiss-2D PAGE, etc.) and other cross-referenced databases (e.g., EMBL/GenBank/DDBJ, OMIM, Medline, FlyBase, ProDom, SGD, SubtiList, etc.), as well as many hyperlinked documents relevant to the field of genomic and proteomic research.

GROUP: Swiss Institute of Bioinformatics (SIB), Geneva, Switzerland

WWW: http://www.expasy.org/

NAME: SRS, Sequence Retrieval System[51]

DESC: The SRS is used to browse the contents of various databases through a web interface. The SRS provides a single point of entry for related searches that integrate more than 250 biological databases. Using these SRS servers users can search the selected database(s) on a number of the factors (entry ID, accession number, description, gene name, keywords, date, organism, NCBI_TaxId, submission date, organelle, plasmid, strain, transposon, tissue, authors, citation, patent, patent date, Medline ref., title, report type, DbName, Dbxref, SeqLength, MolWeight, comment type, comment, feature sequence ID, feature key, feature description, and/or feature sequence).

GROUP: SIB, EBI, Infobiogen, etc.

WWW: There are more than 35 SRS servers available including the EBI SRS
server (http://srs.ebi.ac.uk), the SIB SRS server (http://www.expasy.org/srs5),
and the Infobiogen SRS server (http://www.infobiogen.fr/srs/).

These database retrieval systems are key components of the public protein
databases. The rapid growth of such systems on the Internet increased the public
use of distributed databases and hyperlinked external documents and helped users
not only to access the databases, but to improve the efficiency and effectiveness of
their research processes and workgroup collaboration as well.

4.4   INTEGRATING PROTEIN DATABASES

4.4.1   PROTEIN KNOWLEDGE BASE

Today, proteomics research continually generates additional data sets that differ from
one another. Consequently, a challenge in informatics for proteomics is integrating these
data sets to connect disparate information and developing widely accepted public proteomic databases to share research data and translate research results into knowledge.

Thus, one of the core activities for informatics in proteomics is to integrate all
significant protein information from protein databases into a protein knowledgebase.
The Swiss-Prot is the most widely accepted protein knowledgebase today with
high-quality annotations that can be found in the keyword (KW), comment (CC),
feature table (FT), and database cross-reference (DR) lines of each sequence entry.
Each Swiss-Prot sequence entry is classified by 875 keywords (e.g., 2Fe-2S, 3D-
structure, acetylation, NADP, oxidoreductase, zymogen, etc.). A list of keywords can
be found in the Swiss-Prot Web site (http://www.expasy.org/cgi-bin/keywlist.pl). This
entry contains detailed comments that are organized into 22 different types of topics
(e.g., alternative products, biotechnology, catalytic activity, caution, cofactor, database,
developmental stage, disease, domain, enzyme regulation, function, induction, mass
spectrometry, miscellaneous, pathway, pharmaceutical, polymorphism, ptm, similarity,
subcellular location, subunit, and tissue specificity). The sequence entry has a full set
of features that describe regions or sites of interest in the sequence, posttranslational
modifications, binding sites, enzyme active sites, local secondary structure and other
characteristics. In addition, this entry is linked to a specific reference in the external
databases (e.g., ANU-2D PAGE, AraC-XylS, BLOCKS, CleanEx, CMR, COMPLU-
YEAST-2D PAGE, dbSNP, DDBJ, DictyDb, DIP, ECO2D BASE, EcoCyc, EcoGene,
EMBL, Ensembl, ENZYME, FlyBase, GenBank, GeneCards, GeneCensus,
GeneDB_SPombe, GeneLynx, Genew, GK, GlycoSuiteDB, GO, GPCRDB, Gramene,
HAMAP, HIV, HSC-2D PAGE, HSSP, HUGE, IMGT, InterPro, Leproma, ListiList,
MaizeDB, MAIZE-2D PAGE, MEROPS, MGD, Micado, MIM, ModBase, MypuList,
NRSub, NucleaRDB, PDB, Pfam, PHCI-2D PAGE, PhosSite , PIR, PMMA-2D PAGE,
PRESAGE, PRINTS, ProDom, PROSITE, ProtoMap, ProtoNet, REBASE, SagaList,
SGD, Siena-2D PAGE, SMART, SOURCE, SubtiList, Swiss-2D PAGE, TAIR, TIGR,
TIGRFAMs, TRANSFAC, TuberculList, WorfDB, WormBase, WormPep, ZFIN,
Aarhus/Ghent-2D PAGE, and StyGene).

Although Swiss-Prot is widely used as a protein knowledgebase, it was recognized that scientists had to use PIR, TrEMBL, and other databases for quick access to new protein sequences before they were hand-curated and entered into Swiss-Prot. Hence, the United Protein Database project was initiated, which aims to consolidate development effort to quickly provide a single worldwide protein knowledgebase and to efficiently and effectively handle the increasing amounts of data being generated by large-scale genomics and proteomics projects. This project is described in the next section.

4.4.2 United Protein Database

On October 23, 2002, a three-year, $15 million grant was awarded to the United Protein Database (UniProt) project by the U.S. National Human Genome Research Institute (NHGRI), in cooperation with five other institutes and centers at the National Institutes of Health (NIH). An ultimate goal of the project is to combine the Swiss-Prot, TrEMBL, and PIR protein sequence databases into a single, centralized, universal repository for protein data (i.e., UniProt). The UniProt will be established collaboratively by

The European Bioinformatics Institute (EBI), Hinxton, Cambridge, U.K.
The Protein Information Resource (PIR), Georgetown University Medical Center (GUMC) and National Biomedical Research Foundation (NBRF), Washington, DC, U.S.
The Swiss Institute of Bioinformatics (SIB), Geneva, Switzerland

Combining the resources of Swiss-Prot, TrEMBL, and PIR will help scientists today interpret the tremendous amount of data being generated by the Human Genome Project and proteomics related research. Therefore, an immediate focus of the project is to elevate the annotation of PIR's and TrEMBL's computer-annotated records to the Swiss-Prot standard. Currently, Swiss-Prot contains 125,744 entries (release 41.6 of April 30, 2003), TrEMBL contains 861,482 entries (release 23.8 of April 25, 2003), and PIR-PSD, 283,308 entries (release 76.00 of March 31, 2003). By the end of the grant's three-year span, EBI scientists estimate that the total number of protein sequence entries in the UniProt database should reach well above 2 million.

With the increasing volume and variety of protein sequences and functional information, UniProt, as the central database of protein sequence, will function as a cornerstone for a wide range of scientists active in modern biological research, especially in the field of proteomics. UniProt will be the new public resource and researchers around the world will have free, unrestricted access to a comprehensive and nonredundant source of protein information.

4.4.3 One-Stop Shop for Protein Data

Public protein databases have been steadily accumulating protein sequences and protein structures for more than a decade, submitted from disparate laboratories or reported in published literatures. It is critical to make use of both distinct databases to derive protein functions more intelligently and in a more automated way. There are several projects that emphasize integrating protein sequence and structure databases. One major project is being conducted at the San Diego Supercomputer Center, the Keck Graduate Institute, and the Burnham Institute and they aim to develop a one-stop shop for protein information. On April 4, 2002, the U.S. National Institute of General Medical Sciences (NIGMS) at the National Institutes of Health (NIH) announced that a five-year, $5.4 million grant was awarded to the project to develop a worldwide community resource for systematic protein annotation and modeling. Two ultimate goals of the project are to develop new algorithms to take the protein sequence information and assign putative functional annotation and model structures where no experimental structures exist and to store the results in a database for worldwide community access through the Web.

4.4.4 PROTEIN EXPERIMENT DATABASE

Currently, detailed experimental results of proteomics research are usually not available to worldwide researchers. Some of the experimental results are available publicly, but not in the form of a standardized, centralized data repository. So, to make it possible for further analysis of experimental results by comparing others, it is critical to develop protein experiment databases in a common universal standardized format.

However, the importance of database quality cannot be overstated. Sources of error in databases are varied:

1. Data submission errors (e.g., partially or entirely duplicated sequence entries)
2. Data annotation errors (e.g., inaccuracy, omission, mistakes, or inconsistency of annotated data)
3. Data collection errors (e.g., sequencing error, gel analysis error, sample processing error, instrument's sensitivity issues)
4. Data interpretation errors, for example, a null value can be interpreted into many different meanings, such as
   a. An unknown attribute value
   b. Known, but missing, attribute values
   c. A condition that is "not applicable."

Thus, the role of intelligent computational tools has become more significant than ever to use the massive amounts of proteomics data and information, to derive a better understanding of biology, and to apply that new knowledge into clinical applications.

Public protein databases not only contain data, which consist of facts collected from experimental studies or instruments using genomics and proteomics technology, but information as well; that is, the meaningful interpretation and correlation of data that allows someone to make decisions and derive knowledge. They serve not only as a means to efficiently store massive amounts of data and information, but also as a user interface to provide scientists with user-friendly front-end software for retrieving and analyzing data in databases in order to derive information that may be important to biological knowledge.

REFERENCES

1. George, D.G., Barker, W.C., Mewes, H.W., Pfeiffer, F., and Tsugita, A. The PIR-international protein sequence database. *Nucleic Acids Res.,* 24, 17–20, 1996.
2. Wu, C.H., Yeh, L.S., Huang, H., Arminski, L., Castro-Alvear, J., Chen, Y., Hu, Z., Kourtesis, P., Ledley, R.S., Suzek, B.E., Vinayaka, C.R., Zhang, J., and Barker, W.C. The protein information resource. *Nucleic Acids Res.,* 31, 345–347, 2003.
3. Pruitt, K.D. and Maglott, D.R. RefSeq and LocusLink: NCBI gene-centered resources. *Nucleic Acids Res.,* 29, 137–140, 2001.
4. Bairoch, A. and Boeckmann, B. The Swiss-Prot protein sequence data bank. *Nucleic Acids Res.,* 20, 2019–2022, 1992.
5. O'Donovan, C., Martin, M.J., Gattiker, A., Gasteiger, E., Bairoch, A., and Apweiler, R. High-quality protein knowledge resource: Swiss-Prot and TrEMBL. *Brief Bioinformatics,* 3, 275–284, 2002.
6. Bairoch, A. and Apweiler, R. The Swiss-Prot protein sequence data bank and its supplement TrEMBL. *Nucleic Acids Res.,* 25, 31–36, 1997.
7. Mulder, N.J., Apweiler, R., Attwood, T.K., Bairoch, A., Barrell, D., Bateman, A., Binns, D., Biswas, M., Bradley, P., Bork, P., Bucher, P., Copley, R.R., Courcelle, E., Das, U., Durbin, R., Falquet, L., Fleischmann, W., Griffiths-Jones, S., Haft, D., Harte, N., Hulo, N., Kahn, D., Kanapin, A., Krestyaninova, M., Lopez, R., Letunic, I., Lonsdale, D., Silventoinen, V., Orchard, S.E., Pagni, M., Peyruc, D., Ponting, C.P., Selengut, J.D., Servant, F., Sigrist, C.J., Vaughan, R., and Zdobnov, E.M. The InterPro Database, 2003 brings increased coverage and new features. *Nucleic Acids Res.,* 31, 315–318, 2003.
8. Zdobnov, E.M. and Apweiler, R. InterProScan—An integration platform for the signature-recognition methods in InterPro. *Bioinformatics,* 17, 847–848, 2001.
9. Barker, W.C., Pfeiffer, F., and George, D.G. Superfamily classification in PIR-international protein sequence database. *Methods Enzymol.,* 266, 59–71, 1996.
10. Wu, C., Zhao, S., and Chan, H.L. A protein class database organised with ProSite protein groups and PIR superfamilies. *J. Comput. Biol.,* 3, 547–561, 1996.
11. Bateman, A., Birney, E., Cerruti, L., Durbin, R., Etwiller, L., Eddy, S.R., Griffiths-Jones, S., Howe, K.L., Marshall, M., and Sonnhammer, E.L. The Pfam protein families database. *Nucleic Acids Res.,* 30, 276–280, 2002.
12. Srinivasarao, G.Y., Yeh, L.S., Marzec, C.R., Orcutt, B.C., and Barker, W.C. PIR-ALN: a database of protein sequence alignments. *Bioinformatics,* 15, 382–390, 1999.
13. Attwood, T.K. and Beck, M.E. PRINTS—A protein motif fingerprint database. *Protein Eng.,* 7, 841–848, 1994.
14. Scordis, P., Flower, D.R., and Attwood, T.K. FingerPRINTScan: Intelligent searching of the PRINTS motif database. *Bioinformatics,* 15, 799–806, 1999.
15. Servant, F., Bru, C., Carrere, S., Courcelle, E., Gouzy, J., Peyruc, D., and Kahn, D. ProDom: Automated clustering of homologous domains. *Brief Bioinformatics,* 3, 246–251, 2002.
16. Sigrist, C.J., Cerutti, L., Hulo, N., Gattiker, A., Falquet, L., Pagni, M., Bairoch, A., and Bucher, P. PROSITE: a documented database using patterns and profiles as motif descriptors. *Brief Bioinformatics,* 3, 265–274, 2002.
17. Gattiker, A., Gasteiger, E., and Bairoch, A. ScanProsite: A reference implementation of a PROSITE scanning tool. *Appl. Bioinformatics,* 1, 107–108, 2002.
18. Letunic, I., Goodstadt, L., Dickens, N.J., Doerks, T., Schultz, J., Mott, R., Ciccarelli, F., Copley, R.R., Ponting, C.P., and Bork, P. Recent improvements to the SMART domain-based sequence annotation resource. *Nucleic Acids Res.,* 30, 242–244, 2002.

19. Haft, D.H., Selengut, J.D., and White, O. The TIGRFAMs database of protein families. *Nucleic Acids Res.,* 31, 371–373, 2003.
20. Allen, F.H. The Cambridge Structural Database: A quarter of a million crystal structures and rising. *Acta Crystallogr. B,* 58, 380–388, 2002.
21. Hogue, C.W., Ohkawa, H., and Bryant, S.H. A dynamic look at structures: WWW-Entrez and the molecular modeling database. *Trends Biochem. Sci.,* 21, 226–229, 1996.
22. Berman, H.M., Olson, W.K., Beveridge, D.L., Westbrook, J., Gelbin, A., Demeny, T., Hsieh, S.H., Srinivasan, A.R., and Schneider, B. The nucleic acid database. A comprehensive relational database of three-dimensional structures of nucleic acids. *Biophys. J.,* 63, 751–759, 1992.
23. Berman, H.M., Westbrook, J., Feng, Z., Gilliland, G., Bhat, T.N., Weissig, H., Shindyalov, I.N., and Bourne, P.E. The protein data bank. *Nucleic Acids Res.,* 28, 235–242, 2000.
24. Westbrook, J., Feng, Z., Chen, L., Yang, H., and Berman, H.M. The protein data bank and structural genomics. *Nucleic Acids Res.,* 31, 489–491, 2003.
25. Orengo, C.A., Michie, A.D., Jones, S., Jones, D.T., Swindells, M.B., and Thornton, J.M. CATH—A hierarchic classification of protein domain structures. *Structure,* 5, 1093–1108, 1997.
26. Pearl, F.M., Lee, D., Bray, J.E., Sillitoe, I., Todd, A.E., Harrison, A.P., Thornton, J.M., and Orengo, C.A. Assigning genomic sequences to CATH. *Nucleic Acids Res.,* 28, 277–282, 2000.
27. Kabsch, W. and Sander, C. Dictionary of protein secondary structure: Pattern recognition of hydrogen-bonded and geometrical features. *Biopolymers,* 22, 2577–2637, 1983.
28. Holm, L. and Sander, C. The FSSP database of structurally aligned protein fold families. *Nucleic Acids Res.,* 22, 3600–3609, 1994.
29. Sander, C. and Schneider, R. Database of homology-derived protein structures and the structural meaning of sequence alignment. *Proteins,* 9, 56–68, 1991.
30. Murzin, A.G., Brenner, S.E., Hubbard, T., and Chothia, C. SCOP: A structural classification of proteins database for the investigation of sequences and structures. *J. Mol. Biol.,* 247, 536–540, 1995.
31. Bader, G.D., Donaldson, I., Wolting, C., Ouellette, B.F., Pawson, T., and Hogue, C.W. BIND—The biomolecular interaction network database. *Nucleic Acids Res.,* 29, 242–245, 2001.
32. Xenarios, I., Salwinski, L., Duan, X.J., Higney, P., Kim, S.M., and Eisenberg, D. DIP, the database of interacting proteins: A research tool for studying cellular networks of protein interactions. *Nucleic Acids Res.,* 30, 303–305, 2002.
33. Zanzoni, A., Montecchi-Palazzi, L., Quondam, M., Ausiello, G., Helmer-Citterich, M., and Cesareni, G. MINT: A molecular interaction database. *FEBS Lett.,* 513, 135–140, 2002.
34. Hansen, J.E., Lund, O., Nielsen, J.O., Hansen, J.E.S., and Brunak, S. O-glycbase— A revised database of O-glycosylated proteins. *Nucleic Acids Res.,* 24, 248–252, 1996.
35. Garavelli, J.S. The RESID database of protein modifications: 2003 developments. *Nucleic Acids Res.,* 31, 499–501, 2003.
36. Oh, J.M.C., Hanash, S.M., and Teichroew, D. Mining protein data from two-dimensional gels: tools for systematic post-planned analyses. *Electrophoresis,* 20, 766–774, 1999.

37. Celis, J.E., Ratz, G.P., Madsen, P., Gesser, B., Lauridsen, J.B., Kwee, S., Rasmussen, H.H., Nielsen, H.V., Cruger, D., Basse, B., et al. Comprehensive, human cellular protein databases and their implication for the study of genome organization and function. *FEBS Lett.*, 244, 247–254, 1989.

38. Hoogland, C., Sanchez, J.C., Tonella, L., Bairoch, A., Hochstrasser, D.F., and Appel, R.D. Current status of the Swiss-2DPAGE database of two dimensional gel electrophoresis. *Nucleic Acids Res.*, 26, 334–335, 1998.

39. Appel, R.D., Bairoch, A., Sanchez, J.C., Vargas, J.R., Golaz, O., Pasquali, C., and Hochstrasser, D.F. Federated 2-DE database: A simple means of publishing 2-DE data. *Electrophoresis*, 17, 540–546, 1996.

40. Oh, J.M.C., Brichory, F., Puravs, E., Kuick, R., Wood, C., Rouillard, J.M., Tra. J., Kardia, S., Beer, D., and Hanash, S.M. A database of protein expression in lung cancer. *Proteomics*, 1, 1303–1319, 2001.

41. Tatusov, R.L., Natale, D.A., Garkavtsev, I.V., Tatusova, T.A., Shankavaram, U.T., Rao, B.S., Kiryutin, B., Galperin, M.Y., Fedorova, N.D., and Koonin, E.V. The COG database: New developments in phylogenetic classification of proteins from complete genomes. *Nucleic Acids Res.*, 29, 22–28, 2001.

42. Bairoch, A. The ENZYME data bank in 1995. *Nucleic Acids Res.*, 24, 221–222, 1996.

43. Camon, E., Magrane, M., Barrell, D., Binns, D., Fleischmann, W., Kersey, P., Mulder, N., Oinn, T., Maslen, J., Cox, A., and Apweiler, R. The gene ontology annotation (GOA) project: Implementation of GO in Swiss-Prot, TrEMBL, and InterPro. *Genome Res.*, 13, 662–672, 2003.

44. Kawabata, T., Ota, M., and Nishikawa, K. The protein mutant database. *Nucleic Acids Res.*, 27, 355–357, 1999.

45. Pruess, M., Fleischmann, W., Kanapin, A., Karavidopoulou, Y., Kersey, P., Kriventseva, E., Mittard, V., Mulder, N., Phan, I., Servant, F., and Apweiler, R. The proteome analysis database: A tool for the in silico analysis of whole proteomes. *Nucleic Acids Res.*, 31, 414–417, 2003.

46. Roberts, R.J. and Macelis, D. REBASE—Restriction enzymes and methylases. *Nucleic Acids Res.*, 25, 248–262, 1997.

47. von Mering, C., Huynen, M., Jaeggi, D., Schmidt, S., Bork, P., and Snel, B. STRING: A database of predicted functional associations between proteins. *Nucleic Acids Res.*, 31, 258–261, 2003.

48. Fujibuchi, W., Goto, S., Migimatsu, H., Uchiyama, I., Ogiwara, A., Akiyama, Y., and Kanehisa, M. DBGET/LinkDB: An integrated database retrieval system. *Pac. Symp. Biocomput.*, 683–694, 1998.

49. Wheeler, D.L., Church, D.M., Federhen, S., Lash, A.E., Madden, T.L., Pontius, J.U., Schuler, G.D., Schriml, L.M., Sequeira, E., Tatusova, T.A., and Wagner, L. Database resources of the National Center for Biotechnology. *Nucleic Acids Res.*, 31, 28–33, 2003.

50. Appel, R.D., Bairoch, A., and Hochstrasser, D.F. A new generation of information retrieval tools for biologists: The example of the ExPASy WWW server. *Trends Biochem. Sci.*, 19, 258–260, 1994.

51. Zdobnov, E.M., Lopez, R., Apweiler, R., and Etzold, T. The EBI SRS server—New features. *Bioinformatics*, 18, 1149–1150, 2002.

5 Proteomics Knowledge Databases: Facilitating Collaboration and Interaction between Academia, Industry, and Federal Agencies

Denise B. Warzel, Marcy Winget, Cim Edelstein, Chenwei Lin, and Mark Thornquist

CONTENTS

5.1 INTRODUCTION

Knowledge databases facilitate collaboration between independent researchers by providing access to information and resources otherwise unavailable—the knowledge database forming a linkage between parties. Establishing a knowledge database is complex, particularly in proteomics where data sources are heterogeneous, data formats are incongruent, and the data are related to human subjects. This chapter describes the multiparty collaborative framework and ways to manage it; the difference between a database and a knowledge database; issues related to creating databases containing information linked to human subjects, and techniques for managing the associated IRB issues; references for existing proteomics data resources and Web sites; differences between metadata and data standards and the necessity for both when harmonizing and aggregating heterogeneous data sources and data formats; and references to pertinent metadata and data standards.

5.1.1 KNOWLEDGE DATABASES VS. DATABASES

A database is a collection of information stored on a computer,[1] designed to address the database owner's goals. Data are organized in a specific manner to be accessed with speed and precision, regardless of the database architecture (e.g., relational, hierarchical, flat file) or the software (Oracle, Microsoft Access, SAS). The database owners have specific ideas about the way the data will be entered, analyzed, and retrieved. A central theme is present, often so clear the database name personifies its purpose, such as the Employee Database, Admissions Database, or Inventory Database. Events or actions generate new data recorded over time resulting in the database growing, traditionally in a linear or sequential manner; a direct and clear relationship exists between events.

Knowledge, on the other hand, is the sum of that which is known, an understanding gained through experience, education, awareness, or familiarity.[2] Knowledge is formed multidimensionally by the convergence of memory, background, facts, and data. Thus, a knowledge database implies something bigger, larger, and superior to a database per se. It must provide the user with a particular understanding or scholarship about the theme, drawn from the collective facts and data found therein—wisdom that could not be otherwise attained.

The phrase "Proteomics Knowledge Databases" can be interpreted various ways. For the purpose of this chapter it refers to a new resource created by linking together heterogeneous databases representing the whole of that which is known in proteomics from which understanding can be expanded.

5.1.2 Proteomics Knowledge Database Challenges

Biomedical professionals in various discrete settings conduct proteomics research: in academia, industry, and government agencies (The Parties). Research goals vary from interest in basic science or translational research to interest in technology development and to development of new prevention, diagnostic, and treatment modalities. Each type of researcher and each type of study produces a wealth of data and often the data are stored in databases. These databases are unavoidably heterogeneous, designed to meet individual scientific goals.

It would be difficult if not impossible for one researcher or one study to synthesize or understand the whole of proteomics data without the aid of computers; thus, proteomics is an ideal candidate for a knowledge database. Proteomics research involves the identification and analysis of protein sequences, expressions, functions, and interactions in over 300,000 proteins. Data sources might include information on protein function, protein involvement in enzymatic and other pathways, 3D structure, encoding genes (including polymorphisms, loss-of-function mutations, etc.), technologies to quantify protein amounts or function level (e.g., mass spectrometry), resources for assay development (e.g., monoclonal antibodies), results from various assays (e.g., antibody array data), and from computational tools and methods such as support vector machines, artificial neural networks, and boosting.

A proteomics knowledge database representing the sum of that which is known, endeavoring to impart knowledge, is thus inevitably comprised of an extensive collection of related but still heterogeneous data likely generated by both old and new technologies; multiple parties are needed to successfully create such a knowledge database. Scientific, organizational, and data-related challenges result from this type of project. While proteomics is the common thread between the researchers and their data, the parties involved in creating a knowledge database may be competing for recognition or monetary rewards as well as representing separate research sectors having different scientific goals. Gaining cooperation between independent data owners and harvesting diverse data are central to the effort.

The remainder of this chapter provides insight into ways of addressing these fundamental matters. Topics include organizing and facilitating a multiparty collaboration including IRB issues, implementing metadata and data standards to achieve data harmonization, and identifying proteomics data resources that may provide a source for new proteomics knowledge database efforts.

5.2 MULTIPARTY COLLABORATION

Organizing and facilitating collaboration to create a shared resource for proteomics research requires convening scientists who are traditionally rivals, competitors hoping to be the first to make a novel discovery, win a Nobel Prize, or simply gain a promotion; accordingly these collaborations are often fraught with challenges.

5.2.1 Why Collaborate?

The incentive for The Parties to collaborate is established individually by each party involved in the collaboration and must be felt strongly enough to endure the inevitable challenges of working with other equally talented and enthusiastic researchers. In addition to the fundamental disparity between The Parties are methodology and philosophical and regulatory differences that may not be apparent at the onset. Thus, successful collaboration requires a strong commitment to a collaborative strategy itself, as an entity, calling for cultivation and oversight.

5.2.2 COLLABORATIVE FRAMEWORK

In a complex scientific domain in which researchers frequently compete for recognition and financial rewards, one of the hallmarks of success is when parties come together and form a "strategy or set of strategies, one for each party … such that no party has incentive to unilaterally change their action."[3] This characteristic of collaborative efforts can be likened to the "Nash Equilibrium" game theory. Underlying all such efforts is the drive for individual recognition competing with the drive for collaborative success. In a successful collaboration The Parties must feel that they and each member of the collaboration are essential, that without any one of them, individual and group goals would be impaired. Since a knowledge database aims to be the sum of that which is known, the loss of one of the members of the collaboration cripples the group's ability to achieve its goal.

For example, collaboration is advantageous in biomarker research and development because of its complex multiphase process including discovery, validation, and population study. The need for collaboration in curation of biomarkers has been described as thus:

> Validation is a tedious but necessary step in the curation of a biomarker. There is a need for sufficient preliminary evidence of a prescribed level of sensitivity, specificity, and accuracy. While the excitement of the discovery and its relevance is a strong motivator for biomarker innovators, once the discovery is made and the findings are published, or perhaps even before the findings are published, the next phase of curation begins.

> In order to produce the evidence and validate a biomarker for clinical use, an investigator must have a variety of laboratories perform the assay to test its reproducibility. The involvement of many laboratories in the testing requires larger numbers of samples than can usually be obtained by one institution. The process involves identifying participants and conducting a nested case control study followed by a prospective trial. The samples must be processed and stored appropriately and have a well-developed set of longitudinal clinical data. This data must be consistent, utilizing an established set of normal and abnormal values, and electronically accessible in order to be useful to the investigator and the laboratories. Implicit to the success of collaboration is the need for governance such that these resources are well managed and utilized.

> These additional steps and resources require greater funding than is normally afforded an individual researcher. By joining together in a consortium the potential to gain more funding is improved. The larger pool of researchers, often with proven track records, can heighten grant reviewer's confidence that the proposed validation will be successful.[4]

The nature of public health demands that critical health problems be addressed as quickly as possible. A sacrament of research is the right to publish and be individually recognized for scientific discovery; this must be balanced with the pressing need to accelerate new discoveries in screening, prevention, and treatment modalities. The marriage of information technology and science support the collaborative paradigm, where researchers from across the world can share ideas and information for research and patient care with the touch of a button, enabling collaboration in ways never before possible.

5.2.3 PROCESSES FOR FACILITATING COLLABORATION

As previously stated, collaboration takes commitment usually stemming from the belief that collaboration will result in a better outcome than one obtained from a single investigator. Whether within or across academic institutions, industry, or government agencies, one of the keys to successful collaboration is open and clear communication.

5.2.3.1 Communication

The best way to set the tone for open and clear communication is to have meetings frequently in the early planning stages that include at least one representative from each participating Party. Initially, it is reasonable for meetings to be in the form of conference calls but it is recommended to have a face-to-face meeting early in the planning to help build trust between the collaborators and facilitate communication.

Possible topics to present and discuss at the face-to-face meeting include

The benefits each individual or Party hopes to achieve from the collaboration
A plan of action
The timeline
The expected costs of the project and responsible Parties
The expected deliverables from each Party
Ownership of ideas, specimens, and data
Metadata harmonization
Technology transfer
Project leadership
Differing points of view

Face-to-face meetings should include a combination of formal presentations representing the views of each of The Parties involved and informal discussion time to debate controversial issues and make decisions. Depending on the project, issues surrounding technology transfer and intellectual property may constitute an entire meeting. In order to avoid miscommunication and legal errors on these important topics, it is critical that a legal representative from each of The Parties be involved in these discussions.

5.2.3.2 Infrastructure

To ensure productive and efficient meetings, collaborative projects should have an infrastructure to support planning and running the meetings. Depending on the size of the collaboration the infrastructure might consist of a single administrative assistant to coordinate conference calls, in-person meetings, and take meeting minutes. On the other hand, a large collaboration may require a small team of people that includes administrative, informatics, scientific, and possibly legal expertise. Regardless of the size of the project or the number of people involved in the collaboration, development of draft materials of the discussion items (e.g., draft action plan, timeline, list of deliverables) to hand out during face-to-face meetings will ensure that the meetings are efficient and productive. The purpose of the materials is to provide a starting point for discussion. The infrastructure "team" is responsible for developing and maintaining meeting materials and other study documents.

The materials for the initial meeting should be clearly marked "draft" and the person presenting them should clearly state that the materials were developed only to provide a starting point for discussion. Care should be taken in developing the draft materials so that single solutions are not inserted under potentially controversial topics. Instead, if possible, a list of potential scenarios or solutions (with an indication that the list is not complete) should be presented in order to avoid misunderstanding among meeting attendees or among collaborators that do not attend the meeting but who receive the meeting materials. The meeting documents may serve as first drafts of the study protocol, Manual of Operations, contracts between participating Parties, or other official documents needed for the project.

The importance of organizing meetings to periodically assemble all Parties, providing materials to document the goals and work products and facilitating communications and sharing of resources, cannot be understated. These work products are some of the benefits of the collaborative strategy and the process that helps ensure that individual parties concentrate on the collaborative goals and objectives.

5.2.4 Barriers to Collaboration

While there are important scientific benefits to collaboration, one must acknowledge that there are barriers that must be overcome in order to make collaboration effective. Some of the major barriers are presented below.

5.2.4.1 Information Has Value

First, the information that researchers accumulate has intellectual value in advancing their scientific agenda. Often, researchers will have expended considerable time and resources in collecting their information and want to be sure that they have fully utilized it for their purposes before making it available to others.

Second, information has monetary value to the researchers. Promotion and merit salary increases at research institutions are typically dependent on publication of research findings. The Patent and Trademark Law Amendments Act (a.k.a. the Bayh-Dole Act) encourages biomedical researchers to patent their findings and license them to industry. Making their information publicly available may lead to being preempted by other researchers or may prevent their ability to obtain patent protection on their findings.

The National Institutes of Health (NIH) has recently issued a notice that, effective October 1, 2003, mandates that certain NIH research applications "include a plan for data sharing or state why data sharing is not possible."[5] The release of data must be timely, defined as "no later than the acceptance for publication of the main findings from the final data set."[5] Adoption of the data sharing policy will bring NIH-supported research into line with other government-supported research, such as in space and planetary sciences. However, it remains to be seen how researchers will resolve the new rule with the intellectual property rights provided by the Bayh–Dole Act.

5.2.4.2   Human Subject Information Is Private
and Tightly Regulated

The collection and use of human subject information is governed by Title 45 Code of Federal Regulations (CRF) 46,[7] the federal regulation dealing with protection of human subjects; state regulations; and the Health Insurance Portability and Accountability Act (HIPAA),[8] the privacy rule effective March 2003 that establishes national standards for the protection of health information. The Privacy Rule calls individually identifiable health information "protected health information (PHI)."[9] PHI applies to any "information, including demographic [or payment] information collected from an individual ... [pertaining] to the past, present, or future physical or mental health or condition, of an individual, that identifies the individual or [if] there is a reasonable basis to believe that the information can be used to identify the individual."[10]

Anonymous data, i.e., data from individuals where it is not possible to determine from whom the data came, can be readily shared. However, if there exists a way for an individual's data to be linked to him/her by any person, even if only by a researcher with access to a password-protected computer file containing the link, then the data are not anonymous and the manner of their release must be reviewed and approved by the appropriate Institutional Review Boards (IRBs) and must satisfy the HIPAA requirements. In this case, the degree to which data may be shared is limited by participant consents and whether or not the possibility of the sharing of research data and specimens with commercial entities was addressed. Investigators cannot make such data available in a publicly accessible database unless they receive IRB approval for data sharing, reconsent their participants, or establish mechanisms to prevent access to the database by unapproved entities. Particular attention must be paid to protection of the HIPAA-defined individual identifying elements, which include not just such obvious identifiers as names, addresses, and telephone numbers, but also less obvious identifiers such as zip codes and dates that include more precision than year. Researchers must also ensure that the data shared are the minimum necessary for the intended purpose.

Thus, there can be a substantial time investment in ensuring that data to be shared meet all the regulatory requirements.

5.2.4.3   There Are Costs to Sharing Data

The process of creating a public repository of research data is more complex than putting a study's database on a Web site. First, the study's design, data collection schedule, and detailed data dictionary must be extensively documented, so that individuals accessing the data use them correctly. Most research studies get by with a lesser level of documentation since they rely on the memories of the researchers conducting the study, whereas public repositories' documentation must be complete enough to stand on its own. This documentation must include the metadata that will become the basis for future harmonization activities.

Second, the owners of the repository must receive all appropriate institutional approvals. This includes the IRB and HIPAA approvals described in the previous section, as well as approval for allowing individuals outside the institution to access the computers on which the data reside.

Third, a user-friendly site must be created and tested to ensure that it is easy for users to obtain the information they need.

Finally, the repository, institutional site, and institutional approvals must be maintained; for example, IRB approval must be renewed at least annually. This work requires time and equipment and is frequently of very low priority to the owners of the data. The NIH data sharing initiative, which importantly allows for investigators to include funds for data sharing efforts, can help to reduce the financial burden that arises in the creation of a data repository.

5.2.4.4   There Is No Central Gateway for Dissemination of Data

Many public databases and tools have been made available, as indicated in Table 5.1 and Table 5.2. However, chances are good that within a year of the publication of this book, many of the links provided in that table will no longer work, and the content of the databases that remain will have substantially changed. Further, once the full impact of the Final NIH Statement on Sharing Research Data is felt, there will likely be a larger number of public databases from NIH-sponsored research.

What is needed is a central, permanent portal with indexes and links to all of these databases. The National Cancer Institute (NCI) Cancer Biomedical Informatics Objects (caBIO) web portal (http://ncicb.nci.hih.gov/core/caBIO) uses one approach to addressing this need by utilizing an object model and architecture to help navigate the disparate database formats and access methods.[11] This type of approach will be particularly valuable to encouraging cross-discipline collaboration, since researchers in one domain may be very unfamiliar with the research being performed in other domains and unaware that data relevant to their research has been made public.

5.2.5   Institutional Review Board (IRB) Considerations

There are many challenges facing multisite research trials that use human subjects. It is most important to protect the rights and confidentiality of human subjects. Researchers must consider carefully and respectfully how biological samples are collected, distributed, and utilized for research purposes. Setting up an IRB working group, comprised of the participating Party's IRB representatives is essential to meeting this objective. The charge of this group is to identify issues specific to sharing data from biological samples from human subjects for research purposes.

Considerations identified to be specific to multiple IRB review are listed below:

Timing: Each local IRB has a unique schedule and timeline for reviews.

Consistency: Each IRB has their own set of unique requirements and interpretations of the ever-changing government guidelines; the IRB submission forms are unique and questions asked of investigators may vary between IRBs.

Coordination: All appropriate IRB approvals must be in place before a collaborative study can begin. It is important that required changes or modifications made by any IRB are incorporated into each participating

TABLE 5.1
Pertinent Standards Bodies and Federal Informatics Initiatives

Name	Description	Web Address	Type
ASC X12	The Accredited Standards Committee (ASC) X12 is an American National Standards (ANSI)-accredited organization that develops standards, in X12 and XML formats, for cross-industry electronic exchange of business information.	http://www.x12.org/	Cross-industry data standards
HL7	Health Level Seven (HL7) is an ANSI-accredited Standards Developing Organization (SDO) operating in the healthcare arena. HL7's domain is clinical and administrative data.	http://www.hl7.org	Healthcare metadata standards for electronic transmission
	The Reference Information Model (RIM) is an object model representing the clinical data, life cycles of events, and relationships between the data and HL7 messages fields.	http://www.hl7.org/about/	HL7 Object Model
ISO	International Organization for Standardization (ISO) is a network of national standards institutes from 145 countries working in partnership with international organizations, governments, industry, and business and consumer representatives. ISO is the world's largest developer of standards and although ISO's principal activity is the development of technical standards, ISO standards offer previously defined and accepted standards for data and metadata. The ISO Web site provides access to a number of useful standards; some can be downloaded and others must be ordered in hard copy.	http://www.iso.org http://isotc.iso.org/livelink/livelink/fetch/2000/2489/IHf_Home/Public[yavai-lablestandards.htm	National and international data and metadata standards Freely available JTC1, ISO, and IEC and standards
CDISC	Clinical Data Interchange Standards Consortium (CDISC) is an open, multidisciplinary, nonprofit organization committed to the development of industry standards to support the electronic acquisition, exchange, submission, and archiving of clinical trials data and metadata for medical and biopharmaceutical product development. The CDISC standard includes the following components: Operational Data Modeling (ODM), Submission Data Modeling (SDM), Analysis Data Model (ADaM), and the Laboratory Data Interchange Standards. More information about these and other CDISC standards can be found on the CDISC Web site.	http://www.cdisc.org	Clinical trials data standards

(Continued)

TABLE 5.1
Pertinent Standards Bodies and Federal Informatics Initiatives (continued)

Name	Description	Web Address	Type
CDC	The Centers for Disease Control and Prevention (CDC) is recognized as the lead federal agency for protecting the public health and safety. The agency provides information for researchers and the public. The Data and Statistics section of the CDC home site provides links to current initiatives for scientific, surveillance, health statistics, and laboratory information including information from the Agency for Toxic Substances and Disease Registry (ATSDR).	http://www.cdc.gov	Government agency, healthcare data standards
CMS	Centers for Medical and Medicaid Services (CMS) (formerly HCFA) influences public health data standards and development of information systems. It is responsible for the following: Medicare, Medicaid, State Children's Health Insurance Program (SCHIP), Health Insurance Portability and Accountability Act (HIPAA), and Clinical Laboratory Improvement Amendments (CLIA).	http://cms.hhs.gov/	Government agency influences data standards
DICOM	Digital Imaging and Communications in Medicine (DICOM) Standards Committee exists to create and maintain international standards for communication of biomedical diagnostic and therapeutic information in disciplines that use digital images and associated data. DICOM is a cooperative standard. Every major diagnostic medical imaging vendor in the world has incorporated the standard into their product design and most are actively participating in the enhancement of the standard. Most of the professional societies throughout the world have supported and are participating in the enhancement of the standard as well. DICOM is partners with HL7 and ISO, among others.	http://medical.nema.org/	Data standards for medical imaging
IEEE 1073	The Institute of Electrical and Electronics Engineers 1073 (IEEE1073) series of standards that allow for healthcare providers to plug medical devices into information and computer systems that allow healthcare providers to monitor information from an ICU or through telehealth services on Indian reservations, and in other circumstances.	http://www.ieee1073.org/	Data standards for medical devices and computer systems

LOINC	The purpose of the Logical Observation Identifiers Names and Codes (LOINC®) database is to facilitate the exchange and pooling of results, such as blood hemoglobin, serum potassium, or vital signs, for clinical care, outcomes management, and research. LOINC codes are universal identifiers for laboratory and other clinical observations that solve this problem. The Regenstrief Institute maintains the LOINC database and its supporting documentation.	http://www.loinc.org http://www.regenstrief.org	Laboratory data standards
NCPDP	The National Council on Prescription Drug Programs (NCDCP) is an ANSI-accredited SDO. NCDCP develops standards for ordering drugs from retail pharmacies to standardize information between healthcare providers and the pharmacies. Three standards have been approved: Telecommunications, SCRIPT and Manufacturer Rebate Standard.	http://www.ncpdp.org/	Pharmacy data standards
SNOMED CT®	Created by SNOMED International, a division of the College of American Pathologists (CAP), Clinical Terms® (SNOMED CT®) is a comprehensive and precise clinical reference terminology that can be used to improve the comparability of data. SNOMED CT combines SNOMED RT and Clinical Terms Version 3 (also known as Read Codes V3). The content areas encompass nursing, veterinary, pharmaceutical and cancer and include terminology for clinical observations, procedures, and specimens; provide cross-mappings to ICD-9-CM, ICD-10 (U.K. Edition), NIC, ICD-03, OPCS-4 (U.K. Edition), NANDA, and LOINC®	http://www.snomed.org	Reference terminology resource for clinical data
WHO	The World Health Organization produces a variety of research tools including the WHO family of International Classifications such as International Classification of Diseases (ICD)-10.	http://www.who.int/ research/en	Health data standards such as disease and histopathology

Informatics Initiatives and Resources

NCICB	The NCICB communicates, coordinates, or establishes information exchange standards within the NCI research community. Core informatics tools include vocabulary (EVS), metadata (caDSR), and object models (caBIO) for cancer research.	http://ncicb.nci.nih.gov	Resources for cancer vocabulary and terminology and metadata standards

(Continued)

TABLE 5.1
Pertinent Standards Bodies and Federal Informatics Initiatives (continued)

Name	Description	Web Address	Type
EVS	Enterprise Vocabulary Services (EVS) is part of the NCI caCORE infrastructure. EVS is a set of services and vocabulary.	http://ncicb.nci.nih.gov/core/EVS	Resources for cancer vocabulary and terminology and metadata standards
NCI Metathesaurus	Part of the EVS, the Metathesaurus is an online comprehensive biomedical terminology database that, at the time of this publication, contained 850,000 concepts mapped to 1,500,000 terms by over 4,500,000 relationships. It contains vocabularies from over 70 sources including the National Library of Medicine's UMLS Metathesaurus, as well as a growing number of NCI-specific vocabularies developed by the National Cancer Institute. Vocabularies from GO, LOINC, HL7, and other sources are mapped into the database where they are enriched with thesaurus information and a set of tools to allow semantic browsing, including definitions, synonyms, and broader and narrower concept searches.	http://ncimeta.nci.nih.gov/	Resources for cancer vocabulary and terminology and metadata standards
NCI Thesaurus	The NCI Thesaurus, published by NCI, is a knowledgebase containing the working vocabulary used in NCI data systems. In addition to the NCI concept hierarchy organizing terms into cancer specific semantic concepts, it contains clinical, translational, and basic research as well as administrative terminology.		Resources for cancer vocabulary and terminology and metadata standards
caDSR	Cancer Data Standards Repository (caDSR) supports a broad initiative to standardize the metadata used for cancer research. The caDSR includes tools for creating, editing, browsing, and exporting cancer metadata standards.	http://nci.nih.gov/core/caDSR	Resources for cancer vocabulary and terminology and metadata standards

Name	Description	URL	Purpose
caBIO	caBIO is UML models of biomedical objects to facilitate the communication and integration of information from the various initiatives at NCI such as the Cancer Genome Anatomy Project (CGAP), the Cancer Molecular Analysis Project (CMAP), the Genetic Annotation Initiative (GAI), and clinical trials databases and from publically available databases such as Unigene, HomoloGene, LocusLink, RefSeq, BioCarta, and GoldenPath.	http://nci.nih.gov/core/caBIO	Resources for cancer vocabulary and terminology and metadata standards
NEDSS	The National Electronic Disease Surveillance System (NEDSS) is an initiative that promotes the use of data and information system standards to advance the development of efficient, integrated, and interoperable surveillance systems at federal, state, and local levels. Surveillance System collects and monitors data for disease trends and/or outbreaks so that public health personnel can protect the nation's health. The NEDSS Base System is an example of an NEDSS-compatible system that can be used by a state health department for the surveillance and analysis of notifiable diseases. The NEDSS base system provides both a platform upon which modules can be built to meet state and program area data needs as well as providing a secure, accurate, and efficient way for collecting and processing data.	http://www.cdc.gov/nedss/	Government initiative; data standards
NLM	The United States National Library of Medicine (NLM) located in Bethesda, MD, provides access to publications and databases. The NLM has created a number of web-based research tools and aids, of which MEDLINE/PubMed and MeSH (Medical Subject Headings) are widely known. MeSH is the NLM's controlled subject vocabulary thesaurus and is used for indexing and cataloging medical information. MeSH descriptors, arranged in both alphabetic and hierarchical structure, assist users in searching at various level of specificity. At the time of publication of this book MeSH contained over 20,000 descriptors and over 130,000 supplementary concept records. NLM vocabulary staff have built applications utilizing MeSH to index medical articles in MEDLINE and NLM's Index Medicus®, a monthly guide to over 4,000 journals.	http://www.nlm.nih.gov	Resources for health and scientific vocabulary and terminology

(Continued)

TABLE 5.1
Pertinent Standards Bodies and Federal Informatics Initiatives (continued)

Name	Description	Web Address	Type
NIGMS	The National Institute of General Medical Sciences (NIGMS) is a component of the National Institutes of Health in the U.S. Department of Health and Human Services. By supporting basic biomedical research and training nationwide, NIGMS lays the foundation for advances in disease diagnosis, treatment, and prevention. TargetDB, a result of the Protein Target Database Initiatives in the Structural Genomics Initiative, is a searchable target registration database that was originally developed to provide registration and tracking information for NIH P50 structural genomics centers. TargetDB has now been expanded to include target data from worldwide structural genomics and proteomics projects. To reach the NIGMS Web site and find links to the Structural Genomic Initiative go to the following URL and look under "Major Initiatives."	http://www.nigms.nih.gov/	Proteomics resources
NIST	The National Institute of Standards and Technology (NIST) conducts and supports scientific and engineering research in disciplines ranging from chemistry and physics to information technology. In carrying out this research, NIST works collaboratively with colleagues in industry, academia, and government. Research tools developed at NIST—including measurement methods, standards, data, and various technologies—assist industrial, academic, and government scientists worldwide. To find out more about NIST activities related to researchers, go to the NIST home page and follow the links to "Researchers."	http://www.nist.gov/public_affairs/researchers.htm	Influences and develops standards for industrial, academic, and government research
PHDSC	Public Health Data Standards Coalition (PHDSC). This consortium, officially established in January 1999 as the Public Health Data Standards Consortium, serves as a mechanism for ongoing representation of public health and health services research interests in HIPAA implementation and other data standards-setting processes. The Consortium will facilitate the use of existing national standards and identify priorities for the development of new data standards for public health and health services research. The scope of interest includes all public health data: HIPAA claims-related data (e.g., discharge data); birth and death data; disease registry and surveillance data; immunization data; and birth defects data.	http://www.cdc.gov/nchs/otheract/phdsc/phdsc.htm	Influences public health data standards

TABLE 5.2
Proteomics Databases and Tools

Name	Description	URL and Resource
	DNA/Protein Sequence Searching and Homology Analysis	
Blast® (Basic Local Alignment Search Tool)	Blast® is a set of similarity search programs designed to explore all of the available sequence databases regardless of whether the query is protein or DNA	http://www.ncbi.nlm.nih.gov/BLAST/ NCBI; U.S.
Homology Search	This site provides links to homology search programs such as BLAST — Basic Local Alignment Search Tool (gapped BLAST); SWsrch — Smith-Waterman Searching Programs (Experimental); FastA — Sequence Similarity Searching; SRS BLAST — Sequence Similarity Searching with SRS	http://www.dna.affrc.go.jp/htdocs/homology/homology.html National Institute of Agrobiological Sciences; Japan
BLAST (gapped BLAST) Search Service	Tools that compare a nucleotide query sequence against a nucleotide sequence; an amino acid query sequence against a protein sequence; a nucleotide query sequence translated in all reading frames against a protein sequence database; a protein query sequence against a nucleotide sequence database dynamically translated in all reading frames; the six-frame translations of a nucleotide query sequence against the six-frame translations of a nucleotide sequence database.	http://www.dna.affrc.go.jp/htdocs/Blast/blast.html National Institute of Agrobiological Sciences; Japan
ExPasy Tools	Expert Protein Analysis System. This site provides links to a variety of proteomics informatics tools including tools for protein identification and characterization; DNA protein translation; similarity searches; pattern and profile searches; posttranslational modification prediction; topology prediction; primary structure analysis; secondary structure prediction; tertiary structure; sequence alignment; and biological text analysis.	http://us.expasy.org/tools/ Swiss Institute of Bioinformatics; Switzerland

(Continued)

TABLE 5.2
Proteomics Databases and Tools (continued)

Name	Description	URL and Resource
Advanced BLAST	Advanced BLAST provides most of prokaryotic and eukaryotic DNA and protein databases and all sequence gap penalty matrices. Protein sequence alignments, unlike nucleotide sequences comparisons, allow substitution. The Advanced BLAST scoring matrix determines which substitutions are more likely than others. Advance BLAST has more options than most BLAST search sites, which use common scoring matrices such as BIOSUM50, BIOSUM62, and PAM120.	http://www.ch.embnet.org/software/BottomBLASTadvanced.html Swiss Institute of Bioinformatics; Switzerland
	Protein Structure Analysis	
Block	This tool allows for searching of a protein or DNA sequence against a blocks database. Blocks are multiply aligned ungapped segments corresponding to the most highly conserved regions of proteins.	http://blocks.fhcrc.org/blocks/blocks_search.html Fred Hutchinson Cancer Research Center; U.S.
Dotlet	A diagonal plot tool for sequence comparisons using the dot-matrix method.	http://www.isrec.isb-sib.ch/java/dotlet/Dotlet.html Swiss Institute of Bioinformatics; Switzerland
HLA peptide binding prediction (epitopes finding)	This tool ranks potential 8-mer, 9-mer, or 10-mer peptides based on a predicted half-time of dissociation to HLA class I molecules.	http://bimas.dcrt.nih.gov/molbio/hla_bind/ Computational Bioscience and Engineering Lab (CBEL), Center for Information Technology (CIT), National Institutes of Health (NIH), U.S.
PDB	This is a repository for the processing and distribution of 3D biological macromolecular structure data.	http://www.rcsb.org/pdb/ Rutgers University; U.S.
Pfam	This Web site is a large collection of multiple sequence alignments and hidden Markov models.	US Mirror Site: http://pfam.wustl.edu/ http://www.sanger.ac.uk/Software/Pfam/ Sanger Institute; (Wellcome Trust), U.K.
PFSCAN	Searches all known motifs that occur in a protein sequence. This form lets you paste a protein sequence, select the collections of motifs to scan for, and launch the search.	http://hits.isb-sib.ch/cgi-bin/PFSCAN? Swiss Institute of Bioinformatics; Switzerland

Protscale	Allows computation and representation of the profile produced by any amino acid scale on a selected protein; protein secondary structure properties prediction.	http://www.expasy.org/cgi-bin/protscale.pl Swiss Institute of Bioinformatics; Switzerland
Vast	This is a structure–structure similarity search service that compares 3D coordinates of a newly determined protein structure to those in the MMDB/PDB database.	http://www.ncbi.nlm.nih.gov/Structure/VAST/vastsearch.html NCBI; U.S.

Protein Posttranslational Modification Prediction

FindMod	This tool predicts potential protein posttranslational modifications (PTM) and finds potential single amino acid substitutions in peptides.	http://us.expasy.org/tools/findmod/ Swiss Institute of Bioinformatics; Switzerland
MITOProt	MitoProt calculates the N-terminal protein region that can support a mitochondrial targeting Sequence and the cleavage site. Supports human, mouse, *Caenorhabditis elegans*, and *Neurospora crassa*.	http://www.mips.biochem.mpg.de/cgi-bin/proj/medgen/mitofilter Ludwig-Maximilians-University; Germany Claros, M.G. and Vincens, P. Computational method to predict mitochondrially imported proteins and their targeting sequences. *Eur. J. Biochem.*, 241, 779–786, 1996.
NetNGlyc	The NetNglyc server predicts N-glycosylation sites in human proteins using artificial neural networks that examine the sequence context of Asn-Xaa-Ser/Thr sequins.	http://www.cbs.dtu.dk/services/NetNGlyc/ The Center for Biological Sequence Analysis; Technical University; Denmark
NetOGlyc	The NetOglyc server produces neural network predictions of mucin type GalNAc O-glycosylation sites in mammalian proteins.	http://www.cbs.dtu.dk/services/NetOGlyc/ The Center for Biological Sequence Analysis; Technical University; Denmark
NetPhos	The NetPhos server produces neural network predictions for serine, threonine, and tyrosine phosphorylation sites in eukaryotic proteins.	http://www.cbs.dtu.dk/services/NetPhos/ The Center for Biological Sequence Analysis; Technical University; Denmark
SignalP	The SignalP Web server predicts the presence and location of signal peptide cleavage sites in amino acid sequences from different organisms: Gram-positive prokaryotes, Gram-negative prokaryotes, and eukaryotes. The method incorporates a prediction of cleavage sites and a signal peptide/nonsignal peptide prediction based on a combination of several artificial neural networks.	http://www.cbs.dtu.dk/services/SignalP/ The Center for Biological Sequence Analysis; Technical University; Denmark

(Continued)

TABLE 5.2
Proteomics Databases and Tools (continued)

Name	Description	URL and Resource
Sulfinator	The Sulfinator predicts tyrosine sulfation sites in protein sequences (secretory protein).	http://us.expasy.org/tools/sulfinator/ Swiss Institute of Bioinformatics; Switzerland
TargetP	It predicts the subcellular location of eukaryotic protein sequences. Optionally, TargetP provides a potential cleavage site for sequences predicted to contain a cTP, mTP, or SP.	http://www.cbs.dtu.dk/services/TargetP/ The Center for Biological Sequence Analysis; Technical University; Denmark
Mass Spec/Peptide Fragment Mass Analysis		
PeptideCutter	PeptideCutter predicts potential cleavage sites cleaved by proteases or chemicals in a given protein sequence.	http://us.expasy.org/tools/peptidecutter/ Swiss Institute of Bioinformatics; Switzerland
Peptide-Mass	Peptide-Mass cleaves a protein sequence from the Swiss-Prot and/or TrEMBL databases or a user-entered protein sequence with a chosen enzyme and computes the masses of the generated peptides. The tool also returns theoretical isoelectric point and mass values for the protein of interest. If desired, PeptideMass can return the mass of peptides known to carry posttranslational modifications and can highlight peptides whose masses may be affected by database conflicts, isoforms or splice variants.	http://us.expasy.org/tools/peptide-mass.html Swiss Institute of Bioinformatics; Switzerland
ProFound	Peptide mapping with results of mass proteometry experiments.	http://129.85.19.192/profound_bin/WebProFound.exe The Rockefeller University; U.S.
ProteinProspector	ProteinProspector provides tools for mining sequence databases in conjunction with mass spectrometry experiments.	http://prospector.ucsf.edu/ UCSF; U.S.
Tagident	Tagident is a tool which allows; the generation of a list of proteins close to a given pI and Mw; the identification of proteins by matching a short sequence tag of up to 6 amino acids against proteins in the Swiss-Prot/TrEMBL databases close to a given pI and Mw; and the identification of proteins by their mass, if this mass has been determined by mass spectrometric techniques for one or more species and with an optional keyword.	http://us.expasy.org/tools/tagident.html Swiss Institute of Bioinformatics; Switzerland

(Continued)

Protein Function Knowledgebase Databases

Name	Description	URL/Source
BioCarta	It provides dynamic graphical models for gene and gene interaction. The online maps depict molecular relationships from areas of active research. This Web site also catalogs and summarizes resources providing information for over 120,000 genes from multiple species.	http://www.biocarta.com/genes/index.asp BioCarta; U.S.
CMAP	CMAP contains cancer profiles by chromosomes or 2D arrays; contrasts cancer-to-normal expression ratios displayed by chrosomal locus, using SAGE or EST data; in 2D array format, contrasts expression patterns across multiple experiments, using SAGE data (includes both cancer and normal) or microarray data for NCI60 cell lines.	http://cmap.nci.nih.gov/Profiles NCI, NIH; U.S.
Cope	Cytokines Online Pathfinder Encyclopaedia, including fully integrated MiniCOPE dictionaries on apoptosis, cell lines, chemokines, cytokine topics, hematology, metalloproteinases, virokines, viroceptors, and virulence factors.	http://www.copewithcytokines.de/cope.cgi Ludwig-Maximilians-University, Germany
CSNDB	Cell Signaling Networks Database (CSNDB) is a data and knowledge base for signaling pathways of human cells. It compiles the information on biological molecules, sequences, structures, functions, and biological reactions that transfer the cellular signals.	http://geo.nihs.go.jp/csndb/csn_search.html NIHS; Japan
DRAGON	Database Referencing of Array of Genes Online (DRAGON). This Web site provides free access to a number of bioinformatics tools, allowing a user to annotate a list of genes with information obtained from databases including Unigene, Swiss-Prot, Pfam, and the Kyoto Encyclopedia of Genes and Genomes (KEGG). DRAGON also provides a group of data visualization tools to help interpret microarray or other gene expression data sets.	http://www.dragondb.org Johns Hopkins University, U.S.
Ensembl	Ensembl is a software system that produces and maintains automatic annotation on eukaryotic genomes including a tool to browse the human, mouse, rat, zebrafish, fugu, mosquito, fruit fly, *Caenorhabditis elegans*, and *Neurospora crassa* genomes.	http://www.ensembl.org/ http://www.ensembl.org/Homo_sapiens/ Sanger Institute (Wellcome Trust), U.K.

TABLE 5.2
Proteomics Databases and Tools (continued)

Name	Description	URL and Resource
GeneCard	GeneCards™ is a database of human genes, their products, and their involvement in diseases. It offers concise information about the functions of all human genes that have an approved symbol, as well as selected others.	http://bioinfo.weizmann.ac.il/cards/index.html Weizmann Institute of Science; Israel
GO	Gene Ontology™ is a dynamic controlled vocabulary that can be applied to all organisms even as knowledge of gene and protein roles in cells is accumulating and changing. It also provides annotations of gene products for many organisms, including human.	http://www.geneontology.org/ UC Berkeley; U.S.
INFOBIOGEN	The Atlas of Genetics and Cytogenetics in Oncology and Haematology is a peer-reviewed, online journal and database in free access on the Internet devoted to genes, cytogenetics, and clinical entities in cancer and cancer-prone diseases. The search by themes includes genes; cytogenetic/clinical entities in hematology; solid tumors; cytogenetic/clinical entities; and cancer-prone diseases. Search by chromosomes to find a gene or a chromosome rearrangement is also provided.	http://www.infobiogen.fr/services/chromcancer/ Infobiogen; France; GFCO, ARMGHM, University Hospital of Poitiers – Poitiers University
KEGG	KEGG Regulatory Pathways provides graphical pathway maps, ortholog group tables, and molecular catalogs.	http://www.genome.ad.jp/ Bioinformatics Center, Institute for Chemical Research, Kyoto University, Japan
Mapview	Mapview contains a unified graphical view of maps (genetic and physical) and sequence data for *Homo sapiens*.	http://www.ncbi.nlm.nih.gov/mapview/maps.cgi? NCBI, NIH; U.S.
Membrane protein resources	Membrane protein resources provides links to 3D structures of membrane proteins; Mptopo topology database; Membrane Protein Explorer (MPEx); and Principles of Membrane Protein Folding and Stability.	http://blanco.biomol.uci.edu/MemPro_resources.html UC Irvine; U.S.

Name	Description	URL; Institution
OMIM	Online Mendelian Inheritance in Man (OMIM) is a catalog of human genes and genetic disorders.	http://www.ncbi.nlm.nih.gov/entrez/query.fcgi?db=OMIM&cmd=Limits; Johns Hopkins University; NCBI, NIH; U.S.
PKR	Protein Kinase Resources (PKR) provides information on the protein kinase family of enzymes. This resource includes tools for structural and computational analyses as well as links to related information maintained by others.	http://pkr.sdsc.edu; http://pkr.sdsc.edu/html/index.shtml; University of California (UCSD); SCSD; U.S.
Prowl	Prowl provides amino acids general information; suggested amino acid substitutions; chemical properties; structural properties; and genetic properties.	http://prowl.rockefeller.edu/aainfo/contents.htm; The Rockefeller University; Eli Lilly & Company; U.S.
RefSeq	The Reference Sequence (RefSeq) database aims to provide a biologically nonredundant collection of DNA, RNA, and protein sequences.	http://www.ncbi.nlm.nih.gov/RefSeq/; NCBI, U.S.
SAGE Anatomic Viewer	SAGE Anatomic Viewer displays gene expression in normal and malignant tissues by shading each organ in one of ten colors, each representing a different level of gene expression.	http://cgap.nci.nih.gov/SAGE/AnatomicViewer; CGAP, NCI; U.S.
SEREX	SEREX is a tool within the Cancer Immunome Database (CIDB) that documents the repertoire of antigens eliciting an antibody response in cancer patients.	http://www2.licr.org/CancerImmunomeDB/ (required registration); http://www2.licr.org/CancerImmunomeDB/SerexSearchDB.php; Ludwig Institute for Cancer Research (LICR); UK; Academy of Cancer Immunology; European Cancer Immunome Program, DE
Swiss-Prot	Swiss-Port TrEMBL Advanced queries a protein knowledge in Swiss-Prot and TrEMBL databases. It supports queries by description or by full text.	http://us.expasy.org/sprot/sprot-search.html; Swiss Institute of Bioinformatics; Switzerland
Miscellaneous		
SAGE Data	This Web site uses a new analytical method of reliably matching SAGE tags to known genes. Based on this novel tag to gene mapping, the Web site visualizes human gene expression analysis in tissues or individual libraries using displays that are highly intuitive. The database is comprised of NCI SAGE experiment data and the NCI Sage Genie Tools.	http://cgap.nci.nih.gov/SAGE; (Boon et al. PNAS in press). Ludwig Institute for Cancer Research, U.S.

(Continued)

TABLE 5.2
Proteomics Databases and Tools (Continued)

Name	Description	URL and Resource
EST Data	Tools for searching tissue specific libraries: cDNA Libraries from Tissues; Genes in Tissues — Gene Expression Tissues; SNPs in Tissues	http://cgap.nci.nih.gov/Tissues NCI, U.S.
Genetic Annotation Initiative (GAI)	Utilizing data mining approaches and laboratory methods to perform searches the GAI identifies "candidate" variation in cancer-related genes. Tools include access to the CGAP-GAI Identified Variation in Genes and an SNP finder.	http://gai.nci.nih.gov/ NCI, U.S.
Expression Measurements	Tools for data analysis, data manipulation, and experiment design; information on biological resources including clone sets and nucleic acid reference sets; protocols and information related to performing cDNA and oligo array experiments; and hosts a growing body of publicly available microarray data.	http://dc.nci.nih.gov/ NCI Directors Challenge, U.S.
UniGene	Each UniGene cluster contains sequences that represent a unique gene, as well as related information such as the tissue types in which the gene has been expressed and map location.	http://www.ncbi.nlm.nih.gov/entrez/query.fcgi?db=unigene NCBI, U.S.
HomoloGene	HomoloGene is a resource of curated and calculated orthologs for genes as represented by UniGene or by annotation of genomic sequences.	http://www.ncbi.nlm.nih.gov/HomoloGene/ NCBI, U.S.

Party's documentation and submitted to their IRB for further review before a specific study can proceed.

Monitoring: IRB compliance oversight for all collaborative studies would help confirm IRB approvals are in place.

Guidelines and regulations: Maintaining familiarity with changing human subject regulations and policy guidelines is essential. For example, there are and will continue to be a number of new or proposed federal regulations, policy changes, or recommendations relating to human subject protections, which include the following:

> Department of Health and Human Services (DHHS) requirement for human subject education for key personnel on federally funded grants (June 5, 2000)
>
> DHHS requirement for inclusion of monitoring plans and adverse event reporting for federally funded research (June 5, 2000)
>
> Office for Human Resource Protection (OHRP) procedures for registering IRBs and filing federal wide assurance of protection of human subjects (Dec 3, 2000)
>
> Public Health Service (PHS) policy on instruction in responsible conduct of research (Dec 1, 2000)
>
> Government-wide regulations governing the definition of research misconduct and the handling of research misconduct allegations (December 6, 2000)
>
> Proposed PHS standards for the protection of whistleblowers
>
> National Bioethics Advisory Commission (NBAC) recommendations for research involving human biological materials (July 16, 1999)
>
> NBAC recommendations for the local oversight systems (December 19, 2000)
>
> HIPAA Regulations - 45 CFR 160 and 164, standards for privacy of individually identifiable health information (December 28, 2000)

5.2.5.1 Approaches for Enabling Multiparty IRB Review and Streamlining IRB Process

Proactively addressing the multifaceted and complex IRB issues will enhance the research agenda and provide the foundation for moving research forward. Creating a collaborative platform for investigators and IRB administrators by engaging them in problem solving unique to human subject issues relating to multisite projects will benefit the mission. The following are suggested approaches for enhancing and streamlining the local IRB review process for collaborative studies:

1. Develop an IRB administrator network where IRB issues can be discussed on a regular basis
2. Create a standardized protocol template form for human subjects
3. Pilot a central collaborative review IRB
4. Standardize consent language for collaborative studies
5. Establish an IRB monitoring system to confirm IRB compliance

6. Develop a secure Web site to post IRB specific forms, instructions, and regulations that can be accessed by all participating parties
7. Enhance confidentiality protections
8. Utilize IRB authorization agreements specific to each validation study

It is insightful to have views from the institutional review board representatives, the investigators, the funding agency personnel, and personnel from the OHRP. By proactively identifying possible issues relating to multisite IRB review, you will be able to provide a basis of understanding and education for the collaboration.

A standardized consent form template that can be modified to meet individual party and study needs appears to be a good approach in streamlining the work involved in creating common consent forms that will likely meet with little resistance. By creating a protocol template, investigators work together in developing a single document that incorporates all necessary scientific and human subject considerations while also assuring consistency in all IRB submissions.

In summary, and as proven by the Carotene and Efficacy Trial (CARET), coordinated by the CARET Coordinating Center at the Fred Hutchinson Cancer Research Center in Seattle, Washington, use of a centralized IRB review has shown that the IRB process can be streamlined while maintaining a high level of responsibility for their human subjects.[12] Use of a centralized IRB review will likely streamline the IRB review process for other multisite research projects.

5.3 METADATA AND DATA STANDARDS FOR MULTIPARTY COLLABORATIONS

Harmonizing data is another of the essential components in building proteomics knowledge databases from unrelated databases. In order to ensure semantic equivalence between individual data items, one must be able to understand the data item in its broad context as well as the specific details related to the data item itself. The data characteristics such as its name, length, data type, and definition are among the obvious required attributes; however, it is also necessary to know the details about the protocol or study that produced the data. If data are encoded, a reference document for code translation is needed, as is a list of all the possible values that existed when the item was recorded. Additional information that assists potential users to correctly interpret a given data item include the unit of measure, format, derivation or aggregation rules, the language in which the item was recorded, and the data format. This genus of information is called metadata*. Metadata and data have an important relationship that together transform data

---

* In April 1998, the Federal Geographic Data Committee (FGDC) received a letter from the firm Cislo & Thomas LLP on behalf of their client The Metadata Company advising the FGDC that the word METADATA is a registered trademark (Nos. 1,409,260 and 2,185,504) owned by The Metadata Company. The FGDC forwarded the letter to the Department of the Interior, Office of the Solicitor. The Solicitor responded back to Cislo & Thomas regarding the alleged infringement basically stating that due to the fact that the FGDC was not selling any goods under the word Metadata, that the FGDC was not infringing upon the trademark registration. Furthermore, it was determined that the word "metadata" is widely used within the computer science community and in computer programs without reference to the trademark and therefore the word "metadata" has entered the public domain and is no longer a trademark.

captured in either existing or prospective single party studies or multiparty collaborations into a shared resource.

5.3.1 METADATA VS. DATA

Metadata are data that define and describe other data.[13] The term meta comes from Greek and generally means change, after or beyond as in metamorphosis.[13] In information technology it is commonly used to mean more comprehensive, underlying, or fundamental as in metapsychology or metamathematics. Thus metadata deals with the more critical aspects of data. Use of metadata provides us with the ability to produce and exchange data that are

Machine understandable
Human understandable
Interoperable

While data are the representation of facts about something real, metadata are comprised of all the attributes that convey qualities about the data and should include definitions, permissible values, and data characteristics such as data type, minimum and maximum length, and reference to the standard, if used, to record the fact (e.g., ICD-09 code for a diagnosis or ISO 3166 code for country).

Adequate metadata to make possible the above are necessarily capacious, time consuming, and arduous to produce. The International Standards Organization (ISO) and the International Electrotechnical Commission (IEC) developed the ISO/IEC 11179 Information Technology—Metadata Repositories (MDR) Parts 1–6[14–19] (ISO/IEC 11179) for the specific purpose of facilitating worldwide standardization by providing guidance for identification, development, and description of data elements. In particular, ISO/IEC 11179 Part 3 describes the metadata required to define a data element in an unambiguous manner. Adoption and deployment of metadata standards, such as the ISO/IEC 11179, and data standards for database development by all parties involved in a multiparty collaboration can greatly simplify database implementation and thus smooth the achievement of harmonization and interoperability between databases downstream, even without a specific application in mind.

The benefits of incorporating metadata and data standards are that they

Minimize effort in the management and utilization of clinical and research data without compromising data quality
Enable development of software with broad applications, facilitating additional uses of the data
Allow data integration to generate a holistic view and facilitate analyses across multiple studies
Ease compatibility with other health data standards, information technology standards, naming conventions and systems
Ease the implementation of regulatory requirements such as HIPAA broadly across studies

Enable development of data that are defined, collected, stored and transmitted in identical ways, making data more easily sharable, retrievable, interoperable and useful to all groups

Reduce training, programming, data entry, and associated costs

Knowledge requires understanding generally gained from experience. With regard to databases, metadata are a surrogate for experience. By utilizing metadata, which outwardly and fully describe the data independent of a particular computer system, database, or party, reuse within multiparty collaboration is enhanced, even if each party has never before been exposed to the others' databases or computer systems.

5.3.2 STANDARDS RESOURCES

While metadata provide the information necessary for interoperability, its presence does not ensure semantic equivalence. Adoption and use of standard vocabularies and terminologies for storing data will enhance understanding and interoperability. At the time of publication of this book there are few proteomics-specific metadata or data standards to draw upon; but, there are specific vocabulary standards that can be utilized to enhance semantic understanding and it is the organizing party's responsibility to ensure that these standards are examined and appropriately incorporated into the data design. Metadata and data standards developed by accredited American National Standards Institute (ANSI) Standards Development Organizations (SDO), such as ISO, the Institute of Electrical and Electronics Engineers (IEEE), Health Level 7 (HL7), and Logical Observation Identifier Name Codes (LOINC) are assured to have been through a rigorous and controlled release process and therefore provide a sound starting point for parties looking to create metadata or data standards for use in a new database.

Other reliable sources for standards are those arising from government-funded agencies such as the National Library of Medicine (NLM), who are developing both tools and vocabulary services to support research, of which are the Unified Medical Language System (UMLS) and Medical Subject Headings (MeSH).[20] Two components of the NCI caCore, a key initiative of the NCI Center for Bioinformatics (NCICB) (http://ncicb.nci.nih.gov) are the NCI Enterprise Vocabulary Services (EVS), which provides tools for accessing the UMLS and cancer specific vocabularies, and the Cancer Data Standards Repository (caDSR), which provides tools for creating and accessing cancer-specific metadata. Both of these tools are accessible via the NCICB Web site. Other publicly available standard vocabularies and terminologies include Gene Ontology (GO), LOINC, International Classification of Diseases (ICD), and World Health Organization (WHO).

The role of data standards and metadata cannot be understated. It is worth noting that in March 2003, the Departments of Defense, Veterans Affairs, Health and Human Services announced a commitment to developing standards to facilitate the electronic exchange of clinical health information.[21] These three agencies have joined forces with other federal agencies as part of the Consolidated Health Informatics initiative (CHI). Leading by example, and driven by a desire to improve public health care and decrease costs, the members of the CHI announced that federal agencies will adopt several standards to promote the use of electronic health data systems and programs including electronic health records. The standards that were selected were HL7, National Council on Prescription Drug Programs (NCDCP), Institute of Electrical and Electronics Engineers 1073 (IEEE1073), Digital Imaging Communications in Medicine (DICOM), and LOINC (see Table 5.1).

Another resource for biomedical metadata and data standards is the Centers for Disease Control (CDC), where work is being conducted to develop health data standards for the National Electronic Disease Surveillance System (see Table 5.1). Some standards bodies, such as HL7 and LOINC, are formed by or include participation by influential academic, industry, and governmental agencies where research is conducted. Standards bodies generally provide a current membership list on their Web sites. Therefore, it is helpful to be aware of the dominant standards bodies and resources for identifying and using appropriate existing standard vocabularies in building metadata to record proteomic data.

5.4 WEB RESOURCES AND EXISTING DATABASES

Existing proteomics databases can provide a basis for prospective proteomics knowledge database development. As mentioned previously, while many projects and databases provide proteomic information, each project owner organizes their information somewhat differently, creating databases to meet specific end-user requirements. Owing to their nature these databases are heterogeneous, presenting different data formats and access mechanisms, and thus the creation of a knowledge database from existing information is challenging but, with appropriately designed informatics tools, achievable.

Two technical approaches for unifying heterogeneous databases to form a knowledge resource are found within projects sponsored by the NCI. The NCI's Cancer Bioinformatics Infrastructure Objects (caBIO) model and architecture,[22] also part of the NCI caCORE infrastructure, and the NCI's Early Detection Research Network (EDRN). caBIO is discussed briefly below and the EDRN project is discussed in Chapter 4. Both approaches utilize metadata and software to access disparate data sources harvesting knowledge otherwise beyond reach.

As noted in Table 5.2, information regarding proteomics databases and Web sites available at the time of publication may provide insight into sources for future knowledge database efforts.

5.4.1 caBIO

caBIO unifies disparate data sources by utilizing model-driven software to dynamically perform user-directed queries. caBIO information models describe both the scientific and informatics aspects of database information. A model is something that represents or is patterned after another. In computer programming modeling of databases, concepts, or classes of information (objects) are often documented using Universal Modeling Language (UML). UML draws on specific syntactic and visual presentation to portray the semantic meaning of relationships between objects.[23]

Creating a model describing known relationships between classes of information separately from the description of a specific database is a way to allow post-coordination of disparate data without knowing the specific contents of each data source. Separating the scientific model from the database models in this way allows the caBIO software to access any relational database pertaining to the scientific model by using the semantic meanings conveyed by the UML model. This layer of abstraction provides a way to access data semantically, regardless of the specific relational database in which data is stored.

The caBIO scientific model includes genomic and clinical trials information; the database models are based upon specific NCI and non-NCI data sources. The underpinnings for the genomic and clinical trials model are classes of information such as gene, chromosome, protein, agent, and protocol. Described by metadata and available to the caBIO software, an application named BIOgopher, are class-specific attributes such as "gene name" and "agent source"; data characteristics such as data type and definition; search criteria such as "gene clone name" and "agent NSC number"; and relationships between the classes such as gene "has a" chromosome.

These components, the scientific model and the database models, are described using controlled terminology and vocabularies from the NCICB EVS and represented by metadata in the NCI Cancer Data Standards Repository (caDSR), another part of the NCI caCORE infrastructure. BIOgopher utilized the information contained in the models together with the associated metadata to integrate multiple data sources; portraying them in a single representation to the end-user. Using the relationships and search criteria when examining the various data sources the software reveals biomedical knowledge that was previously undetectable.[24] The data sources for caBIO include the Cancer Genome Anatomy Project (CGAP), the Cancer Molecular Analysis Project (CMAP), the Genetic Annotation Initiative (GAI), clinical trials databases as well as other publicly available data repositories including Unigene, HomoloGene, LocusLink, RefSeq, BioCarta, and GoldenPath (through Distributed Annotation System [DAS]). caBIO provides interfaces via the Internet, allowing anyone to access the knowledgebase either directly through BIOgopher queries or programmatically through Application Programming Interfaces (APIs). The caBIO software is open source and can be downloaded from the NCICB Web site. More information about caBIO, its UML model, and the metadata can be found in a technical document, "caCORE Technical Guide" and in the NCICB caDSR, both part of the NCICB Core and accessible via the NCICB home page (http://ncicb.nci.nih.gov).

5.5 CONCLUSIONS

Knowledge databases represent the sum of that which is known about a specific theme. Creating a knowledge database from existing or new data requires collaboration between multiple parties, often with competing or opposing goals; the databases themselves represent a diverse set of heterogeneous data formats and access methods. An initiative to form a new proteomics knowledge database requires facilitating multiparty collaboration and harmonization of data and metadata. Targeted bioinformatics tools and techniques can create new knowledge sources from existing ones, allowing information to be harvested in new ways and collaborations to occur between diverse parties.

Multiparty collaboration tips:

Strive for strategic equilibrium
Facilitate communication
Provide infrastructure to coordinate conference calls, meetings, and documentation
Set up an IRB working group
Adopt, establish, and maintain metadata and data standards
Recognize barriers
Recognize that information has intellectual and monetary value
Recognize that human subject information is private and regulated
Understand that there are initial and ongoing costs of sharing data
Understand that there is no central gateway monitoring public tools and database changes

Careful planning and coordination and the use of bioinformatics tools can enable the creation of a proteomics knowledge database to facilitate collaboration.

REFERENCES

1. *Merriam-Webster's Collegiate Dictionary,* 10th ed. s.v. "database."
2. *Merriam-Webster's Collegiate Dictionary,* 10th ed. s.v. "knowledge."
3. GameTheory.net, "Nash Equilibrium," http://www.gametheory.net/Dictionary/Nash Equilibrium.html (May 30, 2003).
4. Bunn, P. University of Colorado Cancer Center Director, personal communication, February 2002.
5. U.S. Department of Health and Human Services, National Institutes of Health, Final NIH Statement on Sharing Research Data, NOT-OD-03-032. February 2003; 1.
6. U.S. Department of Health and Human Services, National Institutes of Health, Final NIH Statement on Sharing Research Data, NOT -OD-03-032. February 2003; 2.
7. U.S. Department of Health and Human Services, National Institutes of Health, Office for Protection from Research Risks, Title 45, Part 46, Code of Federal Regulations, 160, 103.
8. Office of Civil Rights, U.S. Department of Health and Human Services, Health Insurance Portability and Accountability Act. Standards for Privacy of Individually Identifiable Health Information—Rules and Regulations. *Federal Register,* 65, 250, 2000.
9. Office of Civil Rights, U.S. Department of Health and Human Services, Summary of the HIPAA Privacy Rule. May 2003.
10. Office of Civil Rights, U.S. Department of Health and Human Services, Health Insurance Portability And Accountability Act Of 1996, Public Law 104-191, 104th Congress Part C, Section 1172:6.
11. Covitz, P.A., Hartel, F., Schaefer, C., De Coronado, S., Fragoso, G., Sahni, H., Gustafson, S., and Buetow, K.H. caCORE: A common infrastructure for cancer informatics. Bioinformatics, 19:2402–12, 2003.

12. Thornquist, M.D., Edelstein, C., Goodman, G.E., and Omenn, G.S. Streamlining IRB review in multi-site trials through single-study, IRB Cooperative Agreements: Experience of the beta-Carotene and Retinol Efficacy Trial (CARET). *Contr. Clin. Trials,* 23, 80-86, 2002.

13. *Merriam-Webster's Collegiate Dictionary,* 10th ed. s.v. "meta."

14. International Organization of Standardization and the International Electrotechnical Commission. International Standard. Information Technology. Metadata Registries (MDR)—Part 1. Framework ISO/IEC 11179–1, 2004.

15. International Organization of Standardization and the International Electrotechnical Commission. International Standard. Information Technology Specification and Standardization of Data Elements—Part 2. Classification of Data Elements. ISO/IEC 11179–2, 2000.

16. International Organization of Standardization and the International Electrotechnical Commission, International Standard. Information Technology. Metadata Registries (MDR)—Part 3. Registry metamodel and basic attributes. ISO/IEC 11179–3, 2003 (E).

17. International Organization of Standardization and the International Electrotechnical Commission. International Standard. Information Technology. Metadata Registries (MDR)—Part 4. Formulation of data definitions. ISO/IEC 11179–4, 2004 (E).

18. International Organization of Standardization. Information Technology. Specification and Standardization of Data Elements—Part 5. Naming and Identification Principles for Data Elements. ISO/IEC 11179–5, 1995.

19. International Organization of Standardization. Information Metadata Registries (MDR)—Part 6. Registration. ISO/IEC 11179–6, 2005.

20. National Library of Medicine (NLM), National Institutes of Health, U.S. Department of Health and Human Services, Library Services. Retrieved from http://www.nlm.nih.gov/libserv.html, May 2003.

21. Department of Health and Human Services, CMS Public Affairs, Federal Government Announces First Federal eGov Health Information Exchange Standards, Press Release. March 21, 2003.

22. Covitz, et al.

23 Rumbaugh, J., Jacobson, I., and Booch, G. *The Unified Modeling Language Reference Manual.* Addison-Wesley, Reading, MA, 1999.

24. Covitz, et al.

6 Proteome Knowledge Bases in the Context of Cancer

Djamel Medjahed and Peter A. Lemkin

CONTENTS

6.1   INTRODUCTION

The origin of most cancers can be often traced to a single transformed cell.[1] The evolution of the disease follows a yet-to-be completely understood pathway of molecular transformations occurring at both genomics and proteomics levels as depicted in Figure 6.1. Most cancers show a significant preponderance to statistically originate from a well-defined part of their respective organs. It is then only normal that investigations to identify biomarkers indicative of the early onset of the disease be focused on these organ-specific regions.

This point was elegantly demonstrated by Page et al.[2] in a careful experiment, where they used magneto-immuno-chemical purification methods to extract pure cell populations and compare the protein expression observed in experimental two-dimensional poly-acrylimide gel electrophoresis (2D PAGE) maps obtained from normal, milk-producing luminal epithelial cells exhibiting a tendency to exhibit carcinomas vs. outer, myoepithelial cells as described in Figure 6.2.

This thorough characterization was achieved by using a combination of enabling technological platforms, some of which are listed in Figure 6.3, which allowed them to flag a number of proteins exhibiting a significant differential expression between the two types of cells and therefore warranting a closer evaluation of their potential

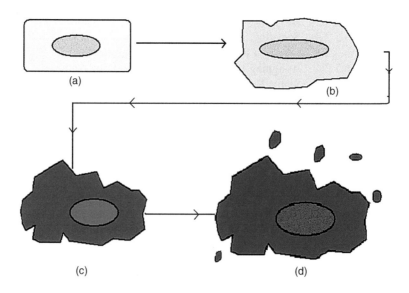

FIGURE 6.1 Illustration of the progressive evolution from a normal cell (a) to precancer (b and c), and finally the cancerous state (d).

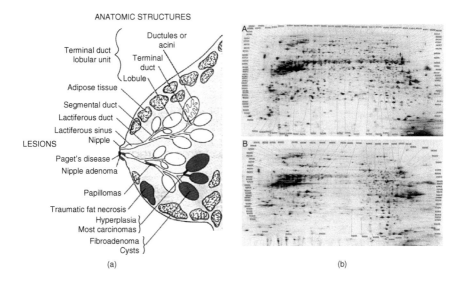

FIGURE 6.2 Dissection of the organ breast. (a) The lobular-alveolar regions show the nature of lesions statistically originating from them. (b) Comparison of protein expression profiles in the inner, epithelial luminal (A) that account for 95% of breast carcinomas vs. (B) outer, myoepithelial of healthy patients. The annotations are those of 51 proteins, which display more than twofold expression change between the two samples.

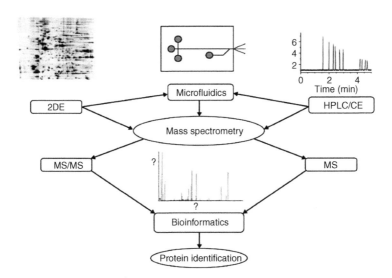

FIGURE 6.3 Technological platforms used in proteomic characterization.

as biomarkers of breast cancer. The time and costs involved in using these techniques can be quite prohibitive, particularly on a large scale.

This led to the initial motivation to address the need that either on a routine basis, or to establish optimal experimental conditions beforehand, one might be interested in predicting the gene products likely to be detected in narrow ranges of isoelectric focusing point (pI) and molecular weight (Mw).

We believe that the initial search for cancer biomarkers can greatly benefit by formulating hypotheses developed from knowledge-based bioinformatic tools. This chapter will describe in some detail two such predictive databases whose development was at least in part motivated by these pressing issues.

6.2 VIRTUAL 2-D: A WEB-ACCESSIBLE PREDICTIVE DATABASE FOR PROTEOMICS ANALYSIS

Over the past three decades and thanks to continuous developments in chemistry,[3] automation, and data collection,[4–6] 2D PAGE[7,8] has evolved from a labor intensive, multiprocess protein separation method to becoming an integral part of most comprehensive proteomics efforts.[9–12] In particular, the advent of immobilized pH gradients[13] in the first dimension has ushered in an era where reproducible, high-resolution iso-electric focusing measurements can routinely be carried out, making it conceivable to predict from the primary sequence the equilibrating positions of proteins within a pH gradient. When solubilized with high concentrations of urea (8.5–10 M), proteins unfold and only the ionizable groups or those amino acids located at the N- or C-terminal amino acids will affect the electrophoretic mobility of the extended conformation. Using a series of well-characterized peptides, Bjellqvist[14] determined the pK values of all the amino acids in similar experimental conditions.

The approach used to determine the isoelectric focusing point and molecular mass of a peptide can then simply be summed up as follows:

1. Scan the primary sequence of the peptide
2. Assign the pK of each contributing amino acid according to Table 6.1
3. Sum up all the mass contributions

The resulting Pi/Mw for the peptide is then given by the ratio of:

$$\{pK_{Cterm} + \Sigma_{int}\, pK_{int} + pK_{Nterm}\}$$

$$Pk_{tot} = (n - 2)$$

and

$$M_{rtot} = \Sigma_i\, M_t i \qquad (6.1)$$

where the pI summation runs over all n contributing, internal amino acids.

TABLE 6.1
Values of Amino Acid Masses and pKs (Determined[14] at High Molar Concentrations of Urea Used in pI MW Computation)

Ionizable Group	PK[a]	Molecular Mass
C-terminal	3.55	
N-terminal		
Met	7.00	132.994
Thr	6.82	102.907
Ser	6.93	88.88
Ala	7.59	72.88
Val	7.44	100.934
Glu	7.70	130.917
Pro	8.36	98.918
Internal		
Asp	4.05	116.89
Glu	4.45	130.917
His	5.98	138.943
Cys	9	104.94
Tyr	10	164.978
Lys	10	114.961
Arg	12	157.989
C-terminal side chain groups		
Asp	4.55	116.89
Glu	4.75	130.917

[a] The pKs of roughly half the internal amino acids fall below pH 6.0, while for the rest they are greater than or equal to 9.0 leading to the segregation of the resulting pIs.

6.2.1 DATABASE MINING

Homo sapiens were the first of several organisms to be examined. The resulting plot of pI versus the molecular mass yields a theoretical 2D PAGE map with a striking bimodal distribution (Figures 6.4 and 6.5). A total of 86,518 inferred or experimentally determined peptides were included in this calculation. One obvious feature of these maps is the presence of a region seemingly devoid of proteins centered around pH 7.4 to 7.5.

The biochemical justification most often advanced in explanation of this observation is that the majority of proteins would tend to naturally precipitate out of solution around the cytoplasmic pH of approximately 7.2. The pI is the pH for which the protein charge is overall neutral. It therefore represents the point of minimum solubility due to the absence of electrostatic repulsion, resulting in maximum aggregation. While this provides an explanation for experimental 2D PAGE maps, we must remember that no such correction was incorporated in the modeling. What then is the basis for the separation of proteins into acidic and basic domains in computed pI/MW charts? In our efforts to answer these questions, we carried out a simulation whereby groups of 1545 peptides varying in length from 50 to 600 AA, in increments of 10, were

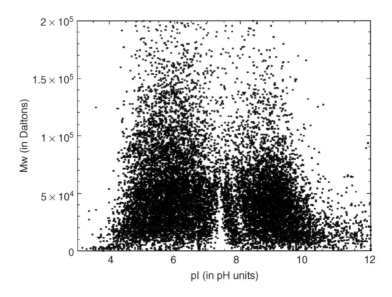

FIGURE 6.4 pI/MW map for *Homo sapiens*. To keep in line with the experimental limits encountered in practice, the pI/MW plot has been confined to less than 2×10^5 kD for the molecular mass and $3.0 < pI < 12.0$ for the isoelectric focusing point. As shown in Figure 6.5, this pattern is by no means unique to *Homo sapiens* and has been reported for other organisms.[21-23]

randomly generated. This brings the total number of simulated sequences to 86,520 vs. 86,518 real peptides extracted from current databases, thereby improving the prospects of any meaningful comparative statistics. As mentioned earlier, the calculation of the pI values is carried out iteratively. The pK of a peptide is calculated by tallying the contributions to the charge from the n-terminus, the c-terminus, and the internal portion of the peptide. As can be observed in Figure 6.6, the resulting simulated pI/MW distribution is strikingly similar to that adopted by the extracted sequences. While this may seem surprising at first, given the total absence of bias in both the lengths and content of the peptides used for the simulation, it is in fact a direct consequence of the constraints imposed by a limited proteomic alphabet of twenty amino acids with distinct pKs, roughly half of which are either acidic or basic (Table 6.1).

In fact, as is reflected in Table 6.1, only seven internal amino acids make non-zero contributions to the pI of the peptide. These seven amino acids are cysteine, aspartic acid, glutamic acid, histidine, lysine, arginine, and tyrosine. It is reasonable to suspect that a high percentage of the variation in the calculated pI values of the simulated data would be modulated by the representation of these seven amino acids as the majority of the contribution to the charge comes from the internal portion of the peptide. To investigate the actual contribution of these seven amino acids in determining an overall pI value, a multiple regression model was developed using the adjusted numbers of these seven amino acids as predictor variables and the pI value as the dependent variable. The adjusted count for an amino acid is equal to

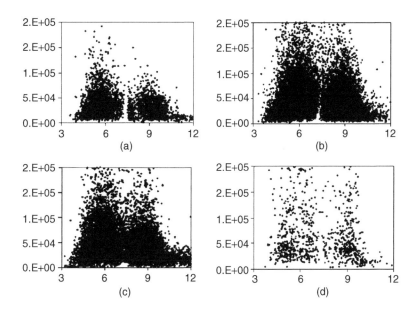

FIGURE 6.5 pI/MW charts for (a and b) *E. coli*, (c) mouse, and (d) *Plasmodium falciparum*.

the actual number of times the amino acid is found in the peptide divided by the length of the peptide. The adjusted counts will be denoted as follows:

    aR = adjusted count for arginine
    aC = adjusted count for cysteine
    aD = adjusted count for aspartic acid
    aE = adjusted count for glutamic acid
    aK = adjusted count for lysine
    aH = adjusted count for histidine
    aY = adjusted count for tyrosine

The regression model in question uses the linear, quadratic, and cubic powers for each adjusted number of the seven amino acids that contribute to the pI calculation when they are part of the interior of the protein. A total of 21 independent variables were employed in the regression analysis. This analysis yields a multiple correlation factor R of .931. The coefficient of determination (the square of the multiple R) gives the proportion of the total variance in the dependent variable accounted for by the set of independent variables in a multiple regression model. For the model in question, .866 is the square of the multiple R. Consequently, 86.6% of the total variation in the pI values was accounted for by the aforementioned seven amino acids. The simulation result confirms the hypothesis that the total number of these seven amino acids is the key factor is explaining the pI value of a peptide.

The predicted pI score in the regression model is denoted as pI′ and it is the dependent (criterion) variable in the regression model. The equation for the regression model is

$$\mathrm{pI'} = a + \Sigma b_i X_i \tag{6.2}$$

where a is the intercept of the model, b_i is the partial slope for the ith predictor in the model, and X_i is the ith predictor in the model. There will be 21 different predictors in the model: 7 linear terms (aR, aC, aD, etc.), 7 quadratic terms (aR^2, aC^2, aD^2, etc.), and 7 cubic terms (aR^3, aC^3, aD^3, etc.). All parameters were estimated by ordinary least squares using the SPSS 8.0 computer package.[15]

The coefficient of determination or R^2 for the model is the proportion of variance of the pI values accounted for by the regression model. It is equal to the sum-of-squares regression divided by the total sum-of-squares:

$$\Sigma \mathrm{pI'} - <(\mathrm{pI})>)^2$$

$$R^2 = \Sigma(\mathrm{pI} - \mathrm{pI'})^2 \tag{6.3}$$

where $<(\mathrm{pI})> = \Sigma \ \mathrm{pI}/N$

Unpredictable bottlenecks associated with Internet traffic and limitations in the size of the files that could be downloaded at any given time from the pI/Mw server force one to typically fragment the proteome of an organism into several smaller files no bigger than 2000 gene product entries. A Perl script was written to address this issue, and, when applied to organism-specific, curated proteome datasets in FASTA format downloaded from the European Bioinformatics Institute's Web site, will output tab-delimited files of the molecular mass, pI, Swiss-Prot accession number and identification for each protein entry. In order to increase the analytical value of Virtual2D to the scientific community, interactivity is built into these plots by implementing the following features (displayed in Figure 6.7).

Possibility of using the database on any JAVA-enabled computer
Pan, zoom, and click features
With an Internet connection, hyperlinks between each data point and popular databases (Swiss-Prot, NCBI, etc.)

6.2.2 COMPARISON WITH EXPERIMENTAL DATA

Computed pI/MW values were compared against those reported experimentally in two cases. In the first example, a high-resolution map for *E. coli* obtained over a narrow pH range (4.5–5.5) was used. Landmarks provided by reference proteins whose characteristics were independently confirmed can be used to calibrate positions over the entire area of the image. pI, molecular masses, and relative intensities can then be determined by interpolation for all detected protein spots (Figure 6.6a). A minimally distorted "constellation" consisting of proteins whose predicted pI/MW values are fairly close to their experimentally determined counterpart, displayed in

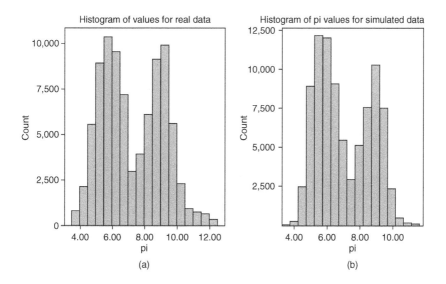

FIGURE 6.6 Side-by side comparison of "pI/MW" histograms for *Homo sapiens*. (a) Computed using amino acid sequences from TrEMBL/Swiss-Prot vs. (b) randomly generated as described in the text.

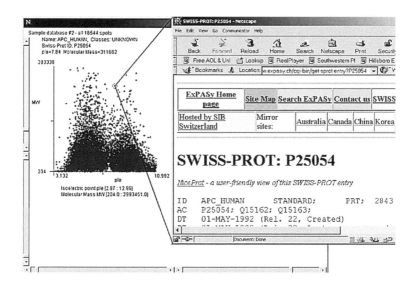

FIGURE 6.7 On-the- fly interaction and identification. By using the controls, one can zoom in on a particular area. Simply moving the mouse over or clicking on any spot will either display a short description or bring up comprehensive information from the hyperlinked Web server of choice. (Protplot uses Java code modified from MicroArray Explorer).

Figure 6.6b can then be used in principle to "warp" (align) the experimental gel onto the theoretical one.

To understand warping in its simplest form, one can imagine dividing up the gel into several regions around each one of these pairs of spots so that for any given region the local experimental landmark will be transformed to its predicted counterpart by a translation specific to that neighborhood (Figure 6.7). Any experimental spot (including the landmark) within region 1, for instance, will undergo the same local translation defined by

$$X_{\text{pred}} = X_{\text{exp}} + \Delta X_1$$

$$Y_{\text{pred}} = Y_{\text{exp}} + \Delta Y_1 \tag{6.4}$$

where ΔX_1 and ΔY_1 are the components of the local translation needed to bring an experimental landmark onto its predicted counterpart. If the spot happens to be in region 3, then

$$X_{\text{pred}} = X_{\text{exp}} + \Delta X_3$$

$$Y_{\text{pred}} = Y_{\text{exp}} + \Delta Y_3 \tag{6.5}$$

and so on.

For those areas without a designated landmark, such as region 2, one can interpolate using the translations from the surrounding neighborhoods

$$X_{\text{pred}} = X_{\text{exp}} + \Delta X_2$$

where

$$X_2 = (\Delta X_1 + \Delta X_3 + \Delta X_6)/3$$

$$X_{\text{pred}} = Y_{\text{exp}} + \Delta Y_2$$

and

$$\Delta Y_2 = (\Delta Y_1 + \Delta Y_3 + \Delta Y_6)/3 \tag{6.6}$$

The outcome of this two-dimensional alignment is not a trivial task as it is a function of several factors including the resolution of the experimental gel (the higher, the better) as well as the number and spatial distribution of landmark reference points. It involves working out the transformations that reflect the local distortions of the gel. Several software packages[16–18] currently existing on the market offer robust and flexible spot detection from many popular image file formats coupled with sophisticated statistical and warping tools.

In the second example, we (arbitrarily) selected and downloaded from Swiss-2D PAGE a map of human colorectal epithelia cells.[19] Figure 6.8 depicts the overlap of observed and corresponding computed pI/MW values for 40 proteins. A quantitative

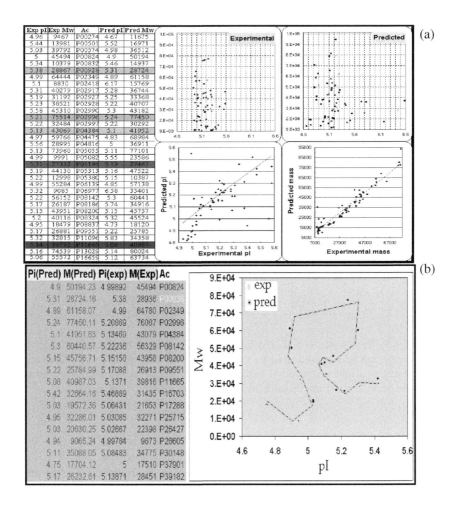

FIGURE 6.8 (a) Comparison of the values of isoelectric focusing points and molecular mass extracted from a high-resolution *E. coli* 2D PAGE map downloaded from Swiss-2D PAGE and those computed in this work. In the two upper charts, a small number of corresponding data points from each set have the same color for a quicker visual inspection. (b) For a small subset of proteins, computed pI/MW values are fairly close to the experimental counterparts, providing a "constellation" of reference points that can be used for warping.

measure of the discrepancy between the two data sets can be obtained by using the relative shift (r.s) of a protein spot between experimental and theoretical values

$$r.s = [(\Delta pI/pI_{exp})^2 + (\Delta Mw/Mw_{exp})^2]^{1/2}$$

where

$$\Delta pI = pI_{exp} - pI_{pred} \text{ and } \Delta Mw = Mw_{exp} - Mw_{pred} \tag{6.7}$$

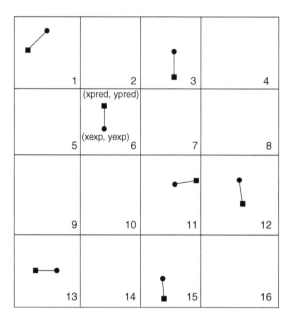

FIGURE 6.9 The warping of a 2D PAGE map on a computed pI/MW chart can be achieved by dividing it in areas surrounding each pair of experimental (●) and predicted (■) landmarks and applying to all the protein spots belonging in a particular neighborhood the necessary local translation to transform the coordinates (Xpred, Ypred) to (Xexp, Yexp).

Despite the broad nominal intervals for pI (4-8 pH units) and MW (0-200kD), more than 66% of the predicted values have a relative shift less than or equal to 0.12 compared to their observed counterpart. However, one must still face the reality of the numerous types of modifications occurring co- and post-translationally that can severely alter the electrophoretic mobility of the proteins affected. As can be seen in Figure 6.9, while relatively small local differences can easily be reconciled, no amount of warping will be able to totally and correctly align a collection of computed pI/MW data points onto a set of experimentally determined protein spots without individually identifying and incorporating the aforementioned corrections in the computation of these attributes.

6.3   TMAP (TISSUE MOLECULAR ANATOMY PROJECT)

By mining publicly accessible databases, we have developed a collection of tissue-specific predictive protein expression maps (PEM) as a function of cancer histological state. Data analysis is applied to the differential expression of gene products in pooled libraries from the normal to the altered state(s). We wish to report the initial results of our survey across different tissues and explore the extent to which this comparative approach may help uncover panels of potential biomarkers of tumorigenesis, which would warrant further examination in the laboratory. For the third

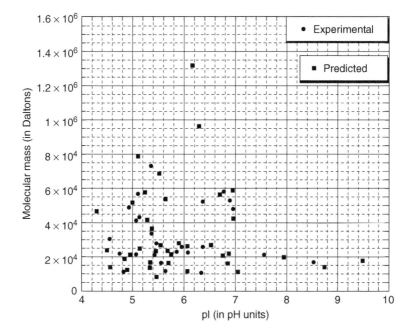

FIGURE 6.10 (**Color insert follows page 204**) Overlap of pI/MW experimental (●) and theoretical (■) values for spots identified in a 2D PAGE map of human colorectal epithelial obtained from Swiss-2D PAGE.

dimension, we computed inferred gene-product translational expression levels from the transcriptional levels reported in the public databases. A number of studies[2,6], have explored the feasibility of molecular characterization of the histopathological state from the mRNA abundance reported in public databases. Many potential tissue-specific cancer biomarkers were tentatively identified as a result of mining expression databases. Thus arose the motivation to explore and catalogue correlations across different tissues as a first step toward comparative cancer proteomics of normal vs. diseased state. One potential clinical application is uncovering threads of biomarkers and therapeutic targets for multiple cancers.

6.3.1 DATA MINING

For each tissue, the CGAP database can be queried by possible histological state, source, extraction, and cloning method. In the initial construction of queries, selecting the option "ANY" from within all of these fields provides an initial overview of the available libraries available. The more restrictive the search, the fewer libraries were selected. Within each library, transcripts are listed along with the number of times they were detected after a fixed number of PCR cycles. Since we were primarily interested in computing protein maps, the gene symbols associated with those ESTs that were clustered to a gene of known function were extracted from UNIGENE.

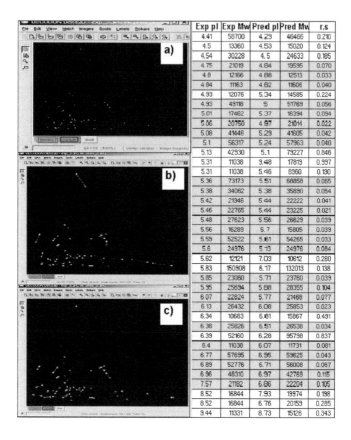

Exp pI	Exp Mw	Pred pI	Pred Mw	r.s
4.41	58700	4.29	46466	0.210
4.5	13360	4.53	15020	0.124
4.54	30228	4.5	24633	0.185
4.75	21019	4.84	19595	0.070
4.8	12166	4.88	12513	0.033
4.84	11163	4.82	11606	0.040
4.93	12076	5.34	14585	0.224
4.93	49118	5	51769	0.056
5.01	17462	5.37	16394	0.094
5.06	30750	4.97	21014	0.022
5.08	41448	5.29	41605	0.042
5.1	56317	5.24	57963	0.040
5.13	42930	5.1	79227	0.846
5.31	11038	3.48	17819	0.997
5.31	11038	5.46	8960	0.130
5.36	73173	5.51	68858	0.065
5.38	34062	5.38	35890	0.054
5.42	21346	5.44	22222	0.041
5.46	22765	5.44	23225	0.021
5.48	27623	5.56	26829	0.039
5.56	16289	5.7	15805	0.039
5.59	52522	5.61	54265	0.033
5.6	24976	5.13	24976	0.064
5.62	12121	7.03	10612	0.280
5.83	150908	6.17	132013	0.138
5.85	23060	5.71	23760	0.039
5.95	25694	5.88	28355	0.104
6.07	22824	5.77	21468	0.077
6.13	26432	6.08	25853	0.023
6.34	10683	6.81	15867	0.491
6.38	25826	6.51	26538	0.034
6.39	52160	6.28	95798	0.837
6.4	11038	6.07	11731	0.081
6.77	57695	6.95	59625	0.043
6.89	52776	6.71	56008	0.067
6.96	48310	6.97	42769	0.115
7.57	21182	6.86	22204	0.105
8.52	16844	7.93	19974	0.198
8.52	16844	6.76	20159	0.285
3.44	11331	8.73	15126	0.343

FIGURE 6.11 (Color insert follows page 204) (a) Overlap of spots identified in 2D PAGE map of human colorectal epithelial cell line (in green) and theoretically computed (in red). (b) Several pairs of corresponding experimentally predicted spots are connected to reflect the translations. (c) A global warping attempts to bring the computed value closer to the corresponding observed member of the pair. While in some cases an almost exact local alignment is achieved, in many instances the differences caused by posttranslation modifications are simply too large to successfully align. This analysis was carried out using a demonstration version of the Delta-2D package.[18]

A Perl script performed the cross-reference checking between the two data sets and output a list of gene symbols and corresponding Swiss-Prot/trEMBL accession numbers (AC). The list of resulting AC was input to the pI/MW tool server, which computed the necessary pI (isoelectric focusing point) and molecular mass (Mw) for the mature, unmodified proteins.[12] In the case of a single library, this information was married to the expression-detection counts in the following manner: The number of hits for each EST was first divided by the sum total of sequences within that library to provide a relative expression for each transcript. Finally, a renormalization was carried out by dividing relative expression levels by the maximum relative expression level. In the event that a tissue search revealed several libraries fulfilling

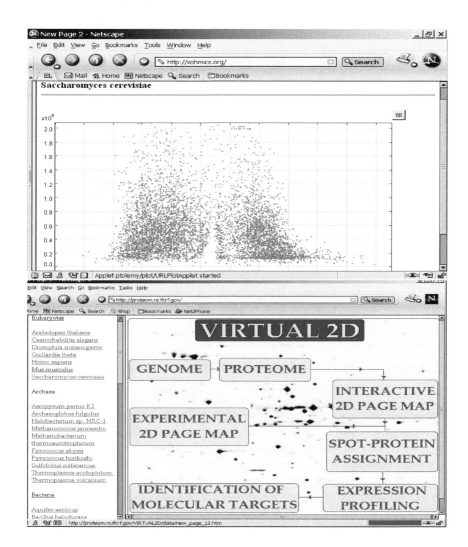

FIGURE 6.12 A snapshot of the screen display of VIRTUAL2D protein expression maps computed for 92 organisms/proteomes using data obtained from the European Bioinformatics Institute[24] can be displayed by clicking on any of the entries on the left.

the requirements of the initial query, to improve the signal-to-noise ratio, the results are first pooled to generate a nonredundant list of entries and a more comprehensive expression map for that tissue and corresponding to that histological state. The databases used are shown in Figure 6.13, and the detailed flow chart is depicted in Figure 6.14.

6.3.2 PROTPLOT

ProtPlot is a Java-based data-mining software tool for virtual 2D gels. It was derived from Opensource MAExplorer project (MAExplorer.sourceforge.net). It may be

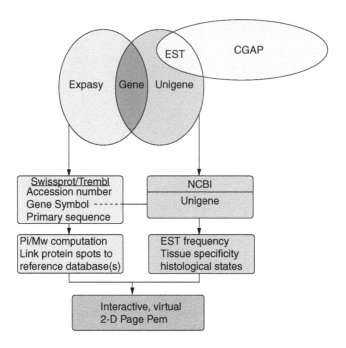

FIGURE 6.13 Overview of the public databases and mining strategy used.

downloaded and run as a stand-alone application. Its exploratory data analysis environment provides tools for the data mining of quantified virtual 2D gel (pIe, Mw, expression) data of estimated expression from the CGAP EST mRNA tissue expression database. This lets one look at the aggregated data in new ways; for example, which estimated "proteins" are in a specified range of (pI,Mw)? Or which sets of estimated "proteins" are up- or downregulated or missing between cancer samples and normal samples? Which sets of "proteins" cluster together across different types of cancers or normals? Here, one may aggregate several different normal and several different cancers as well as specify other filtering criteria.

As is well known, mRNA expression generally does not correlate well with protein expression as seen in 2D PAGE gels.[20] However, some new insights may occur by viewing the transcription data in the protein domain. If actual protein expression data is available for some of these tissues, it might be useful to compare mRNA estimated expression and actual protein expression. This tool may help find those proteins with similar expression and those that have quite different expression. This might be useful in thinking about new hypotheses for protein post-modifications or mRNA posttranscription processing.

ProtPlot generates an interactive virtual protein 2D gel map scatterplot based on a database of derived maximum EST expression over a variety of tissue types from data obtained from the NCI-NCBI CGAP EST database of human cancer, precancer, and cancer mRNA expression (CGAP is the NCI's Cancer Genome Anatomy Project [http://cgap.nci.nih.gov/]. EST is the expressed sequence tag of

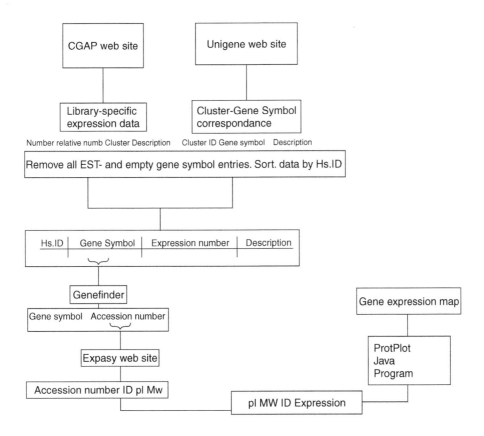

FIGURE 6.14 Flow chart describing in detail steps in the computation of expression maps.

mRNA found in particular tissues). The EST hit rate is a rough estimate of gene expression. These ESTs were mapped to Swiss-Prot (http://www.expasy.ch) accession numbers and Ids; the Mw and pI estimates were computed and used as estimates for corresponding proteins in a pseudo 2D gel.

ProtPlot data is contained in a set of tissue- and histology-specific .prp (i.e., ProtPlot) files described in the data format documentation. These are kept in the PRP directory that comes with ProtPlot when you install it. You will be able to update these .prp files from the ProtPlot Web server http://www.lecb.ncifcrf.gov/TMAP.

6.3.2.1 Using ProtPlot for Data Mining Virtual Protein Expression Patterns

First, one needs to download and install ProtPlot on a local computer. The detailed steps are shown in the following screen shots. This downloads the ProtPlot Java program and the CGAP-derived data set of pseudo 2D gels. If one downloads the

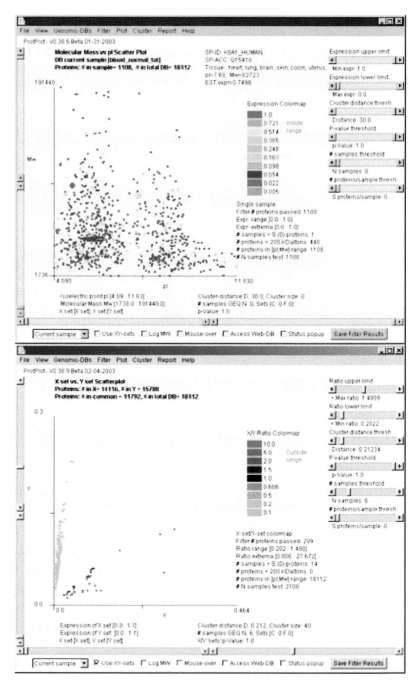

FIGURE 6.15 (Color insert follows page 204) Snapshot of scatterplots from one sample in ProtPlot (top). It is also possible to create (bottom) an (X vs. Y) scatterplot or (mean X set vs. mean Y set) scatterplot when the corresponding ratio display mode is set. The following window shows the (mean X set vs. mean Y set) scatterplot.

(a)

(b)

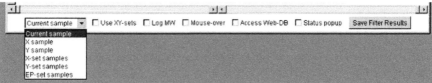

FIGURE 6.16 (a) Tissue and histology selection panel. (b) This may be invoked either from the File menu or the pull-down sample selector at the lower left corner of the main window.

version that includes the Java Virtual Machine (JVM), it will not interact with any other JVM installed.

ProtPlot is started by clicking on the ProtPlot startup icon (Windows, MacOS-X, etc.) or by typing ProtPlot on the command line (Unix, Linux, and other systems).

Once the ProtPlot program is started, it loads the set of PRP files (Figure 6.16a) that were downloaded with the ProtPlot program. The virtual protein data for each tissue is used to construct a master protein index where proteins will be present for some tissues and not for others. The data are presented in a pseudo 2D gel image with the estimated isoelectric point (pI) on the horizontal axis and the molecular mass (MW)

FIGURE 6.17 Snapshot of popup status window.

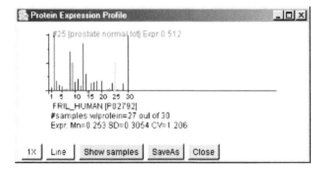

FIGURE 6.18 One can at a glance obtain the expression profile of proteins or groups of proteins across tissues of choice.

(Figure 6.15) on the vertical axis. Sliders on each of the axes allow one to control the minimum and maximum values of pI and Mw displayed and thus the Mw vs. pI scatterplot zoom region one wants to select. By clicking on a spot in the scatterplot, the information on that protein will be displayed. One can also define that protein as the current protein. The current protein is used in some of the clustering methods, protein-specific reports (expression profile report), and the expression profile plot. If one has enabled the popup Genomic-ID Web browser and is connected to the Internet, a Web page from the selected Genomic database for that protein will pop up. One then selects various options from the pull-down menus. Some of the more commonly used options are replicated as check boxes at the bottom of the window.

6.3.2.2 The Scatterplot Display Mode

There are two primary types of pseudo 2D gel (Mw vs. pI) scatterplot display modes (summarized in Table 6.2) of this derived protein expression data: expression mode or ratio mode. The expression data may be for a single sample (the current sample) or the mean expression of a list of samples (called the expression profile, or EP). The ratio data is computed as the ratio of two individual samples called X and Y.

TABLE 6.2

Display Mode	Current Sample	Single X/Y	X Set/Y Set	EP Set
Expression	Yes	No	No	No
Single samples ratio	No	Yes	No	No
X-set and Y-set samples ratio	No	No	Yes	No
Mean Expression	No	No	No	Yes

TABLE 6.3

Filter Name	Current Sample	Single X/Y	X Set/Y Set	EP Set
> 200K Daltons	Yes	Yes	Yes	Yes
Tissue type	Yes	Yes	Yes	Yes
Expression (Ratio) range	Expression	Ratio	Ration	Expression
X/Y (inside/outside) range	No	Yes	Yes	No
(X set, Y set) t-Test	No	Yes	Yes	No
(X set, Y set) KS—Test	No	Yes	Yes	No
(X set, Y set) Missing data	No	Yes	Yes	No
At Most (Least) N samples	No	No	Yes	Yes
AND of saved cluster set	Yes	Yes	Yes	Yes
AND of saved filter set	Yes	Yes	Yes	Yes

Ratio data may alternatively be computed from sets of X samples and sets of Y samples. Generally, one would group a set of samples with similar characteristics together having the same condition (e.g., cancer, normal, etc.). The ratio of X and Y may be single samples, in which case the ratio is computed as

$$\text{Ratio} = (\text{expression } X/\text{expression } Y) \qquad (6.9)$$

where expression X/(expression Y) is the expression of corresponding proteins. Alternatively, one may compute the ratio of the mean expression of two different sets of samples (the X set and the Y set). The X and Y sets may be thought of as experimental conditions and the members of the sets being "replicates" in some sense; e.g., the X set could be cancer samples and the Y set could be normal samples. The ratio of the X/Y sets for each corresponding protein is computed as

$$\text{Ratio} = (\text{mean } X - \text{set expression}/\text{mean } Y - \text{set expression}) \qquad (6.10)$$

Figure 6.15 (bottom) shows a screen shot of one of the (MW vs. pI) scatterplots when the display mode was set to (X set/Y set) ratio mode.

6.3.2.3 Effect of Display Mode on Filtering, Clustering, and Reporting

A particular display mode is selected using the Plot menu commands. When one selects a particular display mode, it will enable and disable Filter, View, Cluster, and Report options depending on the mode. For example, one may only use the t-test or missing XY set test if one is in the XY sets ratio mode. Clustering can only be performed in EP set mode. One may change the display mode using the Plot menu Show Display mode commands. Alternatively, since it is used so often, there is a check box at the bottom of the main window "Use XY sets" that will toggle between the XY sets ratio mode and the previously set mode.

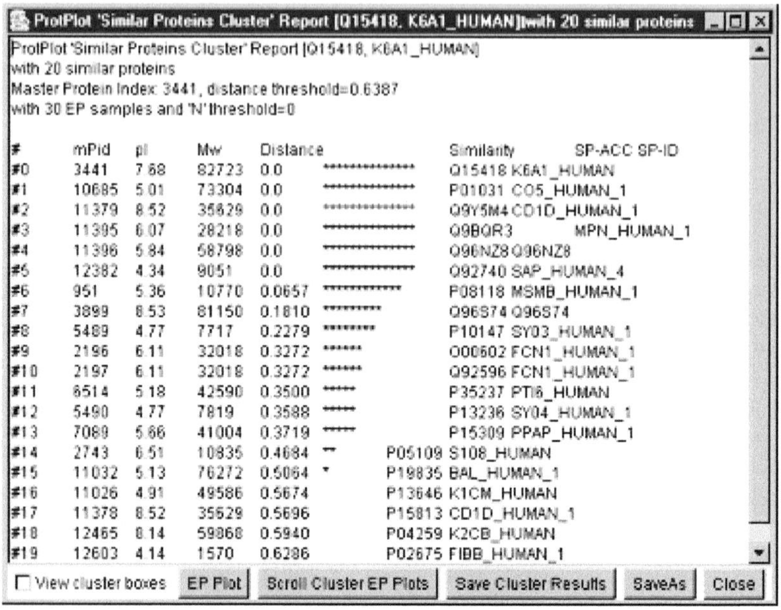

```
┌─────────────────────────────────────────────────────────────────────────────────┐
│ ⧉ ProtPlot 'Similar Proteins Cluster' Report [Q15418, K6A1_HUMAN]with 20 similar proteins  ▄ ▢ ▣ │
├─────────────────────────────────────────────────────────────────────────────────┤
│ ProtPlot 'Similar Proteins Cluster' Report [Q15418, K6A1_HUMAN]                   ▲ │
│ with 20 similar proteins                                                            │
│ Master Protein Index 3441, distance threshold=0.6387                                │
│ with 30 EP samples and 'N' threshold=0                                               │
│                                                                                     │
│ #      mPid    pl    Mw      Distance                  Similarity      SP-ACC SP-ID  │
│ #0     3441    7.68  82723   0.0    ****************    Q15418 K6A1_HUMAN            │
│ #1     10685   5.01  73304   0.0    ****************    P01031 CO5_HUMAN_1           │
│ #2     11379   8.52  35629   0.0    ****************    Q9Y5M4 CD1D_HUMAN_1          │
│ #3     11395   6.07  28218   0.0    ***************     Q9BQR3        MPN_HUMAN_1    │
│ #4     11396   5.84  58798   0.0    **************      Q96NZ8 Q96NZ8                │
│ #5     12382   4.34  9051    0.0    **************      Q92740 SAP_HUMAN_4           │
│ #6     951     5.36  10770   0.0657 *************       P08118 MSMB_HUMAN_1          │
│ #7     3899    8.53  81150   0.1810 **********          Q96S74 Q96S74                │
│ #8     5489    4.77  7717    0.2279 *********           P10147 SY03_HUMAN_1          │
│ #9     2196    6.11  32018   0.3272 *******             O00602 FCN1_HUMAN_1          │
│ #10    2197    6.11  32018   0.3272 *******             Q92596 FCN1_HUMAN_1          │
│ #11    6514    5.18  42590   0.3500 *****               P35237 PTI6_HUMAN            │
│ #12    5490    4.77  7819    0.3588 *****               P13236 SY04_HUMAN_1          │
│ #13    7089    5.66  41004   0.3719 *****               P15309 PPAP_HUMAN_1          │
│ #14    2743    6.51  10835   0.4684 **          P05109 S100_HUMAN                    │
│ #15    11032   5.13  76272   0.5064 *           P19835 BAL_HUMAN_1                   │
│ #16    11026   4.91  49586   0.5674             P13646 K1CM_HUMAN                    │
│ #17    11378   8.52  35629   0.5696             P15813 CD1D_HUMAN_1                  │
│ #18    12465   8.14  59868   0.5940             P04259 K2CB_HUMAN                    │
│ #19    12603   4.14  1570    0.6286             P02675 FIBB_HUMAN_1                ▼ │
├─────────────────────────────────────────────────────────────────────────────────┤
│ ☐ View cluster boxes  [EP Plot]  [Scroll Cluster EP Plots]  [Save Cluster Results]  [SaveAs]  [Close] │
└─────────────────────────────────────────────────────────────────────────────────┘
```

FIGURE 6.19 This window illustrates the scrollable list of EP plots sorted by the current cluster report similarity.

6.3.2.4 Selecting Samples

Samples for the current sample, X sample, Y sample, X set samples, Y set samples, and EP-set samples are selected using a popup check box list chooser of all samples. For example, one may invoke this chooser for the specific tissue sample one wants to view by using the File menu | Select samples | Select Current PRP sample. For $X(Y)$ data, one invokes the choosers using File menu | Select samples | Select $X(Y)$ PRP sample(s). One may switch between single (X/Y) and (X set/Y set) mode using the File menu | Select samples | Use Sample X and Y sets else single X and Y samples (CB) command.

There is an alternative display called the Expression Profile (EP) plot (see Figure 6.18), which displays a list of a subset of PRP samples for the currently selected protein. One may also display the scatterplot on the mean EP data for all proteins. The EP samples are specified using the File menu | Select samples | Select Expression List of samples command.

6.3.2.5 Listing a Report on Sample Assignments

A report of the current sample assignments for the current sample single X sample, single Y sample, X sample set, Y sample set, and EP sample set may be obtained using the File menu | Select samples | List sample assignments command.

6.3.2.6 Assigning the *X* Set and *Y* Set Condition Names

The default experimental condition names for the X and Y sample sets are "X set" and "Y set." One may change these by the File menu | Select samples | Assign X (Y) set name commands.

6.3.2.7 Status Reporting Window

There is a status popup window (Figure 6.17) that first appears when the program is started and reports the progress while the data is loading. After the data is loaded, it will disappear. Toggling the "Status popup" checkbox at the bottom of the window will make it reappear. One may also press the "Hide" button on the status popup window to make it disappear.

6.3.2.8 Data Filtering

The pseudo-protein data is passed through a data filter consisting of the intersection of several tests including pI range, MW range, sample expression range, expression ratio (X/Y) range (either inside or outside the range), t-test comparing the X and Y sample sets, Kolmogorov-Smirnov test comparing the X and Y sample sets, missing proteins test for X and Y sample sets, tissue type filter, protein family filter (to be implemented), and clustering. The filtering options are selected in the Filter menu. Looking at the scatterplot in ratio mode, one may filter by ratio of X/Y either inside or outside of the ratio range. The missing protein test defines "missing" as totally missing and "present" as having at least "N" samples present. Note that the t-test and the missing protein test are mutually exclusive in what they are looking for, so using both results in no proteins found.

6.3.2.9 Saving Filtered Proteins in Sets for Use in Subsequent Data Filtering

One may save the set of proteins created by the current data filter settings by pressing the "Save Filter Results" button in the lower right of the main window. This set of proteins is available for use in future data filtering using the Filter menu | Filter by AND of Saved Filter proteins (CB). Saving the state of the ProtPlot database (Filter menu | State | Save State) will also write out the save protein sets (saved filtered proteins and saved clustered proteins) in the database "Set" folder with ".set" file name extensions. In the Filter menu | State | Protein Sets submenu there are a number of commands to manipulate protein set files. One may individually save (or restore) any particular saved filtered set to (or from) a set file in the Set folder. There are also commands to compute the set intersection, union, or difference between two protein set files and leave the resulting protein set in the saved Filter set.

6.3.2.10 Filter Dependence on the Display Mode

Note that the particular filter options available at any time depend on what the current display mode is. Table 6.4 shows which options are available for which display modes.

TABLE 6.4

Filter Name	Current Sample	Single X/Y	X Set/Y Set	EP Set
Statistics or proteins passing filter	SP-ACC/ID, pI, Mw, expression	SP-ACC/ID, pI, Mw, X/Y, X, Y expr., tissues	SP-ACC/ID, pI, Mw, mnX/mnY, (mn, sd, cv, n) expr. for X and Y sets, tissues. If using t-test then (dF, t-stat, F-stat). If using KS-test then (dF, D-stat)	SP-ACC/ID, pI, Mw, (mn, sd, cv, n) expr. for EP set, tissues
Expression profiles of proteins passing filter	SP-ACC/ID, expr. data EP set	SP-ACC/ID, expr.,data EP set	SP-ACC/ID, expr. data EP set	SP-ACC/ID, expr. data EP set
X and Y sets of missing proteins pasing filter	No	No	SP-ACC/ID, (mn, sd, cv, n) for X and Y sets	No
EP set statistics of proteins passing filter	No	No	No	SP-ACC/ID, (mn, sd, cv, n) for EP set
List of samples in current EP profile	{Nbr, sample-name, expression}	{Nbr, sample-name, expression}	{Nbr, sample-name, expression}	{Nbr, sample-name, expression}
List of all sample assignments	Current, X, Y, X set, Y set, EP set	Current, X, Y, X set, Y set, EP set	Current, X, Y, X set, Y set, EP set	Current, X, Y, X set, Y set, EP set
List of # proteins/sample	{Sample-name, # proteins in sample}	{Sample-name, # proteins in sample}	{Sample-name, # proteins in sample}	{Sample-name, # proteins in sample}
ProtPlot state	State	State	State	State

6.3.2.11 The Data Mining State

The current data mining settings of ProtPlot are called the "state." They may be saved in a named startup file called the "startup state file" in the State folder. The State folder and other folders used by ProtPlot are found in the directory where ProtPlot is installed. Initially there is no startup state file. If one saves the state, then this file is created. As many of these saved state files can be created as desired, one may change the file and thus save various combinations of settings of samples for the current, X, Y, and expression list of samples. The state also includes the various filter, view, and plot options as well as the pI, MW, expression, ratio, cluster distance threshold, number samples threshold, p value threshold sliders, and other settings. The saved Filter and Cluster sets of proteins are also written out as .set files in the Set folder when the state was saved.

Starting ProtPlot by clicking on the ProtPlot startup icon will not read the state file when it starts up. However, if a state is saved, clicking on the state file or a shortcut to the state file will cause it to be read when ProtPlot starts up.

The current state can be saved using either the File | State | Save State command to save it under the current name or the File | State | Save As State command to save it under a new name. The current state may be changed using File | State | Open State file command.

6.3.2.12 The Molecular Mass vs. pI Scatterplot: Expression or Ratio

There are two types of scatterplots: expression for a single sample or the ratio of two samples X and Y. The Plot menu lets one switch the display mode. Ratio mode itself has two types of displays: red (X) + green (Y), or a ratio scale ranging between <1/10 (green) and >10 (red). One may view a popup report of the expression or ratio values for the current protein. If "mouse-over" is enabled, then moving the mouse over a spot will show the name of the protein and its associated data. If mouse-over is not enabled, then clicking on the spot will show its associated data. One may scroll the scatterplot in both the pI and MW axes by adjusting the endpoint scrollbars on the corresponding axes. In addition, one may display the scatterplot with a log transform of MW by toggling the log MW switch.

The popup plots and scatterplot may be saved as .gif image files, which are put into the project's Report folder. Similarly, reports are saved as tab-delimited .txt text files in the Report folder. Because a file name is prompted for, one may browse one's file system and save the file in another disk location.

6.3.2.13 X Sample(s) vs. Y Samples Scatterplot

In X/Y ratio mode (single X/Y samples or X-set/Y-set samples), a scatterplot of the X vs Y expression data can be viewed. Enable the XY scatterplot using the Plot menu | Display (X vs. Y) else (MW vs pI) scatterplot if ratio mode (CB). The scatterplot can be zoomed similar to the MW vs. pI scatterplot. The proteins displayed are those passing the data filter that have both X and Y data (i.e., expression is > 0.0).

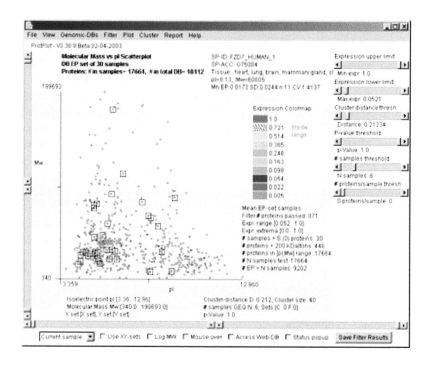

FIGURE 6.20 (**Color insert follows page 204**) The spots marked by boxes belong to the same cluster.

6.3.2.14 Expression Profile Plot of a Specific Protein

An expression profile (EP) shows the expression for a particular protein for all samples that have that protein. The Plot menu | Enable expression profile plot pops up an EP plot window and displays the EP plot for any protein selected. The relative expression is on the vertical axis and the sample number on the horizontal axis. Pressing on the "Show samples" button pops up a list showing the samples and their order in the plot. Pressing on the "n×" button will toggle through a range of magnifications from $1 \times$ through $50 \times$ that may be useful in visualizing low values of expression. Clicking on a new spot in the Mw vs. pI scatterplot will change the protein being displayed in the EP plot. Within the EP plot display, one may display the sample and expression value for a plotted bar by clicking on the bar (which changes to green with the value in red at the top). The EP plot can be saved as a .gif file. One may also click on the display to find out the value and sample. Note: since clustering uses the expression profile, one must be in "mean EP-set display" mode.

6.3.2.15 Clustering of Expression Profiles

One may cluster proteins by the similarity of their expression profiles. First set the plot display mode to "Show mean EP-set samples expression data." The clustering method is selected from the Cluster menu. Currently there is one cluster method;

others are planned. The cluster distance metric is the distance between two proteins based on their expression profile. The metric may be selected in the Cluster menu. Currently, there is one clustering method: cluster proteins most similar to the current protein (specified by clicking on a spot in the scatterplot or using the Find Protein by name in the Files menu). It requires one to specify a) the current protein, and b) the threshold distance cutoff. The threshold distance is specified interactively by the "Distance Threshold T" slider. The Similar Proteins Cluster Report will be updated if either the current protein or the cluster distance is changed.

The cluster distance metric must be computed in a way to take missing data into account since a simple Eucledian distance cannot be used with the type of sparse data present in the ProtPlot database. ProtPlot has several ways to compute the distance metric using various models for handling missing data.

One may save the set of proteins created by the current clustering settings by pressing the Save Cluster Results button in the lower right of the cluster report window. This set of proteins is available for use in future data filtering using the Filter menu | Filter by AND of Saved Clustered proteins (CB). When the state of the ProtPlot database is saved (Filter menu | State | Save State), the set of saved clustered proteins will be saved in the database Set folder. One may restore any particular saved clustered set file.

The EP plot window can be brought up by clicking on the EP Plot button and then clicking on any spot in the scatterplot to see its expression profile. Clicking on the Scroll Cluster EP Plots button brings up a scrollable list of expression profiles for just the clustered proteins sorted by similarity.

One may mark the proteins belonging to the cluster in the scatterplot with black boxes by selecting the View Cluster Boxes check box at the lower left of the cluster report window. This is illustrated in Figure 6.20.

6.3.2.16 Reports

Various popup report summaries are available depending on the display mode. All reports are tab-delimited and so may be cut and pasted into MS Excel or other analysis software. Reports also have a "Save As" button so data can be saved into a tab-delimited file. The default/Report directory is in the directory where ProtPlot is installed. However, it can be saved anywhere on one's file system. The content of some reports depends on the particular display mode. This is summarized in Table 6.5.

6.3.2.17 Genomic Databases

If one is connected to the Internet and has enabled ProtPlot to "Access Web-DB," then clicking on a protein will pop-up a genomic database entry for that protein. The particular genomic database to use is selected in the Genomic-DB menu.

6.4 RESULTS AND DATA ANALYSIS

Figure 6.21 depicts the pI/MW maps computed by our approach for a number of these tissues. They all display the characteristic bimodal distribution that was explained previously as the statistical outcome of a limited, pK-segregated proteomic

TABLE 6.5
Swiss-Prot Accession Numbers (for Those Gene Products Displayed in Figure 6.21, with the Highest and Lowest Cancer/Normal Expression Ratios, Respectively)

Blood		Brain		Breast	
Upregulated	Downregulated	Upregulated	Downregulated	Upregulated	Downregulated
	O00215	P04075	O00184	P02571	O43443
	P01907	P12277	O14498	P05388	O43444
	P01909	P41134	O15090	P12751	O60930
	P05120	P15880	O95360	P18084	O75574
	P35221	P12751	P01116	P49447	P15880
	P42704	P02570	P01118	Q05472	P17535
	P55884	P70514	P02096	Q15445	P19367
	Q29882	P99021	P20810	Q9BTP3	Q96HC8
	Q29890	Q11211	P50876	Q9HBV7	Q96PJ2
	Q99613	P46783	Q9BZZ7	Q9NZH7	Q96PJ6
	Q99848	P26373	Q9UM54	Q9UBQ5	Q9NNZ4
	Q9BD37	P26641	Q9Y6Z7	Q9UJT3	Q9NNZ5

Cervix		Colon		Head and Neck	
Upregulated	Downregulated	Upregulated	Downregulated	Upregulated	Downregulated
	O75331	P00354	O14732	O75770	O60573
	O75352	P02571	P00746	P00354	O60629
	P09234	P04406	P09497	P04406	O75349
	P11216	P04687	P17066	P06702	P30499
	P13646	P04720	P18065	P09211	P35237
	P28072	P04765	P38663	P10321	P49207
	P47914	P09651	P41240	P21741	P82909
	Q02543	P11940	P53365	P30509	Q9BUZ2
	Q9NPX8	P17861	P54259	Q01469	Q9H2H4
	Q9UBR2	P26641	Q12968	Q92597	Q9H5U0
	Q9UQV5	P39019	Q9P1X1	Q9NQ38	Q9UHZ1
	Q9UQV6	P39023	Q9P2R8	Q9UBC9	Q9Y3U8

Kidney		Liver		Lung	
Upregulated	Downregulated	Upregulated	Downregulated	Upregulated	Downregulated
O43257	O60622	P11021	P02792	O95415	O60441
O43458	Q14442	P11518		P01860	O75918
O75243	Q8WX76	P19883		P50553	O75947
O75892	Q8WXP8	P21453		P98176	O95833
O76045	Q96T39	P35914		Q13045	P01160
Q15372	Q9H0T6	P36578		Q15764	P04270
Q969R3	Q9HBB5	P47914		Q92522	P05092
Q9BQZ7	Q9HBB6	Q05472		Q9BZL6	P05413
Q9BSN7	Q9HBB7	Q13609		Q9HBV7	P11016
Q9UIC2	Q9HBB8	Q969Z9		Q9NZH7	Q13563
Q9UPK7	Q9UK76	Q9BYY4		Q9UJT3	Q15816
Q9Y294	Q9UKI8	Q9NZM3		Q9UL69	Q16740

TABLE 6.5
(Continued)

Ovarian		Pancreas		Prostate	
Upregulated	Downregulated	Upregulated	Downregulated	Upregulated	Downregulated
	P02461	P00338	P05451	O00141	O15228
	P02570	P02794	P15085	P08708	O43678
	P04792	P04720	P16233	P19013	P10909
	P07900	P05388	P17538	P48060	P11380
	P08865	P07339	P18621	Q01469	P11381
	P11142	P08865	P19835	Q01628	P98176
	P14678	P20908	P54317	Q01858	Q92522
	P16475	P26641	P55259	Q02295	Q92826
	P24572	P36578	Q92985	Q13740	Q99810
	Q15182	P39060	Q9NPH2	Q96HK8	Q9H1D6
	Q9UIS4	Q01130	Q9UIF1	Q96J15	Q9H1E3
	Q9UIS5	Q15094	Q9UL69	Q9C004	Q9H723

Skin		Uterus	
Upregulated	Downregulated	Upregulated	Downregulated
O14947	O00622		O95432
P01023	P12236		O95434
P02538	P12814		O95848
P06733	P19012		Q08371
Q02536	P28066		Q13219
Q02537	P30037		Q13642
Q13677	P30923		Q9UKZ8
Q13751	P33121		Q9UNK7
Q13752	P36222		Q9UQK1
Q13753	P43155		Q9Y627
Q14733	Q01581		Q9Y628
Q14941	Q9UID7		Q9Y630

alphabet.[12] In addition, one can quickly obtain the most significantly differentially expressed gene proteins by computing the tissue-specific charts of the ratios between normal and cancer states.

A number of proteins detected by the survey described are ribosomal or ribosomal-associated proteins (such as elongation factors P04720, P26641 in colon and pancreas). Their upregulation is consistent with an accelerated cancerous cell cycle. Others may turn out to be effective tissue-specific biomarkers such as phosphopyruvate hydratase (P06733 in skin). A third category will turn out to be druggable targets—molecular "switches" that can be the focus of drug design for therapeutic intervention to reverse or stop the disease.

However, identification of useful potential targets requires additional knowledge of their function and cellular location. Accessibility is an obvious advantage. Such is the case of laminin gamma-2 (Q13753), the second highest differentially expressed

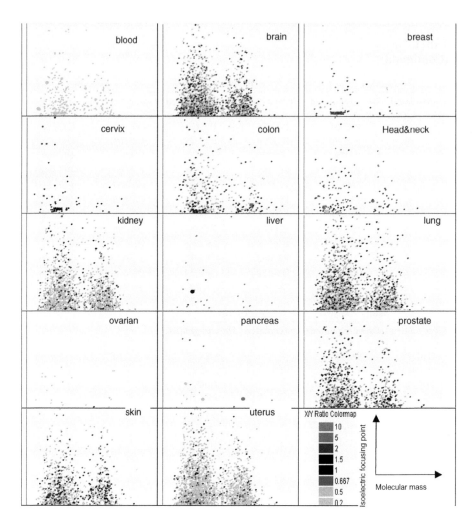

FIGURE 6.21 (Color insert follows page 204) Tissue and histology specific pI/MVv maps surveyed to date. The color code for the scatterplots is the same as in Figure 6.15 for the individual maps, but for ratios (*X/Y*) it is as follows: 10.0, 5.0, 2.0, 1.5, 1.0, 0.666, 0.5, 0.2, 0.1.

protein in skin. It is thought to bind to cells via a high-affinity receptor and to mediate the attachment, migration, and organization of cells into tissues during embryonic development by interacting with other extracellular matrix components.

6.5   CONCLUSION

To date, the charts for 92 organisms have been assembled and are represented within VIRTUAL2D. TMAP results from the survey of 144 libraries from the CGAP public resource to produce more than 18,000 putative gene products encompassing normal, cancerous, and, when available, precancerous states for 14 tissues. These interactive, Web accessible knowledge based proteomics resources are available to the research community to generate and explore in the laboratory hypothesis- driven cancer biomarkers.

REFERENCES

1. Fearon, E.R., Hamilton, S.R., and Volgeinstein, B. *Science,* 238, 193–197, 1987.
2. Page, M.J., Amess, B., Townsend, R.R., Parekh, R., Herath, A., Brusten, L., Zvelebil, M.J., Stein, R.C., Waterfield, M.D., Davies, S.C., and O'Hare, M.J. *Cell Biology,* 96, 22, 12589–12594, 1999.
3. Aebersold, R., Rist, B., and Gygi, S.P. *Ann. NY Acad. Sci.,* 919, 33–47, 2000.
4. Bussow, K. *Trends Biotechnol.,* 19, 328–329, 2001.
5. Fivaz, M., Vilbois F., Pasquali, C., and van der Goot, F.G. *Electrophoresis,* 21, 3351–3356, 2000.
6. Kriegel, K., Seefeldt, I., Hoffmann, F., Schultz, C., Wenk, C., Regitz-Zagrosek, V., Oswald, H., and Fleck, E. *Electrophoresis,* 13, 2637–2640, 2000.
7. O'Farrell, P.H. *J. Biol. Chem.,* 250, 4007–4021, 1975.
8. O'Farrell, P.Z., Goodman, H.M., and O'Farrell, P.H. *Cell,* 12, 1133–1141, 1977.
9. Dihazi, H., Kessler, R., and Eschrich, K. *Anal. Biochem.,* 299, 260–263, 2001.
10. Angelis, F.D., Tullio, A.D., Spano, L., and Tucci, A.J. *Mass Spectrom.,* 36, 1241–1248, 2001.
11. Weiller, G.F., Djordjevic, M.J., Caraux, G., Chen, H., and Weinman, J.J. *Proteomics,* 12, 1489–1494, 2001.
12. Wulfkuhle, J.D., McLean, K.C., Paweletz, C.P., Sgroi, D.C., Trock, B.J., Steeg, P.S., and Petricoin, E.F., III. *Proteomics,* 10, 1205–1215, 2000.
13. Gorg, A., Obermaier, C., Boguth, G., Harder, A., Scheibe, B., Wildruber, R., and Weiss, W. *Electrophoresis,* 6, 1037–1053, 2000.
14. Bjellqvist, B., Sanchez, J.C., Pasquali, C., Ravier, F., Paquet, N., Frutiger, S., Hughes, G.J., Hoschstrasser, and D.F. *Electrophoresis,* 14, 1375–1378, 1993.
15. VanBogelen, A.R., Abshire, Z.A., Moldover, B., Olson, R.E., and Neidhardt, C.F. *Electrophoresis,* 18, 1243–1251, 1197.
16. Lemkin, P.F., Thornwall, G., Walton, K., and Hennighausen, L. *Nucleic Acids Res.,* 22, 4452–4459, 2000.
17. Ptplot is a 2D data plotter and histogram tool implemented in Java that can be accessed at http://ptolemy.eecs.berkeley.edu/java/ptplot/
18. Information on Melanie (Geneva Bioinformatics) can be found at http://www.expasy.ch/melanie/.
19. Information about Z3 is available at http://www.2dgels.com/
20. Ideker et al., *Science,* 292, 929–934, 2001.
21. Bairoch, A. and Apweiler, R. *Nucleic Acids Res.,* 28, 45–48, 2000.
22. PI/MW is part of EXPASY's proteomics tools and can be accessed at http://www.expasy.ch/tools/pi_tool.html.
23. NCBI's Unigene database can be accessed at http://www.ncbi.nlm.nih.gov/U'niGene.
24. The European Bioinformatics Institute Web site can be found at http://www.ebi.ac.uk/

7 Data Standards in Proteomics: Promises and Challenges

Veerasamy Ravichandran, Ram D. Sriram,
Gary L. Gilliland, and Sudhir Srivastava

CONTENTS

7.1   INTRODUCTION

Advances in genome sequencing have created an immense opportunity to understand, describe, and model whole living organisms. With the completion of the
Human Genome Project, the postgenomic era has truly begun. This remarkable
achievement, determining life's blueprint, lays the groundwork for a fundamental
shift in how biological and biomedical research will be performed. However, the
sequence of the human genome, though essential for understanding human genetics,
provides limited insight into the actual working of the cell's functional units—the
proteins—and how cellular systems are integrated to form an entire organism. As
a result, research focus is gradually shifting to the gene products, primarily proteins,
and the overall biological systems in which they act, creating the emerging fields
of systems biology and proteomics (Tyers and Mann, 2003).

The complexity of the proteome often necessitates elaborate sample preparation,
processing and fractionation steps, and multiple experimental platforms. For instance,
samples can be fractionated by 2D gel electrophoresis, liquid chromatography, and
subjected to enrichment by immunoprecipitation, differentially labeled with fluorescent
dyes, isotope-coded "tags" (e.g., ICAT), and analyzed by mass spectrometry or directly
compared using specialized imaging techniques. Absolute or relative quantitation of
proteins that was generally done using radioactivity or amino acid analysis is now
increasingly performed by differential dye labeling (e.g., differential in-gel electrophoresis) or by mass spectrometry. Data analysis involves data capture and validation, data
management, and integration from diverse sources. An important outcome of these
efforts is the understanding of the differences between healthy and diseased tissues and
cells and how the differences in protein expression levels can be correlated with disease.
As the amount of information on the gene products increases, new insights into the
functional interaction of enzymes and other cellular constituents are to be expected.
The complete understanding of cellular function and physiology will require comprehensive knowledge of the complexity of the system-wide protein content of cells.

The free, widespread availability of a large variety of data beyond human genome
sequences, including sequence variation data, model organism sequence data,
organelle-specific data, expression data, and proteomic data, to name a few, is starting
to provide the means for scientists in all disciplines to better design experiments and
interpret their laboratory and clinical results. Due to the complexity of proteins and
their isoforms as well as the dynamic nature of the proteome, enormous amounts
of protein data orders of magnitude larger than that coming from genomics studies
are being generated, making the effective and efficient management of data essential.
The pace of proteomics data generation far outstrips Moore's Law. Having such a
rich source of information is proving invaluable to scientists, whose findings should,
in time, lead to improved and faster strategies for the diagnosis, treatment, and
prevention of genetic diseases.

Along with the rapid data growth has been the development of a wealth of tools for analyzing the expanding data volume. These tools are being applied to extract meaningful information from the data about the system being studied. Currently many options are available and choosing among these is challenging in itself. Despite the combined efforts of biologists, computer scientists, biostatisticians, and software engineers, there is no one-size-fits-all solution for the analysis and interpretation of complex proteomics data. The lack of cohesion between heterogeneous scientific data, resulting from the diverse structure and organization of independently produced data sets, creates an impractical situation for data interoperability and integration. How to handle these data, make sense of them, and render them accessible to biologists working on a wide variety of problems is a challenge facing bioinformatics, an emerging field that seeks to integrate computer science with applications derived from molecular biology. Bioinformatics has to deal with exponentially growing sets of highly interrelated, heterogeneous, complex, and rapidly evolving types of data. The advancement of proteomics high-throughput technologies present challenges for biologists, who have traditionally worked with relatively small data sets and shared results only with others working on similar biological systems.

The enormity and heterogeneity of databases already exceeds our ability to manage and analyze data to produce dependable information in reasonable time frames (Galperin, 2004). The potential impact of improved interoperability derives from the fact that information and knowledge management systems have become fundamental tools in a broad range of commercial sectors and scientific fields. The explosive growth in data will require new collaborative methods and data management tools to locate, analyze, share, and use data and information for research, operations, marketing, and other core business processes. Businesses will continue to invest heavily in knowledge management tools, but the lack of interoperability data standards and related technologies remain persistent and vexing barriers to the integration of diverse knowledge bases and databases.

7.2 DATA ISSUES IN PROTEOMICS

Lack of data standards is the Achilles' heel of data interoperability. The lack of integration, implementation, and use of standards is a barrier to the delivery of optimal biological data (Ravichandran et al., 2004). Even with the dramatic increase in the volume of proteomics data, innovation will be constrained unless technical advances are made in producing critically evaluated data and integrating data sources through data management and data mining techniques. Proteomics data, for example, require systematic data mining, reformatting, annotating, standardizing, and combining of data in a unified computational framework. In the context of time-to-completion pressures and volumes of data, the research community needs certified models that can derive "best" recommended values from critically evaluated experimental data and validated benchmarked predictive methods for any real or proposed measurement. These virtual measurement systems could generate data suitable for immediate use in commercial, scientific, and regulatory applications. Also needed are effective data management standards and techniques (e.g., quality, traceability, or uncertainty estimates) for gathering, integrating, and maintaining information about data accessed from diverse sources. Despite this, a wealth of data exists and is readily available for use in proteomics. Although public proteomics data resources are highly informative individually, the collection of available content would have more utility if provided in a standard and centralized context and indexed in a robust manner. Research and development activities will be much more productive if provided with a wider range of critically evaluated data, virtual measurement methods, and new methods for managing the dramatic increase in research data. The potential benefits are broad.

Biologists usually end up performing a large number of proteomics experiments. All of the parameters in an experiment could be hard to control, describe, or replicate. Attempts at standardization of experiments for the sake of standardization are unlikely to be performed by experimental biologists. Newer technologies can introduce newer types of experiments and standardization issues. There are many ways to represent an experiment and results, which leads to the interoperability problems. For mass spectrometry–derived data, the file format is easier to standardize, but the criteria that are used to statistically "identify" a protein differ greatly and are software dependent (Aebersold and Mann, 2003). For example, in a typical LC-MS/MS experiment, approximately 1000 collision-induced dissociation (CID) spectra can be acquired per hour. Even with the optimistic assumption that every one of these spectra leads to the successful identification of a peptide, it would take considerable time to analyze complete proteomes. Performing a biological experiment does not automatically guarantee validity. For example, a mitochondrial preparation might contain nuclear proteins as contaminants. Marking all the proteins from such a preparation as mitochondrial can lead to erroneous conclusions.

Along with sample preparation, there are various technical challenges in proteome characterization (Verma et al., 2003). Some of the challenges are described below:

Peak picking, cluster analysis, and peak alignment: Data analysis should be conducted with validated algorithms that are able to carry out peak picking, cluster analysis, and peak alignment.

Robustness of technology: High-throughput methods are needed to save time and effort. The current technologies have a mass accuracy of 100 parts per million, with the sensitivity in mass spectrometry in the low femtomolar range. It takes about 40 minutes to run and analyze a sample.

Instrument drift (laser voltage and detector decay): These factors contribute to variations in results. Therefore, standardization of equipment for each analysis is needed.

Protein-chip quality—spot (array) variation, chip variation, and batch variation: Variations in the chip surface and the sample applicator (manual versus automatic) can cause variations in results. Therefore, standardization of the chip surface is crucial.

Calibration—individual and multiple mass spectrometers: Spectrometers should be calibrated frequently using standardized reference material.

Validation: Validation can be performed only after all of the challenges listed above are met. Areas to be addressed include crucial aspects of study groups with diverse populations, sample collection and storage, and longitudinally collected samples.

7.3 DATA STANDARDS

Data standards are essential because they permit cooperative interchanges and querying between diverse, and perhaps dissociated, databases. The ability to interchange data in a seamless manner becomes critically important (Berman et al., 2003). The economic benefits of data interoperability standards are immediate and obvious. Data standards provide well-defined syntax, precise definitions, and examples, as well as data relationships, data type, range restrictions, allowed values, interdependencies, exclusivity, units, and methods.

Standards are generally required when excessive diversity, as in proteomics data, creates inefficiencies or impedes effectiveness. The data should be 1) complete, comprehensive, consistent, reliable, and timely; 2) easily accessible and with effective presentation tools developed to display the data in a user-friendly manner; and 3) available across system boundaries in an interchangeable format. Any system design should maximize the use of standards-based technology, including object-oriented design, modular components, relational database technology, XML, JAVA, and open systems, and the use of standard tools. Metadata should be explicitly represented in the enterprise environment to facilitate application development, data presentation, and data management.

A standard can take many forms, but essentially it comprises a set of rules and definitions that specify how to carry out a process or produce a product (Chute, 1998). For the purpose of this article, we adopt the definition that standards are documented agreements containing technical guidelines to ensure that materials, products, processes, representations, and services are fit for their purpose. Under this definition, there are four broad types of standards.

The first type is the measure or metric standard. This is one used against which to measure; all comparable quantities are measured in terms of such a standard. For example, a test result may have been expressed in two different units (grams/liter and milligrams/milliliter) that are mathematically identical but visually different. Slightly more complex is the case where the units are different, and not mathematically equivalent, for the same test. An example might be grams/deciliter and milligrams/milliliter. A familiar example is the loss of the $125 million Mars Climate Orbiter, due to the inconsistency of the units used.

The second type of standard is process oriented or prescriptive, where descriptions of activities and process are standardized. This type of standard provides the methodology to perform tests and perform processes in a consistent and repeatable way. For example, calibration, validation, and standardization of different instruments in different platforms that perform the same proteomics analysis (MS, for example) are critical for analyzing and comparing the data.

The third type of standard is performance based. In this type of standard, process is not specified, but ultimate performance is. These standards are often based on product experience. For example, analysis and comparison of diverse proteomics data are performance based.

The fourth standard type is based on interoperability among systems. In this type, process and performance are not explicitly determined, but a fixed format is specified. The goal of this type of standard is to ensure smooth operation between systems that use the same physical entity or data. Sharing clinical data through Health Level 7 (HL7) standard exchange format, sharing macromolecule crystallographic data through macromolecular Crystallographic Information File (mmCIF), and sharing of two-dimensional polyacrylamide gel electrophoresis data through markup language that is based on the XML are examples.

Standards are formulated in a number of ways:

1. A single vendor controls a large enough portion of the market to make its product the market standard (example: Microsoft's Windows application)
2. A community agrees on an available standard specification (example: exchange format for macromolecular data exchange)
3. A group of volunteers representing interested parties works in an open process to create a standard (example: data exchange formats for microarray experiments, MIAME)
4. Government agencies such as the National Institute of Standards and Technology (NIST) coordinate the creation of consensus standards (example: physical and data standards).

7.3.1 THE STANDARDS DEVELOPMENTAL PROCESS

The process of creating a standard proceeds through several stages. It begins with an *identification stage*, during which someone becomes aware that there exists a need for a standard in some area and that technology has reached a level that can support such a standard (Ravichandran, 2004). If the time for a standard is ripe, then several appropriate individuals can be identified and organized to help with the *conceptualization stage*, in which the characteristics of the standard are defined: what must the standard do? What is the scope of the standard? What will be its format? In the proteomic area, one key discussion would be on the scope of the standard. Should the standard deal only with the exchange of experimental data, or should the scope be expanded to include other types of data exchange? In the ensuing *discussion stage*, the participants will begin to create an outline that defines content, to identify critical issues, and to produce a time line. In the discussion, the pros and cons of the various concepts are discussed. Usually, few dedicated individuals draft the initial standard; other experts then review the draft. Most standards-writing groups have adopted an open policy; anyone can join the process and be heard. A draft standard is made available to all interested parties, inviting comments and recommendations. A standard will generally go through several versions on its path to maturity, and a critical stage is *early implementation*. This process is influenced by accredited standards bodies, the federal government, major vendors, and the marketplace.

7.3.2 EXAMPLES OF STANDARDS ACTIVITIES IN PROTEOMICS

The International Union of Crystallography (IUCr) appointed a working group in 1991 to develop data standards for crystallographic data to address interoperability problems.

The Crystallographic Information File (CIF) was developed with a dictionary that defines the structure of CIF data files. A Dictionary Description Language (DDL) was developed to define the structure of CIF data files. CIF data files, dictionaries, and DDLs are expressed in a common syntax. Later, IUCr extended CIF to mmCIF (macromolecular Crystallographic Information File) (Westbrook and Bourne, 2000). In addition, in 1998 IUCr recommended about 140 new definitions for adopting newly emerged NMR data. Like many data dictionaries mmCIF is not static. It continues to evolve. The standard representation for Protein Data Bank (PDB) is now mmCIF. By the early 1990s, the majority of journals required a PDB accession code and at least one funding agency (National Institutes of Health) adopted the guidelines published by the International Union of Crystallography (IUCr) requiring data deposition for all structures. Through this community-based effort, the PDB now handles the complex macromolecular data more efficiently. Recently, Protein Data Bank Japan (PDBj), the Macromolecular Structure Database (MSD) group at European Bioinformatics Institute (EBI), and the Research Collaboratory for Structural Bioinformatics (RCSB) have collaborated together to introduce macromolecular data in an XML (Extensible Markup Language) format.

Health Level 7 (HL7) was founded in 1987 to develop standards for the electronic interchange of clinical, financial, and administrative information among independent healthcare-oriented computer systems; e.g., hospital information, clinical laboratory, enterprise, and pharmacy systems. That group adopted the name HL7 to reflect the application's (seventh) level of the Open Systems Interconnection (OSI) reference model. In June of 1994, HL7 was designated by the American National Standards Institute (ANSI) as an ANSI-accredited standards developer. The original primary goal of HL7 was to provide a standard for the exchange of data among hospital computer applications that eliminated, or substantially reduced, the hospital-specific interface programming and program maintenance that was required at that time. The standard was designed to support single, as well as batch, exchanges of transactions among the systems implemented in a wide variety of technical environments. Today, HL7 is considered to be the workhorse of data exchange in healthcare and is the most widely implemented standard for healthcare information in the world.

7.3.3 DATA STANDARDS ACTIVITIES FOR PROTEOMIC DATA

Various approaches have been attempted to unify the diverse and heterogeneous biological data. In order to help the standardization of the microarray data model, Microarray Gene Expression Database Group (MGED) developed a markup language for microarray data called Microarray Markup Language (MAML). As an approach toward standardizing, unifying, and sharing two-dimensional gel electrophoresis data, we are developing a common language—Two-Dimensional Electrophoresis Markup Language (TWODML)—that is based on XML. The Human Proteome Organization's (HUPO) Protein Standards Initiative (PSI) aims to define community standards for data representation in proteomics to facilitate data comparison, exchange, and verification. Currently, PSI is focusing on developing standards for two key areas of proteomics: mass spectrometry and protein–protein interaction data, which will be XML based.

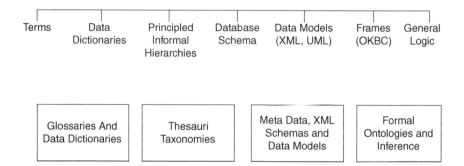

FIGURE 7.1 Techniques for representing proteomics data.

7.4 TECHNIQUES FOR REPRESENTING PROTEOMICS DATA

Proteomics data can be represented in a wide range of formats, as shown in Figure 7.1 (courtesy Michael Gruninger, NIST). At the left end of the spectrum we have glossaries and data dictionaries, which are informal mechanisms for capturing data. Although such schemes provide some organization to data, these are not easily amenable for seamless data exchange. In the center we have XML-based schemas, which provide further organization to the data. XML is becoming a widely accepted language for expressing domain-specific information that can be manipulated using various Web resources. However, XML by itself does not adequately capture semantics of a domain. There are several types of semantics that need to be captured. Here, we use the term to indicate "meaning." Using formal mechanisms, such as logic, we can generate domain-specific ontologies that encode various relationships between data elements. This will aid in the generation of semantically validated data and information models, which can be used for developing self-describing and self-integrating systems.

7.4.1 STANDARD ONTOLOGY

Ontology in a domain defines the basic terms and relationships comprising the vocabulary of a topic area, as well as the rules for combining terms and relationships to define extensions to the vocabulary. Since public proteomics data resources are highly informative individually, the collection of available content would be useful if provided in a standard and centralized context and indexed in a robust manner. The potential impact of improved interoperability derives from the fact that information and knowledge management systems have become fundamental tools in a broad range of commercial sectors and scientific fields. The adoption of common standards and ontologies for the management and sharing of proteomics data is essential. Use of controlled vocabulary is already facilitating analysis of high-throughput data derived from DNA microarray experiments and macromolecular structural information. Gene Ontology (GO) consortium is developing three standards: 1) controlled vocabularies (ontologies) that describe gene products in terms of their associated biological processes, 2) cellular components, and 3) molecular functions in a species-independent manner. The MeSH

Browser is an online vocabulary look-up aid available for use with MeSH (Medical Subject Headings) (Harris et al., 2004). It is designed to help quickly locate descriptors of possible interest and to show the hierarchy in which descriptors of interest appear. The adoption of common ontologies and standards for the management and sharing of proteomics data is essential and will provide immediate benefit to the proteomics community. For example, the use for GO ontologies that is gaining rapid adherence is the annotation of gene expression data, especially after these have been clustered by similarities in patterns of gene expression. The goal of cluster analysis is to reveal underlying patterns in data sets that contain hundreds of thousands of measurements and to present this data in a user-friendly manner.

7.4.2 CLASSIFICATION OF PROTEINS

To deduce possible clues about the action and interaction of proteins in the cell, it is necessary to classify them into meaningful categories that are collectively linked to existing biological knowledge. There have been many attempts to classify proteins into groups of related function, localization, industrial interest, and structural similarities. A proteomics strategy of increasing importance involves the localization of proteins in cells as a necessary first step toward understanding protein function in a complex cellular network. A classification of all the proteins according to their function is necessary for the scientist to get an overview of the functional repertoire of the organism's proteins that will facilitate finding the genes of interest. An example of functional classification of macromolecular protein names is presented in Figure 7.2.

7.4.3 SYNONYMS

A critical requirement for the query selection of proteomics data is the incorporation of comprehensive synonyms for standard vocabularies. A list of synonyms that are internally mapped to the same annotation entry can solve the problem of unmatched synonyms. For example, there are many ways to search for T lymphocyte (T-Lymphocyte, T cell, etc.). An example of synonyms used by authors in depositing structural data for HIV-1 protease in the Protein DataBank:

HIV-I Protease
HIV I Protease
HIV-1 Proteinase
HIV I Proteinase
HIV-1 Proteinase
HIV 1 Proteinase
Human Immunodeficiency Viral (HIV-1) Protease
Human Immunodeficiency Virus (HIV 1) Protease
Human Immunodeficiency Virus (HIV-I) Protease
Human Immunodeficiency Virus (HIV I) Protease
Human Immunodeficiency Viral Protease
Human Immunodeficiency Viral Type 1 (HIV) Protease
Human Immunodeficiency Viral Type 1 (HIV-1) Protease
Human Immunodeficiency Virus Type 1 (HIV 1) Protease

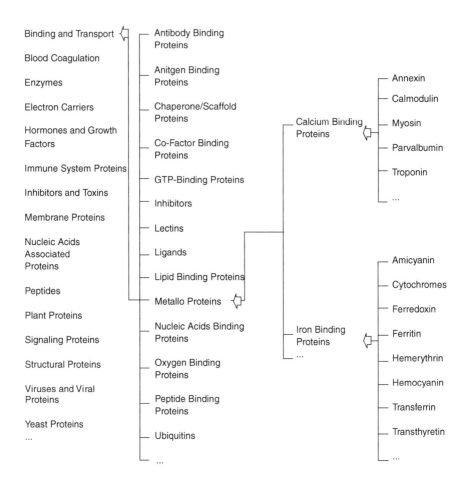

FIGURE 7.2 Functional classification of macromolecular proteins.

Human Immunodeficiency Virus Type 1 (HIV-I) Protease
Human Immunodeficiency Virus Type 1 (HIV I) Protease
Human Immunodeficiency Viral Type 1 Protease
Human Immunodeficiency Viral Type-1 Proteinase
Human Immunodeficiency Viral Type-1 Protease
Human Immunodeficiency Virus Type-1 Protease
Human Immunodeficiency Virus Type-1 Protease
Human Immunodeficiency Virus Type-1 Protease
Human Immunodeficiency Viral Type-1 Proteinase
Human Immunodeficiency Virus Type-1 Proteinase
Human Immunodeficiency Virus Type-1 Proteinase
Human Immunodeficiency Virus Type-1 Proteinase
HIV-1 Protease (Retropepsin)
Retropepsin
EC: 3.4.23.16

TABLE 7.1
Terminology

Standards vs. controls	Standards are used to calibrate equipment or to measure the efficiency of quantitative procedures or methods; for example, known amounts of a variety of proteins used to determine quantitative recovery of a measurement technique (e.g., SELDI-MS). Controls are used to verify that a particular experiment is working properly and may employ the use of standards; for example, "Normal" liver protein preparation from Fischer rat liver homogenate.
Analytical standards	Physical standards used for calibration and validation of methods
Clinical standards	Standards that correlate medical conditions (i.e., disease) with particular measurements (e.g., biomarker determinations)
Standard methods or protocols	Sometimes called Paper Standards, these are used to establish how samples are handled and stored and how assays are run. These standards are particularly important in the clinical diagnostic field.
Data standards	Include the definition of data fields to assure the quality of database searches and data mining, as well as the establishment of data querying protocols and semantics. Appropriate data definitions are crucial for relating similar data collected from different laboratories, protocols, and platforms.
Nomenclature	System of terms that is elaborated according to pre-established naming rules
Terminology	Set of terms representing the system of concepts of a particular subject field
Dictionary	Structural collection of lexical units, with linguistic information about each of them
Vocabulary	Dictionary containing the terminology of a subject field

Such diversity in nomenclature is challenging for any data resource with respect to obtaining a complete answer to a query (Table 7.1).

7.5 BIOLOGICAL DATA EXCHANGE OPTIONS

The popularity of the flat file can be attributable to its simplicity, which facilitates its manipulation by various tools. Flat files consist of columns, each of which represents a parameter, and rows, representing instances of experimental information. This is a limited solution, because it lacks referencing typed values vocabulary control constraints, among other issues. Often fields are ambiguous and their content is contextual. To be interpretable, a "data dictionary"—a document that describes in some depth what each column in each file represents—must accompany the data. A data dictionary is an example of "metadata"—data that describe data. The challenge of standardizing the computational representation of electrophoresis data then reduces to the problem of standardizing the metadata. Below are some techniques that have been developed to address similar situations.

7.5.1 ABSTRACT SYNTAX NOTATION ONE (ASN.1)

ASN.1 is heavily used at the National Center for Biological Information as a format
for exporting the GeneBank data as a means for exchanging binary data with a
description of its structure (McCray and Divita, 1995). ASN.1 is an International
Standards Organization (ISO) standard that encodes data in a way that permits
computers and software systems of all types to reliably exchange both the structure
and the content of the entries. Since ASN.1 files convey the description of their
structure, it offers some flexibility; the client side does not necessarily need to know
in advance the structure of the data. Based on that single, common format, a number
of human-readable formats and tools were developed, such as those used by Entrez,
GenBank, and the BLAST databases. Without the existence of a common format
such as this, the neighboring and hard-link relationships that Entrez depends on
would not be possible. However, ASN.1 software tools do not scale well for very
large data sets. Also, it lacks support for queries. Researchers on ASN.1 Standards
recognized several years ago that there was a requirement (from users of ASN.1) to
have an Extensible Markup Language (XML) representation of the information
structures defined by an ASN.1 specification. ASN.1 now provides a mapping to
XML schema definition, thus allowing ASN.1 to substitute for an XML schema
language such as the XML Schema Definition Language or RELAX NG, with the
added advantage of specification of extremely compact binary encodings in addition
to the XML encodings.

7.5.2 XML FOR PROTEOMICS DATA EXCHANGE

Even though there are many biological data exchange formats available, well-
documented and annotated data with an easily exchangeable data format, such as
an XML format, would help in data mining, annotation, storage, and distribution.
XML was defined by the XML Working Group of the World Wide Web Consortium
(W3C: http://www.w3.org/XML). XML is a markup language for documents containing structured information. Structured information contains both content (words,
pictures, etc.) and some indication of what role that content plays (for example,
content in a section heading has a different meaning from content in a footnote,
which means something different than content in a figure caption or content in a
database table). Almost all documents have some structure. A markup language is
a mechanism to identify structures in a document. The XML specification defines
a standard way to add markup to documents. XML is a simple, very flexible text
format, playing an increasingly important role in the exchange of a wide variety of
data on the Web and elsewhere. Because an XML document so effectively structures
and labels the information it contains, the Web browser can find, extract, sort, filter,
arrange, and manipulate that information in highly flexible ways. XML has been
designed for ease of implementation and for interoperability with the World Wide
Web. Thus XML is an obvious choice for encoding proteomics data syntax.

XML definitions consist of only a bare-bones syntax. When an XML document
is created, rather than use a limited set of predefined elements, the data elements
are created and subjectively assigned names as desired; hence the term "extensible"

in Extensible Markup Language. Therfore, XML can be used to describe virtually any type of document and fits ideally with the requirements for the complex and diverse biological data integration. XML thus provides an ideal solution for handling the rapidly expanding quantity and complexity of information that needed to be put on the Web. A common language, such as XML, should therefore offer power, scalability, adoptability, interoperability, and flexibility with different data types. In order to enhance the interoperability between diverse data, adoption of a universal data exchange language, like XML, to exchange the annotated data would be useful. The number of applications currently being developed by biological communities that are based on, or make use of, XML documents is growing rapidly. This is to facilitate the writing and exchange of scientific information by the adoption of a common language in XML (Table 7.2).

7.5.3 DATA STANDARDS FOR 2D PAGE DATA: MARKUP LANGUAGE FOR ELECTROPHORESIS DATA

Rapid progress has been made in establishing standards for genomic sequence data as well as DNA microarray data. Current attention is on proteomics data standards. Human Proteome Organization's (HUPO) Protein Standards Initiative (PSI) aims to facilitate standards for data representation in proteomics. Currently, the PSI is focusing on developing standards for two key areas of proteomics; mass spectrometry and protein–protein interaction data that will be XML based. We briefly describe here a language for sharing electrophoresis experimental data, Two-Dimensional Electrophoresis Markup Language (TWODML), that is based on the XML. The goal of the TWODML is to

1. Gather, annotate, and provide enough information that may be reported about an electrophoresis based experiment in order to ensure the interoperability of the results and their reproducibility by others
2. Help establishing public repositories and data exchange format for electrophoresis based experimental data
3. Eliminate barriers to data exchange between the electrophoresis data, and permit the integration of data from heterogeneous sources
4. Leverage low-cost XML-based technologies such as XSLT (Extensible Style sheet Language Transformation) and SVG (Scalable Vector Graphics)

The first step in data interoperability is to enforce standards for the electrophoresis data. A much bigger challenge is arriving at what descriptors constitute a "required/minimum acceptable" set with respect to different types of parameters. The minimum information necessary from any 2D PAGE experiment is that associated with the experimental details, in order to ensure first the reproducibility of the experiment, and second the interoperability of the results. By defining the vocabularies in a standard format (e.g., the experimental sample source) the resulting uniformity may permit comparison of data between different systems

TABLE 7.2
Scientific Markup Languages

Markup Language	Purpose	URL
Chemical Markup Language (CML)	Exchange of chemical information	http://www.xml-cml.org
Mathematical Markup Language (MathML)	Exchange of mathematical formula	http://www.w3.org/Math
Bioinformatic Sequence Markup Language (BSML)	Exchange of DNA, RNA, protein sequences, and their graphic properties	http://www.sbw-sbml.org/index.html
BIOpolymer Markup Language (BIOML)	Expression of complex annotation for protein and nucleotide sequence information	http://www.bioml.com/BIOML
Taxonomical Markup Language	Exchange of taxonomic relationships between organisms	http://www.albany.edu/~gilmr/pubxml
Genome Annotation Markup Elements (GAME)	Annotation of biosequence features	http://xml.coverpages.org/game.html
BlastXML	Model NCBI Blast output	http://doc.bioperl.org/releases/bioperl-1.2/Bio/SearchIO/blastxml.html
Ontology Markup Language/Conceptual Knowledge Markup Language (OML/CKML)	Representation of biological knowledge and specifically functional genomic relationships	http://smi-web.stanford.edu/projects/bio-ontology
Multiple Sequence Alignments Markup Language (MSAML)	Description of multiple sequence alignments (amino acids and nucleic acid sequences)	http://xml.coverpages.org/msaml.html
Systems Biology Markup Language (SBML)	Representation and modeling of the information components in the system biology	http://www.cds.caltech.edu/erato/sbml/docs
Gene Expression Markup Language (GEML)	Exchange of gene expression data, Gene Expression Markup Language	http://www.oasis-open.org/cover/geml.html
GeneX Gene Expression Markup Language (GeneXML)	Representation of the Gene Expression Databases datasets	http://xml.coverpages.org/geneXML.html
Microarray Markup Language (MAML)	Integration of microarray data	http://xml.coverpages.org/maml.html
Protein Markup Language (ProML)	Exchange of protein sequences, structures, and families-based data	http://www.bioinfo.de/isb/gcb01/talks/hanisch/main.html
RNA Markup Language (RNAML)	Exchange of RNA information	http://www-lbit.iro.umontreal.ca/rnaml/

(microarray data, macromolecular data, etc.). Hence, in the case of 2D PAGE, the common data elements and data definitions for the required information have been outlined.

The following data elements should be collected in association with their required data categories: sample source, detail about the protein, experimental detail, sample preparation, sample loading, sample separation condition, sample separation, experimental analysis, data analysis, and author information. For example, the sample source information should contain the source record, which specifies the biological and/or chemical source of each molecule in the entry. Sources should be described by both their common and scientific names. Two types of sources will be grouped: the natural source and the genetically modified (recombinant) source. For example, a TWODML format describing a gene used in a recombinant sample source is described as below:

```
<TWODML>

<SAMPLE_SOURCE>

<RECOMBINANT_SOURCE>

<GENETIC_MATERIALS>

<GENE>

<NAME>SNAP-23</NAME>

<PROTEIN_NAME>Snaptosome-associated Protein of 23 kDa

</PROTEIN_NAME>

<SOURCE>

<ORGANISM_SCIENTIFIC>Homo sapiens

</ORGANISM_SCIENTIFIC>

<ORGANISM_COMMON>Human

</ORGANISM_COMMON>

<CELL_LINE>Raji - human B lymphocyte (Burkitt's
Lymphoma)

ATCC number: CCL-86

</CELL_LINE>

</SOURCE>

<GENETIC_VARIANCE>Amino acid 23 is changed from Ser
to Ala

(Ser23Ala)

</GENETIC_VARIANCE>
```

```
<ORIGIN>

<NAME>Dr. V. Ravichandran</NAME>

<ADDRESS>NIST</ADDRESS>

<CONTACT_INFO>vravi@nist.gov</CONTACT_INFO>

</ORIGIN>

<MORE_INFO>GeneBank/EMBL Data Bank accession number:

U55936

</MORE_INFO>

</GENE>
```

In this partial TWODML document, enclosing sets of angle brackets mark data elements. Standardized values for well-defined data elements are embedded within the elements. In addition to well-defined syntax described above, each TWODML document also carries information about data relationships, data types, range restrictions, interdependencies, exclusivity, units, and methods. Free, open-source software is available to validate and parse the data files.

As in the natural source, data items in the genetically modified category record details of the source from which the sample was obtained. Associated data for this category include the gene modified in the source material for the experiment, the genetic variation (transgenic, knockout), the system used to express the recombinant protein, and the specific cell line used as the expression system (name, vendor, genotype, and phenotype). Data items in the natural source category will record details of the sample source. Associated data for this category will include the common name of the organism and its scientific name, the source condition (normal, disease), any genetic variation, sex, age, organ, tissue, cell, organelle, secretion, and cell line information. The cell line and strain should be given for immortalized cells when they help to uniquely identify the biological entity studied.

The TWODML application is being defined by creating an XML schema that defines and names the elements and attributes that can be used in the document and the order in which the elements can appear and constrains the values of elements and attributes. A variety of schema languages are available for XML. The oldest and simplest of these schema languages is the Document Type Definition (DTD). Although DTDs enjoy strong software support, they are inadequate for representing strongly typed or context-sensitive information. Two newer XML schema languages, the W3C's XML Schema Definition Language and ISO's RELAX NG, address these shortcomings of DTDs. Therefore, we plan to specify the TWODML schema using one or both of these languages. Our application will also include one or more Extensible Style Sheet Language (XSL) style sheets. These style sheets will enable electrophoresis XML data to be transformed into other useful formats such as HTML (Hyper Text Markup Language) and SVG.

7.6 UNIFYING PROTEOMICS DATA: COMMON DATA REPOSITORY

There is a necessity for specialized data warehouse and data archival for proteomics to meet the individual needs of the various research communities in order to collect and annotate different kinds of data related to a particular area of interest, to add knowledge of experts to the raw data, and to provide in-depth information complementary to the breadth available in public databases. Proteomics data should be complete, comprehensive, consistent, reliable, and timely. Improvements in productivity will be gained if the systems can be integrated—that is, made to cooperate with each other—to support global applications accessing multiple databases. The major issues are the integration of heterogeneous data sources into a central repository and the systematic establishment of the correlation between the different types of data to allow meaningful comparisons. A framework that supports the integrative analysis of high-throughput biological data should enable analytical work flows that follow the scheme. Several building blocks can be identified, including an information technology infrastructure for data management, an analysis interface for bioinformatics methods, and an annotation module.

Numerous Web resources have been created that focus on many areas that have direct relevance to identifying proteins and assigning protein function. Each resource has its own definitions of its data elements that are perhaps similar to other resources, but *not identical*. Thus, serious problems result, limiting the interactions between these resources. These efforts afford tremendous value to the biological researcher since they, in essence, reduce the massive "sequence space" to specific, tractable areas of inquiry and, by doing so, allow for the inclusion of many more types of data than are found in the larger data repositories. These databases often provide not just sequence-based information, but additional data such as gene expression, macromolecular interactions, or biological pathway information, data that might not fit neatly onto a large physical map of a genome. Most importantly, data in these smaller, specialized databases tend to be curated by experts in a particular specialty. These data are often experimentally verified, meaning that they represent the best state of knowledge in that particular area.

7.6.1 DATA MINING

With the introduction of sophisticated laboratory instrumentation, robotics, and large, complex data sets, biomedical research is increasingly becoming a cross–disciplinary endeavor requiring the collaboration of biologists, engineers, software and database designers, physicists, and mathematicians. Techniques used in other fields can be extremely valuable if we can adapt them to biological problems. The ultimate goal is to convert data into information and then information into knowledge. Before extracting data from an external source, some potential questions one could start with include the following topics: source of the data, reliability of the source, nature of the data, accessibility of the data, and ease of interoperability of the data. When extracting data from external sources, all the available information for a particular data set of interest should be considered. Of course, one may have to gather this information from many different specialized data resources. The problem that many investigators encounter is that larger databases often do not contain specialized information that would be of interest to specific groups within the scientific community. While educational efforts such as this help to address the need for rational ways to approach mining genomic data, additional efforts in the form of providing curated views of the data in specialized databases have been taking place for many years now.

7.6.2 DATA SOURCE

One of the main objectives of a common proteomics data repository is to provide the community with detailed information about a given protein of interest, including qualitative and quantitative properties. This requires more comprehensive data for each of the data items and data groups. One way to achieve this is to encourage the proteomic community to deposit their experimental data, so this will facilitate the entry of proteomic data and associated information through a Web-based common repository. It is clear that having to supply such detail for every single parameter in an experiment data set can be a highly onerous task for data submitters. If a public repository is to encourage submission of experimental data, the designers of such databases must strive to reduce the amount of manual labor required of submitters. Software tools accompanying electronic repositories must provide the equivalent of GeneBank's SEQUIN (http://www.ncbi.nlm.nih.gov/Sequin), BankIt (http://www.ncbi.nlm.nih.gov/BankIt), or Gene Expression Omnibus (GEO: http://www.ncbi.nlm.nih.gov/geo/). Nonetheless, if a proteomic data repository has sufficient data in enough categories then it will be possible to get a meaningful answer to a query. Also, a parallel effort has to be established to gather proteomic data from related published articles and integrate into the repository. This knowledge component of a resource is usually held in scientific natural language as text. Extracting proteomics experimental information could also be done using information retrieval (IR). IR is the field of computer science that deals with the processing of documents containing free text so that they can be rapidly retrieved based on keywords specified in a user's query (Nadkarni, 2002). Data derived from the public repository as well as from the mined data can be stored in a raw data repository. Raw data should be archived in a standard format to ensure the data integrity, originality, and traceability.

7.6.3 DATA ANNOTATION/VALIDATION

For annotation of proteomics data, it may be useful to list some potential questions, such as:

What information (i.e., data fields) is necessary to allow the data to be useful, say, 3, 5, or 10 years from now?
What information is necessary to make comparisons between measurement methods (e.g., SELDI, ICAT, MudPIT, gene chips, protein chips)?

Where is the balance between requesting too much information and having useful data?
How much data should be captured?
In what format should the data be stored?
Who will take responsibility for the data?

Proteomics data and its value need to be examined or validated for its correctness and completeness. The annotation of data elements also requires that all of the related data records within a file are consistent and properly integrated across the group of files. The first phase would be largely automated — annotation resources that will be routinely consulted to provide a complete range of updated, annotated information. Ideally, the information that accompanies the high-throughput data should be seamlessly integrated into the annotation forms. When uploading data into the internal database, this information should also be added to the corresponding forms. Such a mechanism allows highly reliable and efficient annotation, as well as convenient quality control by querying the annotation. The most valuable annotation data for automatic integrative analyses are systematic annotation. Lists of controlled vocabularies (catalogs, ontologies, or thesauri) can be used to avoid problems of interoperability. Continuous annotation of proteomics data will improve the interoperability of cross-platform data sets. Adopting an exchange format, the annotation of proteomics data is also extendable to Distributed Annotation System (DAS). DAS is a client-server system in which a single client integrates information from multiple servers to retrieve and integrate dispersed proteomics data (Stein, 2003). Since DAS adopts the integration through XML, well-annotated proteomics data can be integrated from various specialized resources using XML.

7.6.4 DATA MANAGEMENT AND ARCHIVING

The rapid growth in the volume of proteomics data poses a problem in terms of data management, scalability, and performance. Building of the database structure is the first step toward the structured recording of electrophoresis data in a relational database. This consists of precisely defining data fields and precisely defining the relationships between them, which are represented by links between the tables. Proteomics data are complex to model and there are many different varieties of data with numerous relationships. Data models are the logical structures used to represent a collection of entries and their underlying one-to-one, one-to-many, and many-to-many relationships. The main motivation for creating proteomics data models is usually to be able to implement them within database management systems, usually as a relational database management system (e.g., ORACLE, SQL Server, SYBSASE, MySQL, etc.). Proteomics data model corresponds to a way of organizing the pertinent values obtained on measurement of an experiment. Databases should be modeled to handle the heterogeneous data from various external data sources. New types of proteomics data emerge regularly and this raises the need for updating the whole data semantics and integrating the sources of information that were formally independent. Data analysis generates new data that also have to be modeled and integrated.

7.6.5 DATA DISTRIBUTION

Data are accessed intensively and exchanged very often by users on the Internet. Users can gather their data in an XML format through a Web interface that can be queried. On the other hand, users might also want to view the data. In this case, the data should be presented in a user-friendly format. The data should be available across system boundaries in an interchangeable format. The systems design should maximize the use of standards-based technology including object-oriented design, modular components, relational database technology, XML, Java, open systems, and use of standard tools. An obvious choice is HTML since it is supported by all Internet browsers. Another, perhaps more compelling, format useful for human viewing is SVG (Scalable Vector Graphics)—a standard XML format for describing two-dimensional graphics. Unlike "vanilla" HTML, SVG drawings are dynamic, interactive, and may even be animated.

7.7 FUTURE DIRECTIONS

Unfortunately, knowledge of the protein content of cells (proteomics) is much more complicated than genomics. Protein analyses face many challenges that genomic analyses do not. The chemical properties of the nucleic acid bases are very similar, so separation and purification is relatively easy compared with protein separations, where proteins can have very diverse chemical properties, complicating handling, separations, and identification.

Also, many proteins exist at very low levels in a cell, making it difficult to identify and analyze. In genomics, nucleic acid sequences can be amplified using the polymerase chain reaction (PCR), allowing one to amplify sequences with low copy numbers to levels that permit accurate detection. Comparable amplification methods are not available for proteins. Even when scientists know the amino acid sequence of a protein, they cannot necessarily deduce what the protein does or which other proteins it engages with. The behavior of proteins is determined by the tertiary structure of the molecule, so an assay that is based on protein binding depends on maintaining the native conformation of the protein. This puts constraints on the systems that are used to capture protein targets in affinity-based assays.

Furthermore, protein quantity is not necessarily correlated with function. Proteins can undergo a number of posttranslational modifications that affect their activities and cellular location, such as metal binding, prosthetic group binding, glycosylation, phosphorylation, and protease clipping, among others. RNA splicing can also produce a number of similar proteins that differ in function. So, a complete proteomics analysis must not only measure cellular protein level, but also determine how the proteins interact with one another and how they are modified.

Proteomics data format is complex and difficult to process with standard tools. Data cannot be interchanged easily among different hardware, software, operating systems, or application platforms. Metadata describing the content, format, interpretation, and historical evolution of the proteomics data are not available to either end users or application designers. Not all proteomic data are definitive. For example, identification of a single peptide does not automatically indicate the exact protein or protein isoform that it is derived from. Where there are tools, there are data, and the tools of proteomics continue to grow by the month. Data collection at a volume and quality that is consistent with the use of statistical methods is a significant limitation of proteomics today. The analysis and interpretation of the enormous volumes of proteomic data remains an unsolved challenge, particularly for gel-free approaches.

Proteomics has emerged as a major discipline that led to a re-examination of the need for consensus and a globally sanctioned set of proteomics data standards. The experience of harmonizing the development and certification of validation and data standards by the National Institute of Standards and Technology (NIST) provides a paradigm for technology in an area where significant heterogeneity in technical detail and data storage has evolved. NIST played a crucial role in the early days of the Human Genome Project in supporting and rapidly accelerating the pace of sequencing and diagnostics, essential for the genomic community and, later, the proteomics community. Currently, NIST is developing a program to meet some of the needs of the proteomics community. Although standards can inhibit innovation by codifying inefficient or obsolete technology, and thus increase the resistance to change, standards generally spur innovation directly by codifying accumulated technological experience and forming a baseline from which new technologies emerge. There is an absolute necessity for data standards that collect, exchange, store, annotate, and represent proteomics data.

ACKNOWLEDGMENT

We wish to acknowledge the thoughtful contributions of Dr. Robert Allen (The Johns Hopkins University).

REFERENCES

Aebersold, R. and Mann, M. Mass spectrometry-based proteomics. *Nature,* 422, 198–207, 2003.

Berman, J.J., Edgerton, M.E., and Friedman, B.A. The tissue microarray data exchange specifications: A community-based, open source tool for sharing tissue microarray data. *BMC Med. Inform. Decis. Mak.,* 3, 5, 2003.

Chute, C.G. *Electronic Medical Record Infrastructures: An Overview of Critical Standards and Classifications.* Springer-Verlag, New York, 1998.

Galperin, M.Y. The molecular biology database collection. *Nucleic Acids Res.,* 32, D3–D22, 2004.

Harris, M.A., Clark, J., Ireland, A., Lomax, J., Ashburner, M., Foulger, R., Eilbeck, K., Lewis, S., Marshall, B., Mungall, C., et al. Gene Ontology Consortium. The Gene Ontology (GO) database and informatics resource. *Nucleic Acids Res.,* 32, D258–D261, 2004.

McCray, A.T. and Divita, G. ASN.1: Defining a grammar for the UMLS knowledge sources. *Proc. Annu. Symp. Comput. Appl. Med. Care,* 1, 868–872, 1995.

Nadkarni, P.M. *Pharmacogenomics J.,* 2, 96–102, 2002.

Ravichandran, V., Lubell, J., Vasquez, G.B., Lemkin, P., Sriram, R.D., and Gilliland, G.L. Ongoing development of two-dimensional polyacrylamide gel electrophoresis data standards. *Electrophoresis,* 25, 297–308, 2004.

Stein, L.D. Integrating biological databases. *Nat. Rev. Genet.*, 4, 337–45, 2003.

Tyers, M. and Mann, M. From genomics to proteomics. *Nature,* 422, 193–197, 2003.

Verma, M., Kagan, J., Sidransky, D., and Srivastava, S. Proteomic analysis of cancer-cell mitochondria. *Nat. Rev. Canc.,* 3, 789–795, 2003.

Westbrook, J.D. and Bourne, P.E. STAR/mmCIF: An ontology for macromolecular structure. *Bioinformatics,* 16, 159–168, 2000.

8 Data Standardization and Integration in Collaborative Proteomics Studies

Marcin Adamski, David J. States, and Gilbert S. Omenn

CONTENTS

8.1 COLLABORATIVE PROTEOMICS

8.1.1 THE NECESSITY OF COLLABORATION

Proteomics is often compared to genomics, and proteomes to genomes. There is, however, a fundamental difference between these two. While genomics deals with a (nearly) fixed set of genes in an organism, proteomics deals with a very large number of related proteins with posttranslational modifications. Furthermore, the proteins undergo major changes in level of expression during physiological, pathological, and pharmacological stresses. Gathering and analyzing a vast amount of data about complex mixtures of proteins require large-scale, high-throughput approaches. Thus, proteomics depends upon collaborations across laboratories and centers producing, collecting, and integrating data, and making datasets available to the scientific community.

Standardized operating procedures and protocols for protein separation and identification techniques are still under development. One of the main challenges for modern collaborative proteomics is to establish reliable, efficient techniques for data acquisition and analysis.[1] Standardization has to begin at sample collection, preparation, and handling while allowing comparison of findings between laboratories and technologies and replication of the experimental results.

The various high-throughput technologies presented in Table 8.1[2,3] help to increase both the sensitivity and confidence of findings.[4] When applied to the same sample set, different experimental techniques will not generate exactly the same results. The sampling space is so large and experimental techniques so susceptible to so many biological and technical sources of variation that the number of findings common to all of them may be quite small. Collaborative studies can add confidence to such results.

Collaboration requires data exchange among the parties. Data exchange, in turn, requires a common data transfer protocol. Primary data can be presented in formats ranging from notes in lab books and computer text files to relational database management systems (RDBMS) and advanced laboratory information management systems (LIMS). In earlier years flat or tabular text file format was commonly used. Internet-based protocols have facilitated introduction of the extensible Markup Language (XML) for high-throughput biology (derived from Standard Generalized Markup Language, SGML; http://www.w3.org/XML); it is now accepted as a standard solution for data transfer and exchange (ISO 8879).

In small collaborative projects, data are usually exchanged directly between collaborating laboratories. More complex and especially global collaborations require dedicated data collection centers. These centers are responsible for data collection, conversion, standardization, quality control, and placement in a central repository where all information can be integrated with other databases, warehoused, and accessed.

8.1.2 THE NECESSITY OF DATA STANDARDIZATION

Requirements for data standardization in collaborative studies depend not only on the aims of the project, but also on the size of the collaboration and diversity of

---

TABLE 8.1
High-Throughput Technologies Currently
Used in Proteomics[2,3]

Fractionation Techniques
Two-Dimensional Gel Electrophoresis
Multi-Dimensional Liquid Phase Separations
Reverse Phase Chromatography
Anion-Exchange Chromatography
Cation-Exchange Chromatography
Immunoaffinity Subtraction Chromatography
Size-Exclusion Chromatography
Mass Spectrometry Techniques
Sample Ionizations:
Electro-Spray Ionization (ESI)
Matrix-Assisted Laser Desorption/Ionization (MALDI)
Surface-Enhanced Laser Desorption/Ionization (SELDI)
Instrument Configurations:
Reflector Time-Of-Flight (TOF)
Time-Of-Flight Time-Of-Flight (TOF-TOF)
Triple Quadrupole, Linear Ion Trap
Quadrupole Time-Of-Flight (QTOF)
Ion Trap
Fourier Transform Ion Cyclotron Resonance (FTICR)
Data Analysis Software
Gel Image Analysis
Protein Database Search
Peptide Mass Fingerprint or Peptide Mass Map Analysis
Peptide Sequence or Peptide Sequence Tag Query
Ms/Ms Ion Search Analysis

---

experimental techniques being used. The debate about data exchange standards should be incorporated into discussions preceding and defining the collaboration. Data exchange protocols and repositories should be chosen as the collaboration goals, expected outputs, and funding limitations are defined. Currently many different databases, repositories, and protocols exist. There is no universal solution to satisfy all possible needs. Standards that attempt to cover an area as wide as proteomics usually are not well suited to accommodate more detailed information that may be crucial to particular studies. At the other end of the spectrum are data standards developed specifically to describe particular aspects of studies in fine detail, which may not be well suited for participants using even modestly different techniques. Furthermore, excessively detailed information about certain experimental techniques may be ill suited for data integration because of lack of equivalent data from other experiments. In sum, common data repositories and standardization protocols cannot replace specific Laboratory Information Management Systems (LIMS), but they rather should operate on a precisely defined level of abstraction allowing for data integration without losing any necessary information. "Necessary information" is a key term in the planning of collaborative studies.

For example, consider a multilaboratory collaboration aimed to catalog proteins in a tissue. Participating laboratories are free to choose experimental techniques, each with its own quality control mechanisms. The process of defining the data standards for such a loosely characterized collaboration must find a common denominator for all the results from different laboratories. Database accession numbers for the identified proteins supported by scores or estimates of the level of confidence for the identifications may seem an obvious choice. But there is more than one protein database. Furthermore, databases are living creatures: new entries appear everyday, and, even worse from the point of view of standardization, the existing entries are often revised, or deleted. One solution is to choose one version of a single database as the collaboration standard and to require participants to use it solely, or else convert all nonstandard data received in the data center to the standard. A database version that is state-of-the-art at the beginning of the project may become obsolete by the time the project conclusions are formulated. And how should identifications referring to subsequently revised or eliminated entries be handled? How should data centers compare the confidence levels of identifications when they come from different experimental techniques, including instruments with embedded proprietary search engines? These are just examples of the problems that may suggest that the protein identifiers themselves may not constitute a sufficient level of data abstraction. If not the database accession numbers, then what ought to be used?

For protein identifications based on mass spectrometry techniques, the identified peptides' sequences may provide a better, even sufficient, basis for data integration. Since these sequences can be searched against any protein sequence database, they eliminate many of the shortcomings of accession numbers alone. Furthermore, they allow the data center to calculate the protein sequence coverage, adding additional confidence to the identification. The main drawback of using lists of peptide sequences is that they are actually not a direct result of experiment, but rather a best fit from the domain of available peptides in the search database via the algorithms embedded in the search engine. Use of peptide sequences does not free the results from influence of the original search database. It may also raise problems in situations where the submitted sequence can no longer be found in the database. There is no way of confirming that not-perfectly-aligned peptides can still be explained by their underlying mass spectra; even single amino acid differences put such alignments in question.

Therefore, the next more informative level is the use of the mass spectrometry peak lists. They can be then searched against the selected standard database with a standard search engine and standardized settings. Peak lists are database and search engine independent. As they can be represented by text-based files, they are quite easy to collect and process. But their use adds significant additional workload to the project data integration unit, generally without reducing workload in the data-producing laboratories. The database searches need both hardware and software support. In the laboratory, generating the peak files, organizing them adequately, and sending them to the repository may require no less time and effort than the database searches themselves. Using the peak lists will also generate a whole new set of standardization issues. For example, should only the peak lists supporting a possible identification be collected, or all of them? For tandem mass spectrometry, should peak lists from both stages of the analysis be registered? What settings should be used for obtaining the lists from the spectra?

Finally, instead of the peak lists, the "raw" spectra could be used as the ultimate common denominator. Using them makes a clear cut between the physical experiment and computational analysis. Unfortunately, there is no common format for storing the raw mass spectrometry data. Worse, the data formats are not only instrument specific, but their formats are known only to the manufacturer and usually are not publicly available. Several efforts have been undertaken to build converters able to read many of the formats, or to build a common format in which the instruments could export their data. Converters are now available for the ThermoFinnigan Xcalibur, Waters Mass Lynx, Sciex/ABI Analyst, and Bruker HyStar. A standard converter cannot read all the existing formats and must be updated each time the instrument generic data format is changed by the manufacturer. The raw mass spectra files are rather large, even for current standards; spectra from a simple analysis can easily reach several gigabytes. The files are in binary format, which means that they cannot be directly embedded into text format protocols like flat text or XML. Additionally, use of the raw spectra increases the workload to the data integration unit and demands more hardware support.

Our hypothetical project may require additional supportive data. For example, if proteins or peptides have been separated by gel electrophoresis, perhaps it would be useful to record gel images annotating spots that have been picked for analysis. If so, which format should be used? Could a simple list of the spot gel coordinates include a sufficient amount of information? And how to standardize and then integrate the data when collaborating laboratories use different separation techniques?

In single-laboratory research, data standardization and integration issues do not appear to play a major role, at least not until efforts to compare published findings. In high-throughput collaborative studies, standardization and integration may be no less important than the actual experiments. Underestimating their importance can add an extra burden to participants in the collaboration, distract the group from the actual research goals, and delay formulation of the project conclusions.

8.1.3 INFORMATION FLOW IN COLLABORATIVE STUDIES

8.1.3.1 Data Exchange Strategies

Generally, the strategy for the data exchange in collaborative studies can be arranged with one of the following two topologies.

In the first, called the peer nodes topology, all participants perform experiments generating the data. All participants also collect data generated by others and carry out the data integration and analysis (Figure 8.1a). In this strategy, there is no dedicated data collection and integration center. Each participant exchanges results with all other participants. This strategy requires $n \cdot (n - 1)$ data transfers plus n data integrations (where n stands for the number of participants). The data exchange complexity of this strategy is on the order of n^2. Even when the number

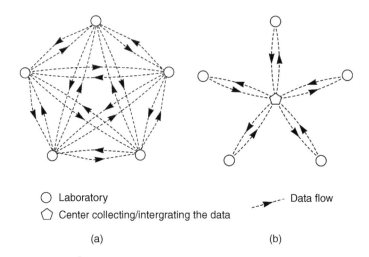

○ Laboratory
⬠ Center collecting/intergrating the data

╌╌➤╌ Data flow

(a)   (b)

FIGURE 8.1 Data exchange strategies (a) peer nodes—strategy without dedicated data collecting center — each laboratory exchanges data with all others laboratories; (b) central node — strategy with dedicated data collection center limiting complexity of the data transfers and integrations.

of collaborating parties is relatively small, the number of necessary data transfers is large and rises quickly with increasing the number of participants. In a ten-participant collaboration there are 90 data transfers—each participant sends their own data in 9 copies. Any divergence in data format from one contributor affects all other participants. These drawbacks of the peer nodes topology limit its use to very small collaborations (typically, two or three participants) and should be avoided in bigger alliances.

In the second strategy, called the central node topology, the participants who generate the data send them to a central, dedicated data collection and integration center (Figure 8.1b). The center collects and processes the data and sends integrated results back to the participants. In this collaboration each contributor sends the data only once to the data collection center. This strategy requires $2n$ data transfers plus single data integrations. Data exchange complexity of this strategy is linear or in order of n. A potential divergence in data format from one participant affects only communication between that participant and the data collection center. The price for the much smaller number of data transfers, with more error proofing and simplified data integration, is the cost of the dedicated data collection center. This strategy should be chosen for virtually any collaboration involving more than two groups. In practice, there can be more than one data collection center. In this case the collaborators submit their data once to one of the centers. The centers may either partially integrate the information and then synchronize it between themselves as in the peer nodes strategy, or send it to a higher ranking center for the final integration (Figure 8.2).

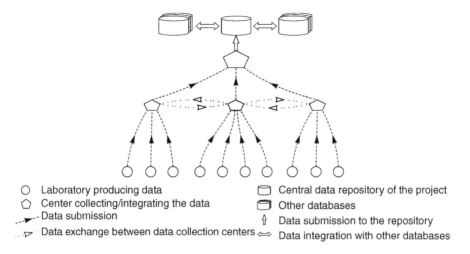

○ Laboratory producing data   ⬒ Central data repository of the project
⬠ Center collecting/integrating the data   ⬓ Other databases
➤ Data submission   ⇕ Data submission to the repository
▽ Data exchange between data collection centers ⇔ Data integration with other databases

FIGURE 8.2 Multilevel data exchange strategy. Data submitted from laboratories to one of the first level data collection centers are partially integrated there and/or sent to the next level integration center. The top level data collecting/integrating center deposits the data in a central repository and performs integration with data from other off-project databases.

8.1.3.2   Data Exchange Technologies

Exchange of data among collaborating parties can be arranged with use of physical media (diskettes, CDs, DVDs, tapes, memory chips) or with use of Internet protocols like email, file transfer protocol, or dedicated on-line submission sites equipped with a Web browser interface.

Use of the physical media may seem outdated nowadays, but disks still remain the most trustworthy way of storing and exchanging information. Clearly, usage of such media is associated with physically sending the disks from one site to another, with unavoidable delays. However, disk capacity, especially the 4.7 GB capacity of the DVD, offers a new potential for exchanging huge amounts of information generated in proteomics studies. Although such data may be transferred over Internet connections, the disks are particularly desirable when not all of the collaborating parties have the ability to transfer huge amounts of data over the Internet.

Sending data sets as attachments through email systems may be the simplest means for data exchange. It is easy to use and widely available. The drawbacks include problems with sending and receiving big files, ineffective encoding of binary files, and recent infestation by various viruses, which have greatly complicated free exchange of binary email attachments. Unfortunately, the limitations apply as well to files archived with popular programs like *.zip* or *.gzip*. The received emails usually have to be processed by hand, which involves lots of time and introduces human errors. Data exchange through the email system may be advisable only for small collaborations and for small amounts of data transfers. Some of the mentioned drawbacks can be overcome by creating an automated data exchange system based on email protocol, in which case use of the file transfer protocol (ftp), designed specifically for data transfers, is a better solution.

Ftp is an Internet protocol that allows for transferring huge amounts of data, downloading or uploading files into specific directory locations, and restoring broken connections (a priceless feature in big data transfers). Ftp was developed in the early days of the Internet and is still the most popular and widely used protocol for file transfers. Both text and GUI (Graphical User Interface) clients are freely available. Properly configured ftp servers constitute fast, efficient, and secure data exchange solutions.

The data exchange technologies described above need some human interaction at the data collection center. A perfect exchange technology should allow for automatic data processing and submission of relatively big files, record additional information about submitted files, handle data resubmission, send back confirmations, and, broadly speaking, interact with the submitters. Many of these features can be achieved with on-line submission sites. Such sites work over the Internet and present their interfaces using regular Web browsers. Access to the site should be restricted to the participants and enforced with passwords. The on-line submission sites may constitute the best and most effective solution for data exchange. They provide simple and common user interfaces, keep track of all the data transfers, and offer additional services like on-line, in-repository data lookup for participants. The only problem with this solution is that usually the sites have to be tailored specifically for the collaboration needs. They have to be built, tested, and then maintained by qualified personnel.

8.1.3.3   Data Collection and Data Integration in Collaborative Studies

The data collecting process can be divided into six consecutive steps:

1. Registration (recording of the submitted files)
2. Processing of the files into the project database
3. Verification of data consistency against specific rules
4. Feedback to the submitting laboratory to permit revision of verified data
5. Data integration (often a complicated, multilevel process)
6. Open-source data warehousing accessible to the scientific community

The data integration process can be divided into two conceptually distinct levels: shallow and deep. Shallow integration transforms schematics structures, maps schema elements, and attributes names across data sources. Deep integration fuses information from multiple sources, frequently containing overlapping or contradictory information, and presents it in a single coherent form. The necessary level of data integration should be defined in formulating the collaboration, since this decision affects requirements for the data standards, defines tasks for the data integration center, and often influences designs of the experiments themselves.

In proteomics studies, the data may require integration on both integration levels. The collaboration may involve participants generating the same level of information, performing experiments that reveal the same kind of findings or whose results complement one another. For example, functional studies usually involve integration of data coming from genomics, transcriptomics, and proteomics experiments. In proteomics studies alone, it is not unusual to integrate findings from mass spectrometry protein identification experiments with quantitative data from experiments like antibody microarrays, protein microarrays, or immunoassays. The data integration process can be divided into the following five consecutive steps:

1. Object identification: A set of distinct entities is established. Objects having multiple accession numbers from different data sources are identified and pulled together. Ambiguous identifiers (identifiers referring to more than one distinct entity) have to be identified and cleared up; some experiments may lack needed identifiers. The main problems at this step are associated with data integration across databases or even across versions of a single database, and the need to describe the same object in different contexts.
2. Data fusion: All the information gathered from multiple data sources is merged; each entity references a union of the gathered information preserving its provenance.
3. Data interpretation: Data validation and interpretation of the overlapping, possibly contradictory information from different sources, with assignment of levels of uncertainty.
4. Integration: Integration of the project findings with existing public databases; e.g., protein catalog, protein–protein interaction databases, and article databases.
5. Inference: Drawing conclusions from the project findings, dealing with incomplete and uncertain information, generally based on statistical analysis of the data, and calculating statistical significance of the findings.

8.1.4　HUPO—The Human Proteome Organization

The main challenges standing before modern proteomics include cataloging the entire human proteome and identifying protein biomarkers of physiological, pathological, and pharmacological changes that can be used for diagnosing diseases at early stages, predicting and monitoring the effects of treatments, and guiding preventive interventions. These long-term goals can benefit from international collaborations of researchers, vendors, and government and private sector founders. The international Human Proteome Organization (HUPO) initiative was launched in 2001 with explicit objectives to accelerate development of the field of proteomics, to stimulate major research initiatives, and to foster international cooperation involving the research community, governments, and the private sector.[5,6] This global cooperation is aimed to catalog and annotate the entire human proteome. That includes several initiatives around organ systems and biological fluids, as well as development of proteomics resources.

The first HUPO initiative focused on biological fluids is the Plasma Proteome Project (PPP). The long-term scientific objectives of the Project include

1. Comprehensive analysis of plasma protein constituents in normal humans in a large cohort of subjects
2. Identification of biological sources of variation within individuals over time and assessment of the effect of age, sex, diet and lifestyle, diseases, and medications
3. Determination of the extent of variation in plasma proteins within populations in various countries and across various populations from around the world[7]

The project was launched in 2002 with a pilot phase aimed at producing preliminary results, defining operational standards for the main part of the project, and assessing possible difficulties. An additional important aim was the development of an efficient method of data acquisition, storage, and analysis in a big collaborative proteomics experiment. More than 40 laboratories in 13 countries and several technical committees are actively engaged in the PPP, with leadership in the U.S.

The HUPO Liver Proteome Project (LPP) and Brain Proteome Project (BPP) have objectives similar to objectives of the PPP: characterization of protein expression profiles in the human liver and brain, construction of protein interaction maps, and studies of liver and nervous system pathologies. HUPO also has a major antibody production project (see http://www.hupo.org).

Developing and adopting standardized approaches that facilitate analysis of proteomics data generated by different laboratories are an important informatics-related effort of HUPO. The HUPO Proteomics Standards Initiative (PSI) was launched with the aim of defining community standards for data representation in proteomics to facilitate data exchange, comparison, and validation (see next sections).

8.2 STANDARDIZATION INITIATIVES IN PROTEOMICS

Standardization of the data is an indispensable step in collaborative proteomics studies, crucial to successful integration of the results. A good data standard should facilitate access to the data, being extensible enough to exchange data with a high level of detail. Many users require only simple access to the data via an easy data search interface. Requirements of others may be more complex and should not be limited by unavoidable simplicity of the common interface. Wherever possible, the data should be generated in the standard form straight away in the producing laboratories, rather than converted in the data collection centers. Such work flow helps to avoid conversion errors and facilitate quality control, but may generate additional work for laboratories that do not have sufficient informatics support. The problem is especially important in high-throughput mass spectrometry studies. As noted above, the instruments produce huge amounts of data not yet standardized among the manufacturers; usually the output data are encoded in unpublished format and can be processed only with dedicated software.

8.2.1 HUPO-PSI: Proteome Standards Initiative

The HUPO-PSI was established in April 2002 as a working group of the HUPO.[8–12] The HUPO-PSI, based at the European Bioinformatics Institute (EBI) in Cambridge, U.K., aims to define community standards for data representation in proteomics, to overcome the current fragmentation of proteomics data, and to facilitate data comparison, exchange, and verification. The initiative focuses its effort on two main aspects of proteomics studies: mass spectroscopy data and protein–protein interactions data, as well as on defining a common work flow for proteomics results integration.

The protein–protein interaction community currently has a relatively small number of databases at its disposal; e.g., BIND,[13] IntAct,[14] DIP,[15] MINT,[16] HPRD,[17] and PS MI (see below). There is a desire to integrate the existing databases. The HUPO-PSI, with the support of major protein interaction data providers, has already proposed a community standard data model (PSI-MI) for the representation and exchange of protein interaction data.[18] As noted above, mass spectrometry data are available at many levels of abstraction, from raw data through peak lists to peptide and protein identification, and there is significant variation of methods used to process these data. There is currently no repository established for mass spectrometry data. Data published in the literature represent only a summary of the huge amount of underlying information collected in these studies.

The current agenda for the mass spectrometry workgroup includes controlled vocabulary for m/z data exchange; standard XML format for search engine results; converters to transform supported vendor proprietary peak list formats to standard XML format; search engines to be supported; navigation and visualization tools; PSI MS data formats, and MIAPE data model. The agenda for the related Molecular Interactions workgroup includes a PSI MI XML schema; controlled vocabularies; procedures for maintenance of controlled vocabularies; controlled vocabulary for external databases; unified identifiers for interactors; and data exchange options.

Finally, there is a group addressing work flows for integrating proteomics: sample preparation; coordination with The Microarray Gene Expression Data Society (MGED); 2D gel electrophoresis — data capture, data storage; modeling columns and other separation techniques; integrating mass spectrometry methods and results; and supporting bioinformatics analyses.

8.2.1.1 MIAPE and PEDRo

Several efforts have been undertaken to create a standard data exchange protocol for mass spectroscopy results and standard tools to capture data and metadata from proteomic experiments. One of them, a proteomics integration model called MIAPE (minimum information to describe a proteomics experiment), is being developed by the EBI under supervision of the HUPO-PSI. The general idea of this model is based on PEDRo (the proteomics experiment data repository) developed at the University of Manchester in the U.K.[19] PEDRo has been aimed to describe and encompass central aspects of a proteomics experiment and is platform independent. Different implementations can be derived from it for use in different software environments. It is intended to capture all the relevant information from any proteomics experiment, such as details of sample sources, experimenters, methods and equipment employed in analyses, and results.

The schema of the PEDRo can be naturally divided into five sections (Figure 8.3):

1. Experiment: Captures rationale and hypothesis, work plan, applied methods, and expected results.
2. Sample Generation: Holds information about the sample itself and its origin (entity *SampleOrigin*). *SampleOrigin*, among others, includes data

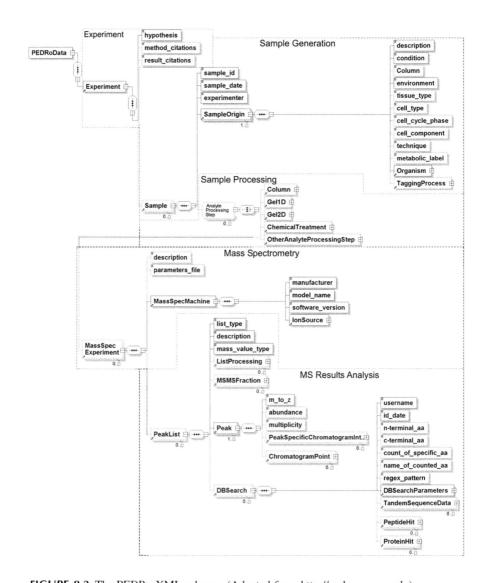

FIGURE 8.3 The PEDRo XML schema. (Adapted from http://pedro.man.ac.uk.)

describing sample labeling (entity *TaggingProcess*), allowing for differential expression studies such as differential gel electrophoresis (DiGE), or isotope-coded affinity tag (ICAT) mass spectrometry.

3. Sample Processing: Holds information describing the sample separation steps. The relation between the sample generation and processing sections can be described as follows: A sample defined in the sample generation section is put through the first step of separation (e.g., depletion, digestion, gel chromatography, etc.) described in the sample processing section. Results of this separation step (e.g., a fraction, band, spot) can be rerouted to the next step of the separation and finally used as input data for the mass spectrometry experiment (described in next section).

4. Mass Spectrometry: Keeps data about spectrometry analyses, including stages of the experiment, equipment, and parameters utilized. It also includes elements containing the attribute-value structures, facilitating the creation of new experiment stages.

5. MS Results Analysis: Holds information about the results of the mass spectrometry experiment carried out on the samples defined and processed in the two first sections. It includes the peak lists, database search hits, and all necessary supporting data.

Although the model currently does not permit storing unprocessed (raw) chromatograms (spectra), it allows for storage of sets of self-referenced peak lists. That means that it may be a raw, not human-edited, peak list with a set of associated peak lists representing human-processed spectra. The second-stage spectra from tandem mass spectrometry are stored in the same manner. The model allows for storing multiple database search results for each peak list, collecting data about peptide hits, protein hits, database used, search software, parameters, and scores.

The PEDRo model assumes existence of a central data repository and data entry tools distributed among data-producing laboratories. The central data repository is implemented with use of an SQL (structured query language) relational database server. Laboratories do not enter their data directly into the repository. Instead they use the data entry tool that encodes the data and produces an XML file, which is then submitted to the repository. The data entry tool has a user-friendly graphical interface to allow for easy data import, permit creation of data templates (e.g., for description of similar experiments), and ensure data integrity and compatibility with the PEDRo model. When the XML file is submitted to the repository, a validation tool checks its correctness, and the data are entered into the repository's database. Once in the repository, the data become publicly available. The availability of particular parts of the data (or of whole repository) may be restricted to a defined group of users or be accessible in unrestricted manner. Access to the data in the repository is possible by compliant query, search, and analysis tools.

Although the PEDRo model requires a fairly substantial amount of data to be captured, much of the information should be readily available in the laboratory. Substantial parts of the data will also be common to many experiments and so will only have to be entered once and saved as a template. One of the main advantages of adopting such a model is that all the data sets will contain information sufficient to quickly establish the origin and relevance of the data set, facilitating nonstandard searches and subsequent data integration.

8.2.1.2 mzData

From the data standardization point of view, the mass spectrometry part of proteomics experiments presents a very complex problem. An XML standard data format called mzData is currently being developed by PSI together with major mass spectrometer producers. The format will represent spectra in the context of the experimental setup and the analyzed system (Figure 8.4).

The *mzData* element is the root element of the mzData document file. The *mzData* element contains two parts — *desc*, a descriptive section of the data, and *raw*,

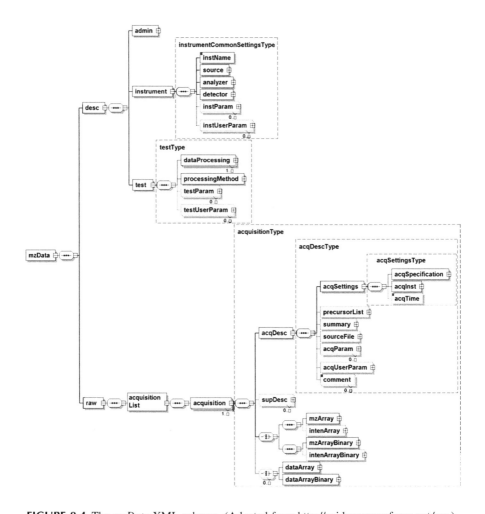

FIGURE 8.4 The *mzData* XML schema. (Adapted from http://psidev.sourceforge.net/ms.)

holding the actual peak lists; *desc* comprises the elements *test* and *instrument*, describing the mass spectrometry experiment and the spectrometer itself.

Included in the *testType* are (a) *dataProcessing*, information about how the XML document has been generated; (b) *processingMethod*, a general description of the peak processing method, which can be overridden by additional description for particular peak list acquisition (see later in text element *acqDesc*); and (c) *testParam* and *testUserParam*, additional information about the experiment.

The spectrometer is further described under *instrumentCommonSettingsType*: (a) *instName*, name of the instrument; (b) *source*, information about the ion source; (c) *analyzer*, mass analyzer data; (d) *detector*, detector data; and (e) *instParam* and *instUserParam*, additional information about the instrument.

The *raw* element contains a list of instrument acquisitions—the *acquisition* elements. Each acquisition can be described in addition by *acqDesc* element and its subelements from the *acqDescType*: (a) *acqSettings*, which stores data on settings for the specific acquisition, such as specifications for combining raw scans into a peak list (element *acqSpecification*) or instrument acquisition settings (element *acqInst*), which hold information about the mass spectrometry level (first- or second-stage MS); (b) *precursorList*, list of precursor ions for the current acquisition; (c) *sourceFile*, stores name and type of the original source file used to generate the particular document; (d) *summary*, of information for acquisition; and (e) *acqParam* and *acqUserParam*, additional information about the acquisition.

The actual acquisition data in the form of peak lists can be stored in two formats: (1) directly readable text values in two arrays, *mzArray* and *intenArray*, storing mass over charge ratios and ion intensities, respectively; and (2) base64 encoded binary float point mass over charge ratios and ion intensities in tables *mzArrayBinary* and *intenArrayBinary*, respectively.

Users can choose the format of the data storage. The text format, with numbers stored as text representation, can be easily read and edited by human personnel, but it generates bigger files than the binary float point data encoded with use of the base64 algorithm.

The *acquisition* element can directly store the raw machine file in the XML document. If the raw file is not binary, or is supplemented by nonbinary data, it can be stored in the *dataArray* element. When the raw file is in binary format, it can be base64 encoded and stored in the *dataArrayBinary* element.

mzData allows for a very detailed description of virtually any mass spectrometry experiment. It allows for inclusion of binary data in a text-based XML document. The experiment and acquisition description and parameters are either hard-built into the data structure or controlled by a set of vocabularies. The vocabularies have been divided into two subsets—one where the names are globally controlled, and the other where the user can define his or her own collection. This solution reinforces global data integrity while permitting users to introduce their own specific elements. mzData has been designed as a data export format, which should be built into mass spectrometer software by the instrument vendor. As such it may be easily introduced and updated with periodic instrument software updates. In addition, it does not require the vendor to publish the internal instrument data format and eliminates a time lag necessary for third-party developers to create a software converter able to turn the data format from an internal instrument into common use.

8.2.1.3 HUP-ML

Other initiatives have been undertaken to facilitate data standardization in proteomic studies. An XML-based, proteomic-oriented markup language, HUP-ML (Human Proteome Markup Language), has been developed in the Proteomics Research Center, NEC Corporation in Tsukuba, Japan (http://www1.biz.biglobe.ne.jp/~jhupo/HUP-ML/hup-ml.htm). HUP-ML has been designed to be capable of describing the experimental results as well as the experiment protocols and conditions. It can describe sample origin and preparation, methodology, 2D electrophoresis gel image/LC results, spot identification, mass spectrometry experiment parameters and peak lists, protein search results and features, and protein 3D structure.

HUP-ML is supported by a helpful visual data entry tool called HUP-ML Editor. The editor is available as a desktop application and as a web-based interface. It can be used to enter all the text descriptions, tables, and graphical gel images with spot annotation.

By its concept, design, and used technology (XML standard) HUP-ML is very similar to MIAPE. Both standards have similar abilities.

8.2.1.4 PSI MI: Molecular Interactions

Identification and characterization of proteins constitute only a part of the proteomics research. The other major goal of proteomics is the complete description of the protein interaction network underlying cell and tissue physiology. It is probable that most of the proteins encoded in the human genome are large, multidomain molecules that participate in molecular interactions with other proteins, DNA, RNA, carbohydrates, and other molecules. There are more protein–protein interactions than sequences.[20] Several interaction databases have been developed, such as DIP (Database of Interacting Proteins), BRITE (Biomolecular Relations in Information Transmission and Expression) (http://www.genome.ad.jp/brite), CSNDB (Cell Signaling Networks Database), and BIND (Biomolecular Interaction Database). To access and manipulate the data in a more efficient way, as well as to exchange data between different repositories, a common data exchange format has to be used. Such a format has been jointly developed by members of the HUPO-PSI and is supported by major protein interactions data providers, including BIND, MINT, IntAct, HPRD, and DIP. Molecular Interaction (MI) XML format PSI MI has been proposed.[18] The PSI MI format is a database-independent exchange protocol. It has been developed using a multilevel approach. Currently the format is on its first level (Level 1) and provides basic elements suitable for representing the majority of all currently available protein–protein interactions data. It allows for representation of binary and more complicated multiprotein interactions, classification of experimental techniques, and conditions. Level 1 currently describes only protein–protein interactions and does not contain detailed data on interaction mechanisms or full experimental descriptions. These data, as well as descriptions of other molecules interactions, will be encompassed in more detail in subsequent PSI MI levels.

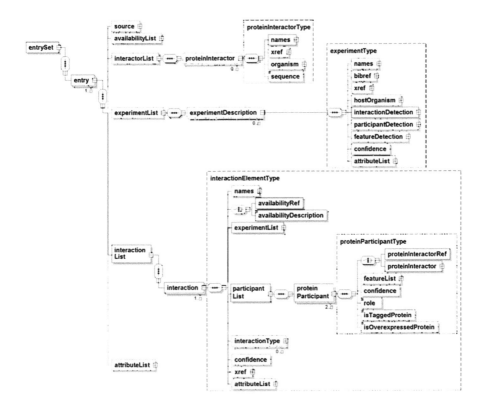

FIGURE 8.5 The PSI MI XML schema. (Adapted from http://psidev.sourceforge.net/mi/xml/doc/user.)

The root element of the PSI MI XML document file (see Figure 8.5) is the *entrySet*, which contains one or more entries. The *entry* element describes one or more protein interactions and contains a *source* element (normally the data provider) and several list elements: *availabilityList, experimentList, interactorList, interactionList*, and *attributeList*. The *availabilityList* provides statements on the availability of the data. The *experimentList* contains descriptions of experiments — the *experimentDescription* elements; each *experimentDescription* describes one set of experimental parameters. The *interactorList* describes a set of interaction participants — the *proteinInteractor* elements. Currently only protein–protein interactions are supported. In the future, the schema will be extended to accommodate protein interactions with other elements (e.g., small molecules). The *interactionList* contains one or more interaction elements constituting the core of the PSI MI structure. Each interaction contains a description of the data availability, a description of the experimental conditions under which it has been determined, and a confidence attribute. The information about the proteins participating in the interactions is stored in the *participantList* as *proteinParticipant* elements. Each *proteinParticipant* contains a description of the molecule given either by reference to an element in the *interactorList*, or directly in a *proteinInteractor* element. Additional elements of the *proteinParticipant* element describe the specific form of the molecule in which it participated in the interaction. The *featureList* describes sequence features of the protein. The *role* element describes this particular protein role in the experiment (e.g., bait or prey). The *attributeList* elements are placeholders for semistructured additional information the data provider might want to transmit. The *attributeList* elements contain tag-value pairs and provide an easy mechanism to extend the PSI MI format.

The PSI MI format can be used in two forms: compact and expanded. In the compact form, all interactors, experiments, and availability statements are given once in the respective list elements—*interactorList*, *experimentList*, *availabilityList*—and then referred to from the individual interactions. The compact form allows a concise, nonrepetitive representation of the data. In the expanded form, all interactors, experiments, and availability statements are described directly in the interaction element. As a result, each interaction is a self-contained element providing all the necessary information. The compact form is suitable mainly for transmitting large data sets, whereas the expanded form creates larger files but is more suitable for a single-experiment description and for conversion to displayed data.

The PSI MI format has been designed for data exchange by many data providers. Therefore, in addition to the standard data format, the meaning of the data items has to be consistent and well defined. To address the problem, PSI MI uses sets of controlled vocabularies instead of free text description wherever possible. For the Level 1 of the format, five sets of vocabularies have been created, which describe the interaction type, sequence feature type, feature detection, participant detection, and interaction detection elements. All the terms' definitions are supported by literature references whenever appropriate. The controlled vocabularies have a hierarchical structure with higher level terms being more general than lower level terms, which has advantages for both annotation and querying of the data.

8.2.2 SASHIMI

Outside the HUPO-PSI initiative, other efforts have been undertaken to create a common standard specifically for mass spectrometry data. The most notable is a project called SASHIMI at the Institute for Systems Biology in Seattle, WA, U.S. (extensive information available at: http://sashimi.sourceforge.net). The goals of the project are to provide the scientific community with mzXML,[21] an XML-based, open, standard file format to represent tandem mass spectrometric data, linked with cutting-edge, free, open-source software tools for the downstream analysis. The adoption of an open standard will provide programmers with an easy way to access this kind of information, thus facilitating development and distribution of software in this field. Additionally, the use of a platform-independent representation will ease the exchange of data sets between collaborators and ultimately allow for the creation of public data repositories.

To address difficulties presented by the diversity of data formats from existing spectrometers, as well as the introduction of new machines' data formats into preexisting data structures, an XML-based common file format for mass spectrometry data has been developed. The mzXML arranges the data in the meaning of mass spectrometry

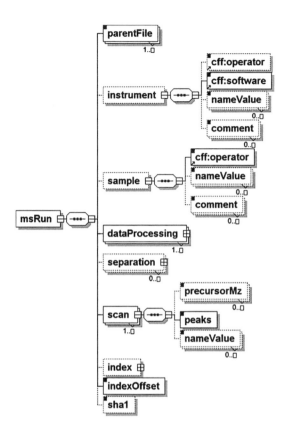

FIGURE 8.6 The mzXML XML schema. (Adapted from http://sashimi.sourceforge.net.)

runs (Figure 8.6). The parent element, *msRun*, stores results of a single MS run. Its children elements, *instrument, sample*, and *separation*, describe the mass spectrometer, the sample, and, if applicable, the sample separation process, respectively. The other two elements, *parentFile* and *dataProcessing*, represent the raw data generated by the instrument. They store, respectively, a chronological list of all the files used to generate a given instance of the document and a list of all the modifications applied to the data. Each run of the spectrometer (element *msRun*) has to have one or more elements *scan* representing analysis of a single precursor peptide (element *precursorMz*). There can be multiple instances of this element per each *scan* element, to account for fragment ion spectra with more than one ion precursor. The other subelement of the *scan*, the *peaks* element, represents the actual peak list (again, there can be more than one such element per each *scan*). The *index* element allows for nonsequential access to the data in *msRun*. This extremely valuable element is not an essential part of the schema, but facilitates much faster data access and reduces the amount of code necessary to build mzXML-based analysis tools. The mzXML incorporates elements with closely defined structures as well as the name-value attributes (called *nameValue* in the schema) allowing for personalization of the documents, while referring to a centralized common schema. It can also be used to introduce new data items necessary to describe information from new mass spectrometers.

The mzXML is supported by a set of tools developed under the SASHIMI project. Validation of the mzXML document against its schema is possible with use of a tool called ValidateXML. The XML document can be accessed with use of the RAP (Random Access Parser), or RAMP (Random Access Minimal Parser; RAP with minimal functionality only), which support the nonsequential data access utilizing the *index* element. Export of the data from mzXML to other formats can be done with use of the mzXML2Other tool. This program can convert mzXML documents into input files for programs like SEQUEST (.dta), Mascot (.mfg), Protein-Lynx (.pkl), or a tab-delimited text format. The other tool, a graphical mzXML viewer, offers a common way to display data generated on different instruments. It can display various properties of the LC-MS run, such as the total ion current (TIC), the base peak chromatogram (BPC), and the MS and MS/MS spectra. It can also load and simultaneously visualize more than one file at the time.

In conclusion, the mzXML is an open-source standard that allows for registering mass spectrometry results in an instrument-independent file format. The mzXML structure has been designed to facilitate easy integration of new instrument data into the preexisting structure. mzXML is supported by a series of open-source tools, allowing for validating and accessing the documents. Viewing of the registered data is possible with the mzXML viewer. The data may also be exported in input formats for several search engines. This feature allows for creation of a common data analysis pipeline, which utilizes several different mass spectrometers and a common analytical procedure (e.g., one framework for protein identification or quantification).

8.2.3 TWODML

A project called TWODML (two-dimensional electrophoresis markup language) has been under development in the Biotechnology Division of the National Institute of Standards and Technology in Gaithersburg, MD, U.S. It is specifically aimed at standardizing the representation of data from 2D PAGE (two-dimensional polyacrylamide gel electrophoresis) experiments.[22] The objectives of the TWODML are to

1. Gather, annotate, and provide enough information that may be reported about an electrophoresis-based experiment in order to ensure the interoperability of the results and their reproducibility by others
2. Help establish public repositories and a data-exchange format for electrophoresis-based experimental data
3. Eliminate barriers to data exchange between the repositories and permit integration of the data from heterogeneous sources

To achieve these goals an XML-based data model has been created (Figure 8.7). The model's main structures describe all the aspects of the 2D PAGE experiment: sample source, preparation and loading, experimental details, sample separation,

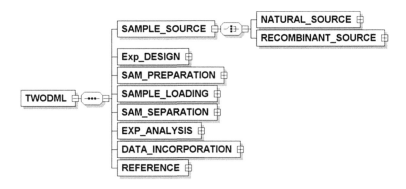

FIGURE 8.7 TWODML XML schema. (Adapted from Ravichandran, V., et al., *Electrophoresis*, 25, 297–308, 2004.)

sample and data analysis, and details about identified proteins. The sample sources are divided into two groups:

1. Natural source: Described by organism, sex, age, condition, strain, organ, tissue, cell, organelle, secretion, etc.
2. Genetically modified (recombinant): Described by type of expression system, genetic materials, growth medium, growth condition, method of transfection, etc.

Sample preparation describes details about cell lysis method, cell lysis buffer, anti-protease used, fractionation, etc. Sample separation describes details about the sample on-gel separation — first dimension, second dimension, and running conditions.

The project also proposes creation of a common electrophoresis data repository. Information in the repository can be shared with any other system supporting TWODML. The data in the repository may be also accessed with use of a web-based data query interface.

8.3 PILOT PHASE OF THE HUPO-PPP: A CASE STUDY

The Human Proteome Organization (HUPO) has launched several major initiatives. As noted above, these are focused on the plasma proteome, the liver proteome, the brain proteome, protein standards/bioinformatics, and certain technologies, including large-scale antibody production. Overall, HUPO aims to accelerate the development of the field of proteomics and to stimulate and organize international collaborations in research and education.

The pilot phase of the Plasma Proteome Project (PPP) was launched in 2002 and focused on the following objectives:

1. Assessment of the analysis resolution and the sensitivity
2. Assessment of the time involved, and volumes of samples required with various separation and analytical technologies

3. Clarification of influences of variability arising from the choice of serum versus plasma
4. Guidance for specimen handling
5. Determination of the place of methods to deplete the most abundant plasma proteins[23]

In the initial planning meetings, interdisciplinary groups of experts from academia, government, and industry proposed a pilot phase to address the following ten scientific issues:

1. Sensitivity of various techniques to deal with the huge dynamic range of concentrations of proteins and peptides in the circulation.
2. Technical aspects of specimen collection, handling, storage, and thawing, aiming for standardization.
3. Evaluation of methods of depleting or prefractionating the several most abundant proteins.
4. Comparisons of the advantages and limitations of analyses of serum vs. plasma, and alternative anticoagulation methods for plasma (EDTA, heparin, citrate).
5. Enumeration and categorization of proteins visualized and identified, with special attention to posttranslational modifications and tissue of origin.
6. Comparison of results obtained from separation of intact proteins vs. separation of peptides from digested proteins.
7. Comparison of use of gel-based vs. liquid-phase multidimensional separation methods.
8. Evaluation of parameters for high-throughput links with mass spectrometry.
9. Comparisons of MALDI vs. direct MS (SELDI) methods.
10. Assessment and advancement of specific peptide-labeling methods.

8.3.1 SETTING UP THE COLLABORATION

In order to compare the attributes of various technology platforms, it is essential to have reference specimens available for use with each platform. The range of the possible options extended from a potential single individual to the vast American Red Cross donor pool. After extensive discussions, the following set of reference specimens was chosen:

Set 1: U.K. NIBSC lyophilized citrated plasma, previously prepared as a reference specimen for hemostasis and thrombosis studies for the International Society for Thrombosis and Hemostasis/Standards Committee.

Set 2: BD Diagnostics sets of four reference specimens for each of three ethnic groups: Caucasian-American, African-American, and Asian-American. Each pool consisted of 400 ml of blood each from one male and one postmenopausal female healthy donor, collected in a standard donor setup, after informed consent, sequentially into ten 10-ml tubes each with appropriate concentrations of K-EDTA, lithium heparin, or sodium citrate for plasma and without clot activator for serum.

Set 3: The Chinese Academy of Medical Sciences set of four reference specimens. This pool was prepared, after review by the CAMS Ethics Committee and informed consent by donors, according to the BD protocol, including tests for viral infections.

A standard HUPO PPP Questionnaire was sent to all established proteomics laboratories whose investigators had expressed interest in participating, either at the workshops and HUPO World Congress or after learning about the Plasma Proteome Project through colleagues, the HUPO Web site (www.hupo.org), or press coverage. Forty-seven laboratories in 14 countries committed to participate in the PPP. Of these laboratories, 28 are in the United States, including 17 academic, 6 U.S. federal, and 5 U.S. corporate; 19 laboratories are in other countries, including 7 in Europe, 1 in Israel, 9 in Asia, and 2 in Australia. Of these labs, 41 requested the U.K. NIBSC specimens, 43 the BD Caucasian-American specimens, 18 the BD African-American and Asian-American specimens, and 18 the Chinese Academy of Sciences specimens. With regard to different kinds of technology platforms, 31 indicated that they would run 2D gels, 29 liquid chromatography separations, 25 protein digestion first, 30 various MALDI/MS or MS/MS, and 15 direct MS/SELDI. Combinations of technologies will help move down the dynamic range of concentrations, which spans about 9 orders of magnitude from albumin (40 mg/ml) to PSA or cytokines (pg/ml).

8.3.2 Data Standardization and Data Exchange Protocols

From the data integration point of view, the pilot phase of the HUPO-PPP can be defined as follows: 47 laboratories worldwide were invited to identify proteins present in up to 17 specimens from 5 donor groups. Laboratories used different techniques and protocols, but the last experimental step was almost always done with the use of mass spectrometers. To evaluate protein abundance, the main effort was supported by independent immunoassays of selected proteins in the reference specimens.[7] In addition to answers to the ten scientific issues raised in the planning phase (see above), the PPP will generate a draft "human plasma proteome as of 2004" as an output from this collaboration.

The investigators agreed to collect the following sets of information in a standardized format, as proposed by the Project Technology and Resources Committee:

Experimental Protocols: Sufficiently detailed to allow the work to be reproduced and considered as acceptable for publication in the *Journal of Biological Chemistry* or *Proteomics*

Protein Identification Data: The protein accession number, name, database (and version) used, sequences of the identified peptides, and estimate of confidence for the identification, plus any supporting information about posttranslational modifications (from 2D gel coordinates and estimated pI and MW), and estimates of relative protein abundance in the specimen

Summary of Technologies and Resources: Estimates of the time and capital and operating costs of the analyses

Participants agreed that the proteins will be identified using a single, common database. The International Protein Index (IPI), developed by the European Bioinformatics Institute (EBI) to incorporate content from several other sources with extensive annotation, was chosen. However, output tied to MS instruments searching other databases like SwissProt[24] and NCBI-nr were also acceptable.

Data collection centers were established at EBI and at the University of Michigan to gather and integrate data received from the participating laboratories. The project data exchange committee designed two data exchange standards: (1) an XML format based on the PEDRo concept and using the PEDRo data entry tool, and (2) a set of Microsoft Excel templates for gathering the protein identification data, resources used, and experimental protocols (in Microsoft Word template). Subsequently, the standard was extended to accommodate m/z peak lists from direct MS/SELDI experiments for a separate subgroup analysis.

The XML format was used by only two of the laboratories, and the PEDRo tool was still in a prolonged development stage. Thus, Excel/Word templates became the overwhelmingly preferred means of data exchange. As a result, the University of Michigan Web-based submission site became the main data collection and integration center. Excel/Word templates were updated to improve their processing, especially with error-proofing abilities.

Analysis of the preliminary results brought to the fore a major problem with data integration and validation based exclusively on protein accession numbers. Participating laboratories did not only use different search databases, but also different algorithms for picking up protein identifications from the database search results. The estimation of confidence of the identification, based on search scores and laboratory binary (high/low) judgment, was inconsistent. To address these problems, the data exchange protocols were enhanced to include peak lists supporting submitted identifications in specially designed text format, and raw files in spectrometer manufacturer format to be subjected to independent PPP analyses. Because of the size, sometimes exceeding several gigabytes, the raw mass spectrometer files had to be burned on CD or DVD disks and sent to the data collection center at the University of Michigan for copying and distribution.

8.3.3 Data Integration

From the initial group of 47 laboratories, 20 laboratories submitted MS or MS/MS data with protein identifications. Altogether, the 20 laboratories submitted 128 protein identification datasets (different specimens and sometimes multiple methods with the same specimens in various labs) — 122 Excel and 6 XML documents. These laboratories also submitted 70 documents with protocols and 27 with resources descriptions. Sixteen laboratories sent disks with raw spectra or peak lists, which generally required conversions at the data integration center. An additional 10 laboratories submitted SELDI findings, 2 laboratories submitted microarray results, and several labs' results were still pending at the time of writing.

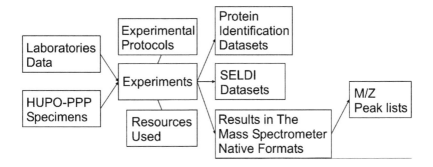

FIGURE 8.8 Block diagram of the HUPO-PPP data repository.

The data integration process was divided into three steps:

1. Receipt of Data: For three days the data remained unprocessed and available for the submitting laboratory to review the submission and, if needed, update the submitted documents.

2. Transfer of data into an intermediate database, done automatically for each Web-submitted document and manually for emailed XML submissions. The data in the intermediate structure presented an exact copy of the data from the documents, without any transformations or integration, available for the integration procedures, to check correctness of the structure of the submitted document. The cleaned data were rewritten in a consistent format of accession numbers, database names, experimental categories converted into a set of controlled vocabularies, and reformatted peptide sequence lists (Figure 8.8). The 20 laboratories' highly redundant 128 data sets had a total of about 50,000 protein identifications from several different databases: about 85% IPI, 14% NCBI-nr, and the rest from RefSeq, SwissProt, Genbank, and Brookhaven databases. Sometimes laboratories used different versions of the databases. Fortunately, 97% of the identifications were supported with sequence lists of the identified peptides.

3. Creation of PPP Standard Database: Mapping the sequence lists into the project standard database revealed about 9,000 distinct IPI proteins, corresponding to about 5,500 ENSEMBL gene identifications. To further analyze the quality of the identifications, a designated group within the PPP is independently repeating the database searches with the submitted peak lists and standard database, search engine, and search criteria. Another independent group is using the raw spectra to reanalyze the peak generation part of the analyses and to compare the results from database searches using several different commonly employed search engines and databases. This work will bring additional confidence to the protein identifications and presumably a major reduction in the number of distinct proteins. The results will also be expressed in terms of a sensitivity analysis with different cut points for confidence levels to call an identified protein a HUPO PPP "hit."

The integrated project results were made available to the participants with use of a specially written Web-based tool that allows for querying the project database in Structured Query Language (SQL) and returns results in HTML, XML, or text format. The results have been converted from the relational structure (Figure 8.9)

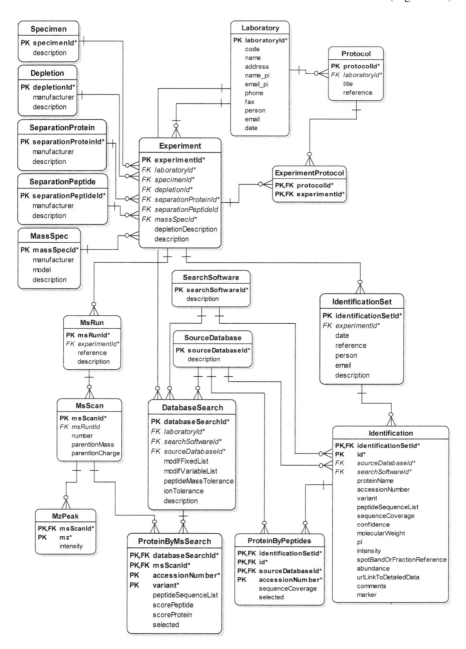

FIGURE 8.9 Entity relationship (ER) diagram of the HUPO PPP data repository.

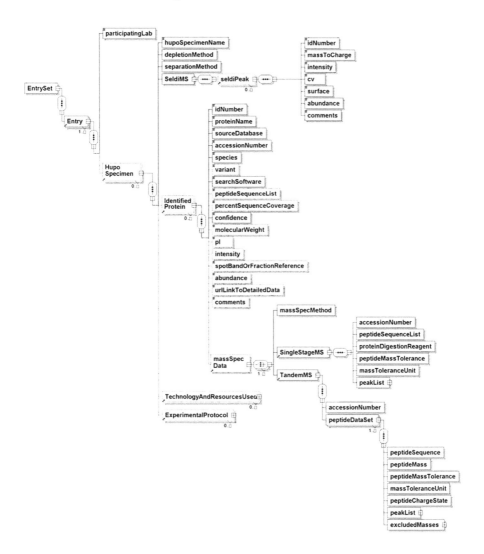

FIGURE 8.10 XML schema of the HUPO PPP data repository. (Adapted from http://psidev.sourceforge.net/ppp/pilotPhase.)

into a nested XML document (Figure 8.10), which will be publicly available at the EBI Web server and the HUPO Web site. Investigators and bioinformatics colleagues will utilize these integrated data resources and primary data for an intensive Jamboree Workshop in Ann Arbor, MI, U.S., in June 2004.

8.3.4 DISCUSSION

The pilot phase of the HUPO PPP provides a very good example of data integration problems and ways to deal with them in collaborative proteomics studies. As a pilot phase of the first project executed under the HUPO, and one of the first large-scale proteome projects, it encountered many learning situations. Two problems generated the most confusion and additional work for participants and data collection/integration centers: overestimation of laboratories' ability to use XML data format, and underestimation of the importance of the peak lists collection.

Thanks to early identification of the problems, both of the issues were successfully overcome. The early identification was possible because of the preliminary submission step and prompt analysis of the initial results. However, since the necessary changes were introduced during the project's operations, the collection/integration center had to deal thereafter with data formatted according to both the old and the new protocols. Additionally, the new protocol came too late for some of the participants, who were not able to completely fulfill its requirements.

Substantial bioinformatics capabilities are essential for collaborative studies. Many variables that contribute to incongruent reports of protein identifications can be traced to the nonstandardized use of proprietary search engines and the multiplicity of databases constantly being updated. Nonstandard procedures for recruitment of specimen donors (overcome in this project by preparation of reference specimens), opportunity for choice among reference specimens (permitted in this project), specimen collection and handling, and depletion or prefractionation of proteins in specimens before MS analysis all contribute complexity and uncertainty. In addition, methods aimed at detecting subproteomes (phosphoproteome, glycoproteome; liver, brain, other organ proteomes) may introduce other variables when seeking potential corresponding detection of the same proteins in the circulation. Flexibility of bioinformatics approaches will be essential for integration of data from many such sources and for comparison with results from non-MS methods, including microarrays.

The HUPO PPP pilot phase is generating valuable information for the choice of standard operating procedures in support of disease-oriented studies and population-based epidemiologic and clinical trials studies with archived and newly collected human specimens.

REFERENCES

1. Huber, L.A. Is proteomics heading in the wrong direction? *Nat. Rev. Mol. Cell. Biol.*, 4, 74–80, 2003.
2. Zhu, H., Bilgin, M., and Snyder, M. Proteomics. *Annu. Rev. Biochem.*, 72, 783–812, 2003.
3. Aebersold, R. and Mann, M. Mass spectrometry-based proteomics. *Nature*, 422, 198–207, 2003.
4. Hanash, S. Disease proteomics. *Nature*, 422, 226–232, 2003.
5. Hanash, S. and Celis, J.E. The Human Proteome Organization: A mission to advance proteome knowledge. *Mol. Cell. Proteomics*, 1, 413–414, 2002.
6. Hanash, S. HUPO initiatives relevant to clinical proteomics. *Mol. Cell. Proteomics*, 3, 298–301, 2004.
7. Omenn, G.S. The Human Proteome Organization Plasma Proteome Project pilot phase: Reference specimens, technology platform comparisons, and standardized data submissions and analyses. *Proteomics*, 4, 1235–1240, 2004.

8. Orchard, S., Hermjakob, H., and Apweiler, R. The proteomics standards initiative. *Proteomics,* 3, 1374–1376, 2003.

9. Orchard, S., Kersey, P., Zhu, W., Montecchi-Palazzi, L., Hermjakob, H., and Apweiler, R. Progress in establishing common standards for exchanging proteomics data. *Comp. Funct. Genom.,* 4, 203–206, 2003.

10. Orchard, S., Hermjakob, H., Julian, R.K., Runte, K., Sherman, D., Wojcik, J., Zhu, W., and Apweiler, R. Common Interchange Standards for Proteomics Data: Public Availability of Tools and Schema. Report on the Proteomic Standards Initiative Workshop, 2nd Annual HUPO Congress, Montreal, Canada, 8–11th October 2003. *Proteomics,* 2004, 4, 490–491.

11. Orchard, S., Zhu, W., Julian, R.K., Hermjakob, H., and Apweiler, R. Further advances in the development of a data interchange standard for proteomics data. *Proteomics,* 3, 2065–2066, 2003.

12. Orchard, S., Kersey, P., Hermjakob, H., and Apweiler, R. The HUPO Proteomics Standards Initiative meeting: Towards common standards for exchanging proteomics data. *Comp. Funct. Genom.,* 4, 16–19, 2003.

13. Bader, G.D. and Hogue, C.W. BIND—A data specification for storing and describing biomolecular interactions, molecular complexes and pathways. *Bioinformatics,* 16, 465–477, 2000.

14. Hermjakob, H., Montecchi-Palazzi, L., Lewington, C., Mudali, S., Kerrien, S., Orchard, S., Vingron, M., Roechert, B., Roepstorff, P., Valencia, A., Margalit, H., Armstrong, J., Bairoch, A., Cesareni, G., Sherman, D., and Apweiler, R. IntAct: An open source molecular interaction database. *Nucleic Acids Res.,* 32 (Database issue), D452–D455, 2004.

15. Xenarios, I., Rice, D.W., Salwinski, L., Baron, M.K., Marcotte, E.M., and Eisenberg, D. DIP: The database of interacting proteins. *Nucleic Acids Res.,* 28, 289–291, 2000.

16. Zanzoni, A., Montecchi-Palazzi, L., Quondam, M., Ausiello, G., Helmer-Citterich, M., and Cesareni, G. MINT: A Molecular INTeraction database. *FEBS Lett.,* 513, 135–140, 2002.

17. Peri, S., Navarro, J.D., Amanchy, R., Kristiansen, T.Z., Jonnalagadda, C.K., Surendranath, V., Niranjan, V., Muthusamy, B., Gandhi, T.K., Gronborg, M., Ibarrola, N., Deshpande, N., Shanker, K., Shivashankar, H.N., Rashmi, B.P., Ramya, M.A., Zhao, Z., Chandrika, K.N., Padma, N., Harsha, H.C., Yatish, A.J., Kavitha, M.P., Menezes, M., Choudhury, D.R., Suresh, S., Ghosh, N., Saravana, R., Chandran, S., Krishna, S., Joy, M., Anand, S.K., Madavan, V., Joseph, A., Wong, G.W., Schiemann, W.P., Constantinescu, S.N., Huang, L., Khosravi-Far, R., Steen, H., Tewari, M., Ghaffari, S., Blobe, G.C., Dang, C.V., Garcia, J.G., Pevsner, J., Jensen, O.N., Roepstorff, P., Deshpande, K.S., Chinnaiyan, A.M., Hamosh, A., Chakravarti, A., and Pandey, A. Development of human protein reference database as an initial platform for approaching systems biology in humans. *Genome Res.,* 13, 2363–2371, 2003.

18. Hermjakob, H., Montecchi-Palazzi, L., Bader, G., Wojcik, J., Salwinski, L., Ceol, A., Moore, S., Orchard, S., Sarkans, U., von Mering, C., Roechert, B., Poux, S., Jung, E., Mersch, H., Kersey, P., Lappe, M., Li, Y., Zeng, R., Rana, D., Nikolski, M., Husi, H., Brun, C., Shanker, K., Grant, S.G., Sander, C., Bork, P., Zhu, W., Pandey, A., Brazma, A., Jacq, B., Vidal, M., Sherman, D., Legrain, P., Cesareni, G., Xenarios, I., Eisenberg, D., Steipe, B., Hogue, C., and Apweiler, R. The HUPO PSI's molecular interaction format — A community standard for the representation of protein interaction data. *Nat. Biotechnol.,* 22, 177–183, 2004.

19. Taylor, C.F., Paton, N.W., Garwood, K.L., Kirby, P.D., Stead, D.A., Yin, Z., Deutsch, E.W., Selway, L., Walker, J., Riba-Garcia, I., Mohammed, S., Deery, M.J., Howard, J.A., Dunkley, T., Aebersold, R., Kell, D.B., Lilley, K.S., Roepstorff, P., Yates, J.R., III, Brass, A., Brown, A.J., Cash, P., Gaskell, S.J., Hubbard, S.J., and Oliver, S.G. A systematic approach to modeling, capturing, and disseminating proteomics experimental data. *Nat. Biotechnol.,* 21, 247–254, 2003.

20. Marcotte, E.M., Pellegrini, M., Ng, H.L., Rice, D.W., Yeates, T.O., and Eisenberg, D. Detecting protein function and protein–protein interactions from genome sequences. *Science,* 285, 751–753, 1999.

21. Pedrioli, P. G., Eng, J. K., Hubley, R., Vogelzang, M., Deutsch, E.W., Raught, B., Pratt, B., Nilsson, E., Angeletti, R.H., Apweiler, R., Cheung, K., Costello, C. E., Hermjakob, H., Huang, S., Julian, R.K, Kapp, E., McComb, M. E., Oliver, S. G., Omenn, G., Paton, N. W., Simpson, R., Smith, R., Taylor, C. F., Zhu, W., and Aebersold, R. A common open representation of mass spectrometry data and its application to proteomics research. *Natl. Biotechnol.,* 22, 1459–1466, 2004.

22. Ravichandran, V., Lubell, J., Vasquez, G.B., Lemkin, P., Sriram, R.D., and Gilliland, G.L. Ongoing development of two-dimensional polyacrylamide gel electrophoresis data standards. *Electrophoresis,* 25, 297–308, 2004.

23. Omenn, G.S. Advancement of biomarker discovery and validation through the HUPO plasma proteome project. *Dis. Markers,* 18, 1–4, 2004.

24. Bairoch, A. and Boeckmann, B. The SWISS-PROT protein sequence data bank: current status. *Nucleic Acids Res.,* 22, 3578–3580, 1994.

9 Informatics Tools for Functional Pathway Analysis Using Genomics and Proteomics

Chad Creighton and Samir M. Hanash

CONTENTS

9.1 INTRODUCTION

This chapter presents an overview of the informatics tools available in the form of databases of pathway and biological network information, as well as software tools and methods by which one could integrate this information with global expression profiling experiments. Global gene expression profiling data, both at the RNA level (e.g., oligonucleotide array data) and at the protein level (e.g., two-dimensional gel electrophoresis or protein microarray derived data) represents a vast resource for exploration, or "data mining," to uncover new insight into the complex biological systems under study. Yet the process of extracting information from such sources may be daunting. A useful strategy to extract meaningful information from a potentially overwhelming sea of expression data is to effectively link the data to external sources of biological information, such as gene annotation or the biomedical liter-ature. In particular, there is currently substantial information available pertaining to biological pathways that has utility in the interpretation of gene expression data. Moreover, numerous databases on biological pathways have recently become avail-able that allow the results of global profiling experiments to be linked to these pathways to understand the observed differential expression of genes and proteins. Pathways and other biological networks may be represented as graphs, with nodes as the genes or proteins and edges as protein–protein or protein–DNA interactions. In functional pathway analysis, graph theoretic methods and graph visualization tools can be used to look for interesting patterns between pathway data and proteomic or genomic data; for example, a graph may display a large number of previously established connections between a set of genes or proteins found to be coregulated in a particular study.

9.2 PROFILING GENE EXPRESSION AT THE RNA AND PROTEIN LEVELS

Although several technologies are currently available for global profiling of gene expression at the RNA level, the DNA microarray approach has had the most impact on biomedical research. Applications of DNA microarrays include uncovering unsus-pected associations between genes and specific biological or clinical features of disease that are helping devise novel molecular-based disease classifications. In rela-tion to cancer, most published studies of tumor analysis using DNA microarrays have either examined a pathologically homogeneous set of tumors to identify clinically relevant subtypes; e.g., responders vs. nonresponders, or pathologically distinct sub-types of tumors of the same lineage; e.g., high-stage vs. low-stage tumors to identify molecular correlates, or tumors of different lineages to identify molecular signatures for each lineage. Very few studies have analyzed microarray data primarily from a functional pathway point of view. An important challenge for microarray analysis is to understand at a functional level the significance of associations observed between subsets of genes and biological features of the samples analyzed. Another challenge is to determine how well RNA levels of predictive genes correlate with protein levels. A lack of correlation may imply that the predictive property of the gene(s) is inde-pendent of gene function. For example, comparisons of mRNA and protein levels for the same tumors reported for lung cancer demonstrated that only a small percentage of genes had a statistically significant correlation between the levels of their corre-sponding proteins and mRNAs.[1]

Given the dynamic nature of the proteome and the occurrence of numerous posttranslational modifications, notably phosphorylation, that is functionally relevant, there is a compelling need to profile gene expression at the level of the proteome. Numerous alterations may occur in proteins that are not reflected in changes at the RNA level. The technologies for proteome profiling such as two-dimensional gels do not currently readily allow the profiling of the entire proteome or a cell or tissue. However, profiling of protein subsets is currently feasible. There is substantial interest in developing microarrays or biochips that allow the systematic analysis of thousands of proteins and eventually proteomic-scale profiling.[2]

Unlike DNA microarrays that provide one measure of gene expression, namely RNA levels, there is a need to implement protein microarray strategies that address the many different features of proteins, including determination of their levels in tissues and cells and their selective interactions with other biomolecules, such as other proteins, antibodies, drugs, or various small ligands. The compelling need for protein chips has led numerous biotechnology companies to devise novel strategies for producing biochips that have utility for biomedical investigations. Proteomic profiling studies that have utilized protein microarrays are beginning to emerge. As a model to better understand how patterns of protein expression shape the tissue microenvironment,[3] Knezevic et al. analyzed protein expression in tissue derived from squamous cell carcinomas of the oral cavity through an antibody microarray approach for high-throughput proteomic analysis.[3] Utilizing laser capture microdissection to procure total protein from specific microscopic cellular populations, they demonstrated that differences in expression patterns of multiple proteins involved in signal transduction within epithelial cells reproducibly correlated with oral cavity tumor progression.

9.3 COMPLEXITIES OF BIOLOGICAL PATHWAY ANALYSES

Biological pathways in the traditional sense are thought of as a set of biological reactions that are carried out in a series of sequential steps (e.g., protein A activates protein B; protein B then binds to protein C; protein B and protein C together then target protein D, protein E, and protein F), while gene expression profiling gives a "snapshot" of the global state of the expression activity of the cell and hence does not capture the sequential aspect of pathways. Likewise, proteome profiling such as with 2D gels provides a snapshot of the levels of different proteins in a cell population or a tissue. Therefore, it may be expedient in functional pathway analysis to represent pathways, where the order of events is specified, as biological networks, where the interactions between genes and proteins are represented (e.g., as edges in a graph), but not in the order in which these interactions are thought to occur sequentially. Another important point to consider in data analysis using pathway information is that pathways operate at multiple levels that cannot be captured with any one analytical tool currently available. For example, signal transduction pathways involve the phosphorylation and dephosphorylation (e.g., the "activation" and "deactivation," respectively) of protein products. Oligonucleotide array data, on the other hand, measure mRNA levels, which often do not correlate well with measured protein levels, presumably because of processes such as regulation of translation of mRNA to protein, as well as targeted protein degradation that causes the correspondence between the two to be something other than 1:1. Although protein levels may be measured directly using 2D gels, this type of data will not reveal such information as the phosphorylation state of a given protein, nor will it reveal the activity state of enzymes. When integrating gene expression data with pathway information, a clear understanding of what the measured values actually represent is needed for correctly interpreting the results from such an analysis.

9.4 THE KYOTO ENCYCLOPEDIA OF GENES AND GENOMES (KEGG)

One of the first publicly available Web databases of pathway information was the Kyoto Encyclopedia of Genes and Genomes, or KEGG.[4] The major component of KEGG is the PATHWAY database that consists of graphical diagrams of biochemical pathways including most of the known metabolic pathways and some of the known regulatory pathways, over 150 pathways in all, as of August 2003. The KEGG/PATH-WAY reference diagrams can be readily integrated with genomic and proteomic information. A KEGG pathway is represented as a graph, with nodes as the proteins that participate in the pathway and edges as protein–protein interactions involved in certain steps of the pathway. Figure 9.1 shows a KEGG graph diagram of the cell cycle pathway in *Homo sapiens*. Using the KEGG tools, a user may enter a set of genes or proteins (such as a set of genes that appear significant in a given microarray experiment) to see which nodes or elements of any given pathway are represented by these genes; such nodes will appear highlighted in the graph. One current limitation of the KEGG database is its primary focus on well-defined metabolic pathways and its lack of

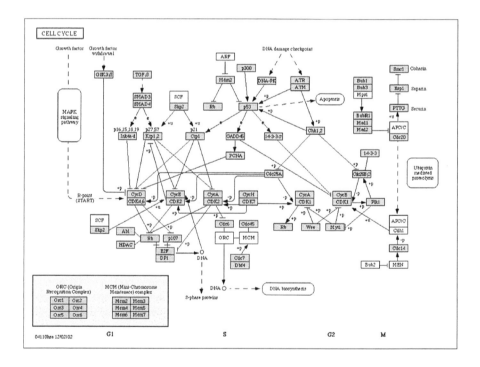

FIGURE 9.1 (Color insert follows page 204) KEGG/PATHWAY graph representation of the cell cycle pathway in *Homo sapiens*. Nodes in the graph represent gene products; edges represent interactions between gene products. Nodes highlighted in pink represent an example set of genes of interest as entered by the user (Web site available at http://www.genome.ad.jp/kegg-bin/ mk_point_ multi_html).

information on the more complicated signal transduction pathways that play a critical role in human diseases such as cancer. One recent study[5] that integrated the KEGG pathway information with data from a set of lung cancer mRNA profiling experiments readily demonstrated that processes of increased cell division and metabolism were characteristic of more advanced tumors but did not readily yield more novel findings.

9.5 GENE MICROARRAY PATHWAY PROFILER

In the study of complex biological systems, pathways not currently represented in KEGG may need to be considered. GenMAPP (Gene Micro-Array Pathway Profiler) is a freely available program for viewing and analyzing expression data on "microarray pathway profiles" (MAPPs) representing biological pathways or any other functional grouping of genes.[6] As of January 2003, GenMAPP included over 50 MAPP files depicting various biological pathways and gene families. Similar to the KEGG tools, GenMAPP can automatically and dynamically color code the genes on a MAPP linked to a gene-expression data set according to criteria supplied by the user. In addition to the MAPP files, GenMAPP includes gene annotation information as described by the Gene Ontology (GO) consortium.[7] The GO consortium has defined a controlled vocabulary consisting of over 2,000 terms for describing a given gene in terms of its molecular function, biological process involvement, or cellular component. Biological pathways in which a gene may participate are also included in that gene's GO annotation. Given a set of genes of interest obtained from an analysis of expression data, the GenMAPP program will identify GO terms that appear overrepresented in the gene set. GO terms that appear significantly enriched for a given set of genes may provide clues as to the processes or pathways that the genes as a whole may represent.

9.6 PROTEIN INTERACTION DATABASES

TRANSPATH is an online Web database on signal transduction and gene-regulatory pathways.[8] As of August 2003, the TRANSPATH database (version r4.2) contained 15,346 protein–protein interactions on 12,262 molecules and 2,604 genes. TRANSPATH focuses on the interactions between genes and proteins, rather than on using predefined pathways as KEGG or GenMAPP does. In this way, a TRANSPATH signaling pathway could be considered as more of a signaling network of biological interactions rather than as a pathway that presumes the order in which the interactions take place. Given any gene or protein of interest, the user can retrieve a list of associated interactions, and given any two molecules of interest, the database will show an ordered list of molecules connecting the two via a series of interactions. The TRANSPATH database is not freely available and a subscription is required for access, although a small demonstration portion of the database is available for browsing.

There are numerous databases of protein–protein interactions other than TRANSPATH that are freely available for querying and downloading, including the Biomolecular Interaction Database, or BIND[9] and the Database of Interacting Proteins, or DIP.[10] In both of these databases, the organism most represented is

Saccharomyces cerevisiae (yeast), for which an abundance of protein–protein interaction data has been generated using high-throughput techniques such as yeast
two-hybrid. As of August 2003, over 15,000 protein–protein interactions were catalogued in the BIND database, most of which were generated from a set of 4,825
yeast proteins (about 75% of the yeast proteome). Curation of such large interaction
data sets makes possible the creation of detailed maps of biomolecular interaction
networks. The sum total of all possible biomolecular interactions within an organism
is often referred to as the "interactome," which, along with the genome and the
proteome, currently represents a vast area for scientific exploration and discovery.
One area of bioinformatics research is to use graph theoretic methods to identify
highly connected subnetworks within a global interaction network. Such densely
connected regions of the interactome may represent molecular complexes. This NP
hard graph search problem is an ongoing research topic in computer science,
although software programs using heuristic algorithms have been developed and are
freely available.[9] Figure 9.2 shows a biomolecular network uncovered in a recently
published analysis of yeast interaction data.[11]

For research groups that study *S. cerevisiae* using genomic or proteomic profiling, the protein–protein interaction data made available by databases such BIND
and DIP can be a valuable resource for the analysis of expression data. One recent
study by the Institute for Systems Biology of the galactose utilization (GAL) pathway

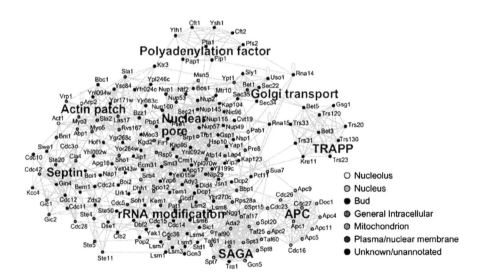

FIGURE 9.2 (Color insert follows page 204) Visual representation of a molecular complex
in protein interaction networks found using the *k*-core method of analyzing yeast protein
interaction data. The above network is a six-core. In a six-core, each node has at least six
edges connected to it. SAGA, Spt-Ada-Gcn5-acetyltransferase transcriptional activator–
histone acetyltransferase complex; TRAPP, transport protein particle complex. Proteins are
colored according to GO (Gene Ontology) cellular component. (Reproduced from Bader,
G.D. and Hogue, C.W. *Nature Biotechnol.*, 20, 991–997, 2002. With permission.)

in yeast[12] illustrates this point. The study analyzed 20 systematic perturbations of the GAL pathway using DNA microarrays, quantitative proteomics, and databases of known physical interactions. In each perturbation, a different gene with an important role in the GAL pathway was knocked out. An interaction network of 2,709 protein–protein interactions and 317 protein–DNA interactions was compiled. For each perturbation experiment from which mRNA and protein expression profiles were obtained, nodes in the network were colored to represent changes in mRNA or protein level corresponding to the observed effects of the perturbation (similar to what has been described above for the KEGG and GenMAPP tools). Figure 9.3 shows some of the graphical results presented from the study. The freely available Cytoscape software package allows users to carry out the same type of analysis as described above using their own expression and interaction datasets. The Linux version of Cytoscape also implements a method for screening a molecular interaction

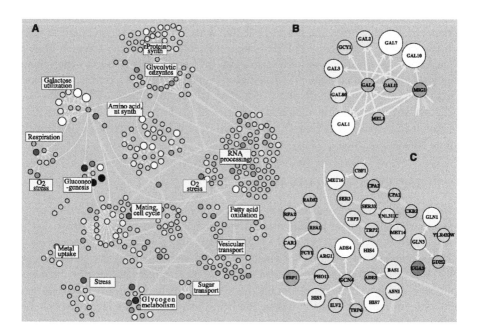

FIGURE 9.3 (Color insert follows page 204) Integrated physical-interaction networks from systematic perturbations of the GAL pathway in yeast. Nodes represent genes, a yellow arrow directed from one node to another signifies that the protein encoded by the first gene can influence the transcription of the second by DNA binding (protein–DNA), and a blue line between two nodes signifies that the corresponding proteins can physically interact (protein–protein). Highly interconnected groups of genes tend to have common biological function and are labeled accordingly. (A) Effects of the *gal4+* gal perturbation are superimposed on the network, with *gal4* colored red and the gray-scale intensity of other nodes representing changes in mRNA as in Figure 9.2 (node diameter also scales with the magnitude of change). Regions corresponding to (B) galactose utilization and (C) amino acid synthesis are detailed at right. (Reproduced from Ideker, T., et al., *Science*, 292, 923–934, 2001. With permission.)

network to identify active subnetworks; i.e., connected regions of the network that show significant changes in expression over particular subsets of conditions.[13]

The integration of protein–protein and protein–DNA interaction data with genomic or proteomic data can allow for analysis of global expression patterns at the level of biological networks in order to find patterns indicative of direct or indirect interactions between genes. Many recent gene expression profiling studies investigate the molecular biology of cancers in humans, and there is much interest in identifying the pathway or network perturbations within the cell that lead to tumorigenesis and disease progression. No large data sets of high-throughput inter-action data are currently available for *Homo sapiens*, although several groups have developed ambitious plans to this effect. As of August 2003, the DIP protein interaction database contained on the order of 1000 interactions for some 700 human proteins, which still represents a very small fraction of the human interactome. To carry out, using human gene expression data, the type of network analysis demon-strated using yeast data, more human protein interaction data may be required. Most of the human protein–protein interactions in DIP and BIND are largely manually curated from the biomedical literature. Databases of literature articles and abstracts, most notably PubMed, are a tremendous resource of human protein interaction information, as these catalogue the results of countless experiments investigating the interaction effects between individual proteins. Extracting protein interaction data from the literature, however, is a nontrivial task. Without the use of computer programs for automatic text processing, the process of manually sifting through any articles detailing a protein or set of proteins of interest, is enormously time consuming, though likely more accurate.

9.7 THE KINASE PATHWAY DATABASE

The Kinase Pathway Database[14] uses a natural language processing (NLP) algorithm to automatically extract protein interaction information from article abstracts. As of August 2003, over 480,000 abstracts have been mined in order to uncover some 26,000 human protein interactions being indicated in the literature. In all, the Kinase Pathway Database contains more than 47,000 protein interactions that span the major model organisms. Protein–protein interactions in the Kinase Pathway Database may be *direct* or *indirect*. A direct interaction being found between two proteins entails that a direct physical interaction may take place between the two (e.g., protein A binds to protein B to form a complex). An indirect interaction entails that the two proteins have a functional relationship in which the activity of one protein may be modulated by the other (e.g., protein A may bind upstream of the gene encoding protein B to downregulate the expression of protein B). Estimates of the accuracy of the database's NLP algorithm were made by manually checking some 500 abstracts from which interaction information was mined, which showed a false-positive error rate of only around 6% in reporting a functional relationship between two proteins. However, this reported accuracy does not evaluate the correctness of describing a protein–protein interaction as being direct or indirect, which may be worse. The Kinase Pathway Database team is actively improving upon their NLP algorithm, and monthly updates are being made to the database.

9.8 OTHER DATABASES UNDER DEVELOPMENT

It is likely that additional databases specialized on an organismal basis or on biological pathway or other bases that address unmet needs will emerge in the near future. For example, a database dedicated to humans, under construction, is the Human Protein Reference Database, or HPRD, which relies on a "brute force" approach to literature information extraction. The HPRD development team includes biologists who spend much of their time reading the literature for relevant information for each protein currently catalogued in the database. As of August 2003, over 10,000 interactions have been entered in HPRD for some 2750 human proteins. The HPRD project represents a tremendous effort in human time. As a result it overcomes the inherent limitations of the Kinase Pathway Database, which relies on automatic natural language processing algorithms, which are not perfect. Thus, the HPRD interactions are inherently more likely to be correct in representing direct physical interactions between proteins as described in the literature. In addition, the HPRD has defined a set of human protein interaction networks surrounding well-known signaling pathways. Each of these networks is comprised of all of the interactions that apply to the key elements of the given pathway. Figure 9.4 shows a graphical

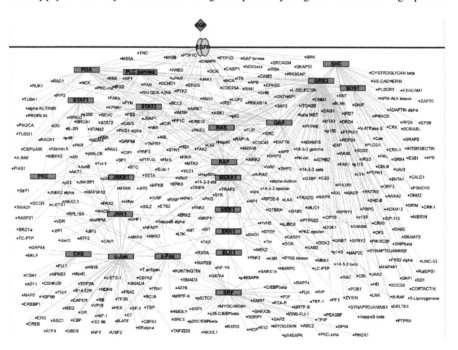

FIGURE 9.4 Protein–protein interaction network surrounding the EGFR signaling pathway, as represented by the Human Protein Reference Database (HPRD). Nodes in the graph represent proteins; edges represent interactions between proteins. Nodes that appear enlarged over the other nodes represent the core components of the EGFR pathway. The protein–protein interactions catalogued in the HPRD that directly involve each of these core pathway components are also represented in the graph.

representation of the EGFR interaction network derived from the EGFR signaling pathway.

One key to functional pathway analysis is the integration of genomic or proteomic data with protein interaction data to uncover significant networks of apparently active interactions. Refining analysis methods for this type of network analysis remains an active area of bioinformatics research, which includes questions such as how to define patterns of interactions that are both statistically significant and biologically meaningful. The biological interpretation of network patterns can also present a challenge, although, graph visualization tools can greatly aid this step of the analysis. Many software tools for drawing graphs are available, including the Pajek program (originally designed for the analysis of social networks), which has been used extensively in analyses of yeast two-hybrid data. At this point in time, programming and database skills are a valuable asset for performing network analyses of expression data. As network analysis methods become more developed, network analysis software made for ease of use by the biology research community at large will likely follow.

9.9 CONCLUSION

The field of global profiling of gene expression, while still in its infancy, is gradually evolving toward integration of multiple sets of data that combine global measurement of RNA and protein levels as well as assessment of protein functional states; e.g., phosphorylation. An important justification for such an undertaking is to extract functionally relevant information from such large data sets. Although there are numerous challenges associated with functional analyses of gene expression, methods/tools and databases are becoming available that facilitate functional pathway analysis.

REFERENCES

1. Chen, G., Gharib, T.G., Huang, C.C., Thomas, D.G., Shedden, K.A., Taylor, J.M., Kardia, S.L., Misek, D.E., Giordano, T.J., Iannettoni, M.D., Orringer, M.B., Hanas, S.M., and Beer, D.G. Proteomic analysis of lung adenocarcinoma: Identification of a highly expressed set of proteins in tumors. *Clin. Canc. Res.,* 8, 2298–2305, 2002.
2. Phizicky, E., Bastiaens, P.I., Zhu, H., Snyder, M., and Fields, S. Protein analysis on a proteomic scale. *Nature,* 422, 208–215, 2003.
3. Knezevic, V., Leethanakul, C., Bichsel, V.E., Worth, J.M., Prabhu, V.V., Gutkind, J.S., Liotta, L.A., Munson, P.J., Petricoin, E.F., III, and Krizman, D.B. Proteomic profiling of the cancer microenvironment by antibody arrays. *Proteomics,* 1, 1271–1278, 2001.
4. Kanehisa, M., Goto, S., Kawashima, S., and Nakaya, A. The KEGG databases at GenomeNet. *Nucleic Acids Res.,* 30, 42–46, 2002.
5. Creighton, C., Hanash, S., and Beer, D. Gene expression patterns define pathways correlated with loss of differentiation in lung adenocarcinomas. *Federation of European Biochemical Societies,* 540, 167–170, 2003.
6. Doniger, S.W., Salomonis, N., Dahlquist, K.D., Vranizan, K., Lawlor, S.C., and Conklin, B.R. MAPPFinder: Using gene ontology and GenMAPP to create a global gene-expression profile from microarray data. *Genome Biol.,* 4, R7, 2003.

7. Ashburner, M., Ball, C.A., Blake, J.A., Botstein, D., Butler, H., Cherry, J.M., Davis, A.P., Dolinski, K., Dwight, S.S., Eppig, J.T., Harris, M.A., Hill, D.P., Issel-Tarver, L., Kasarskis, A., Lewis, S., Matese, J.C., Richardson, J.E., Ringwald, M., Rubin, G.M., and Sherlock, G. Gene ontology: Tool for the unification of biology. The Gene Ontology Consortium. *Nat. Genetics,* 25, 25–29, 2000.
8. Schacherer, F., choi, C., Gotze, U., Krull, M., Pistor, S., and Wingender, E. The TRANSPATH signal transduction database: A knowledge base on signal transduction networks. *Bioinformatics,* 17, 1053–1057, 2001.
9. Bader, G.D., Betel, D., and Hogue, C.W. BIND: The biomolecular interaction network database. *Nucleic Acids Res.,* 31, 248–250, 2003.
10. Xenarios, I., Salwinski, L., Duan, X.J., Higney, P., Kim, S.M., and Eisenberg, D. DIP, the database of interacting proteins: A research tool for studying cellular networks of protein interactions. *Nucleic Acids Res.,* 30, 303–305, 2002.
11. Bader, G.D. and Hogue, C.W. Analyzing yeast protein–protein interaction data obtained from different sources. *Nature Biotechnol.,* 20, 991–997, 2002.
12. Ideker, T., Thorsson, V., Ranish, J.A., Christmas, R., Buhler, J., Eng, J.K., Bumgarner, R., Goodlett, D.R., Aebersold, R., and Hood, L. Integrated genomic and proteomic analyses of a systematically perturbed metabolic network. *Science,* 292, 929–934, 2001.
13. Ideker, T., Ozier, O., Schwikowski, B., and Siegel, A.F. Discovering regulatory and signalling circuits in molecular interaction networks. *Bioinformatics,* 18, S233–S240, 2002.
14. Koike, A., Kobayashi, Y., and Takagi, T. Kinase pathway database: An integrated protein–kinase and NLP-based protein–interaction resource. *Genome Res.,* 13, 1231–1243, 2003.

ADDITIONAL REFERENCES

1. Kyoto Encyclopedia of Genes and Genomes (KEGG), http://www.genome.ad.jp/kegg/kegg2.html.
2. Gene MicroArray Pathway Profiler (GenMAPP), http://www.GenMAPP.org.
3. The Gene Ontology (GO) Consortium, http://www.geneontology.org/.
4. TRANSPATH, http://www.biobase.de/pages/products/transpath.html.
5. Biomolecular Interaction Database (BIND), http://bind.mshri.on.ca/index.phtml.
6. Database of Interacting Proteins (DIP), http://dip.doe-mbi.ucla.edu/.
7. Cytoscape (bioinformatics software platform for visualizing molecular interaction networks and integrating these interactions with gene expression profiles), http://www.cytoscape.org.
8. Kinase Pathway Database, http://kinasedb.ontology.ims.u-tokyo.ac.jp/.
9. Human Protein Reference Database (HPRD), http://hprd.org/.
10. Pajek (program for large network analysis), http://vlado.fmf.uni-lj.si/pub/networks/pajek/.

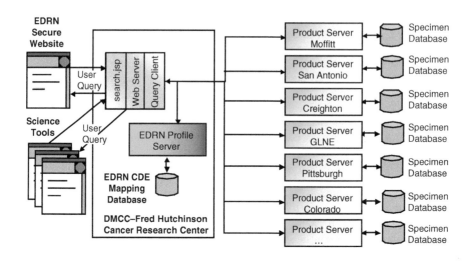

FIGURE 3.7 ERNE scalable system architecture.

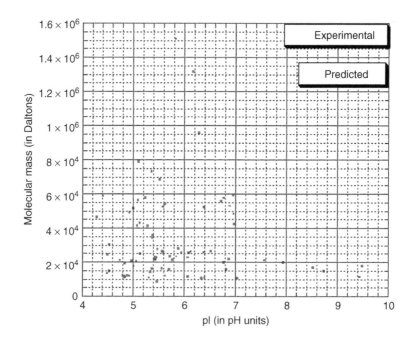

FIGURE 6.10 Overlap of pI/MW experimental (●) and theoretical (■) values for spots identified in a 2D PAGE map of human colorectal epithelial obtained from Swiss-2D PAGE.

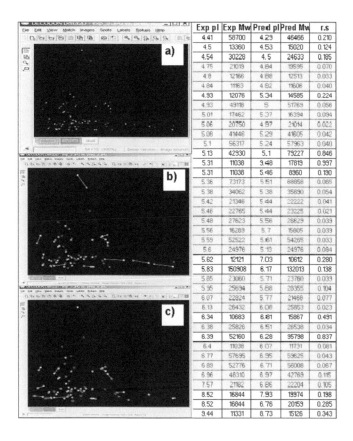

Exp pI	Exp Mw	Pred pI	Pred Mw	r.s
4.41	58700	4.29	46466	0.210
4.5	13360	4.53	15020	0.124
4.54	30228	4.5	24633	0.185
4.75	21019	4.84	19595	0.070
4.8	12166	4.88	12513	0.033
4.84	11163	4.82	11606	0.040
4.93	12076	5.34	14585	0.224
4.93	49118	5	51769	0.056
5.01	17462	5.37	16394	0.094
5.06	20750	4.97	21014	0.022
5.08	41448	5.29	41805	0.042
5.1	56317	5.24	57963	0.040
5.13	42930	5.1	79227	0.846
5.31	11038	3.48	17819	0.997
5.31	11038	5.46	8360	0.190
5.36	73173	5.51	68858	0.065
5.38	34062	5.38	35890	0.054
5.42	21346	5.44	22222	0.041
5.46	22765	5.44	23225	0.021
5.48	27623	5.56	26829	0.033
5.56	16289	5.7	15805	0.039
5.59	52522	5.61	54265	0.033
5.6	24976	5.13	24976	0.064
5.62	12121	7.03	10612	0.280
5.83	150908	6.17	132013	0.138
5.85	23060	5.71	23760	0.033
5.95	25694	5.88	28355	0.104
6.07	22824	5.77	21488	0.077
6.13	26432	6.08	25853	0.023
6.34	10683	6.81	15867	0.491
6.38	25826	6.51	26538	0.034
6.39	52160	6.28	95798	0.837
6.4	11038	6.07	11731	0.081
8.77	57695	6.95	59625	0.043
6.89	52776	6.71	56008	0.067
6.96	48310	8.97	42789	0.115
7.57	21182	6.86	22204	0.105
8.52	16844	7.93	19974	0.198
8.52	16844	6.76	20159	0.285
9.44	11331	8.73	15126	0.343

FIGURE 6.11 (a) Overlap of spots identified in 2D PAGE map of human colorectal epithelial cell line (in green) and theoretically computed (in red). (b) Several pairs of corresponding experimentally predicted spots are connected to reflect the translations. (c) A global warping attempts to bring the computed value closer to the corresponding observed member of the pair. While in some cases an almost exact local alignment is achieved, in many instances the differences caused by posttranslation modifications are simply too large to successfully align. This analysis was carried out using a demonstration version of the Delta-2D package.[18]

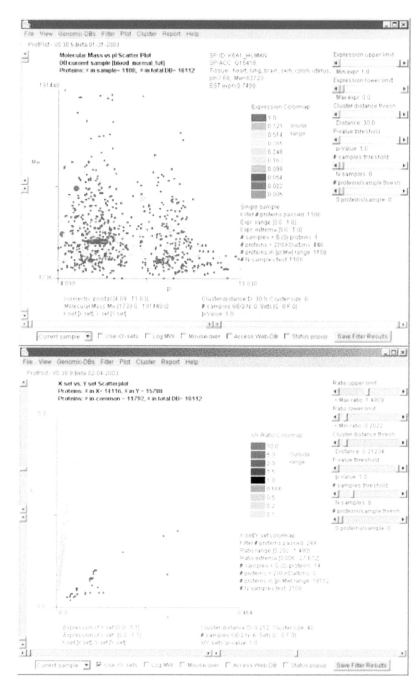

FIGURE 6.15 Snapshot of scatterplots from one sample in ProtPlot (top). It is also possible to create (bottom) an (*X* vs. *Y*) scatterplot or (mean *X* set vs. mean *Y* set) scatterplot when the corresponding ratio display mode is set. The following window shows the (mean *X* set vs. mean *Y* set) scatterplot.

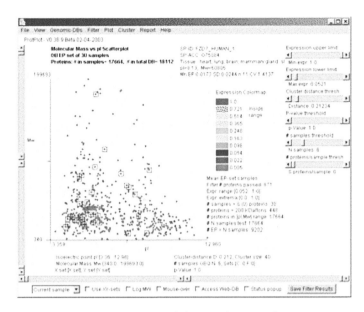

FIGURE 6.20 The spots marked by boxes belong to the same cluster.

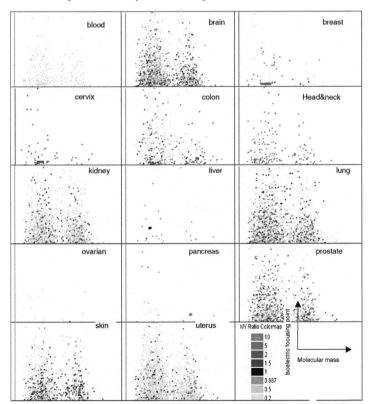

FIGURE 6.21 Tissue and histology specific pI/MVv maps surveyed to date. The color code for the scatterplots is the same as in Figure 6.15 for the individual maps, but for ratios (X/Y) it is as follows: 10.0, 5.0, 2.0, 1.5, 1.0, 0.666, 0.5, 0.2, 0.1.

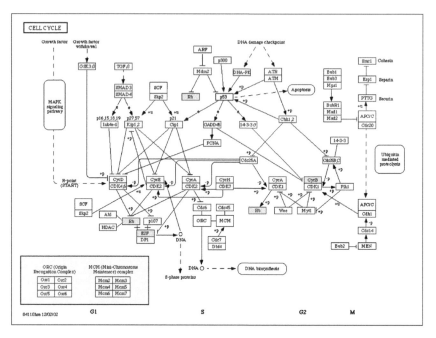

FIGURE 9.1 KEGG/PATHWAY graph representation of the cell cycle pathway in *Homo sapiens*. Nodes in the graph represent gene products; edges represent interactions between gene products. Nodes highlighted in pink represent an example set of genes of interest as entered by the user (Web site available at http://www.genome.ad.jp/ kegg-bin/ mk_point_ multi_html).

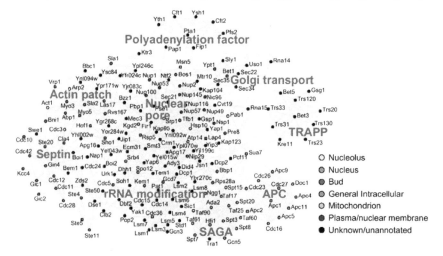

FIGURE 9.2 Visual representation of a molecular complex in protein interaction networks found using the *k*-core method of analyzing yeast protein interaction data. The above network is a six-core. In a six-core, each node has at least six edges connected to it. SAGA, Spt-Ada-Gcn5-acetyltransferase transcriptional activator–histone acetyltransferase complex; TRAPP, transport protein particle complex. Proteins are colored according to GO (Gene Ontology) cellular component. (Reproduced from Bader, G.D. and Hogue, C.W. *Nature Biotechnol.*, 20, 991–997, 2002. With permission.)

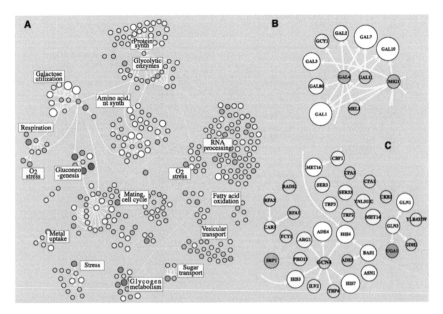

FIGURE 9.3 Integrated physical-interaction networks from systematic perturbations of the GAL pathway in yeast. Nodes represent genes, a yellow arrow directed from one node to another signifies that the protein encoded by the first gene can influence the transcription of the second by DNA binding (protein–DNA), and a blue line between two nodes signifies that the corresponding proteins can physically interact (protein–protein). Highly interconnected groups of genes tend to have common biological function and are labeled accordingly. (A) Effects of the *gal4* + gal perturbation are superimposed on the network, with *gal4* colored red and the gray-scale intensity of other nodes representing changes in mRNA as in Figure 9.2 (node diameter also scales with the magnitude of change). Regions corresponding to (B) galactose utilization and (C) amino acid synthesis are detailed at right. (Reproduced from Ideker, T., et al., *Science*, 292, 923–934, 2001. With permission.)

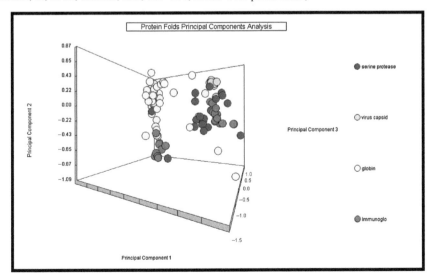

FIGURE 10.3 A PCA plot of proteins with different folds.

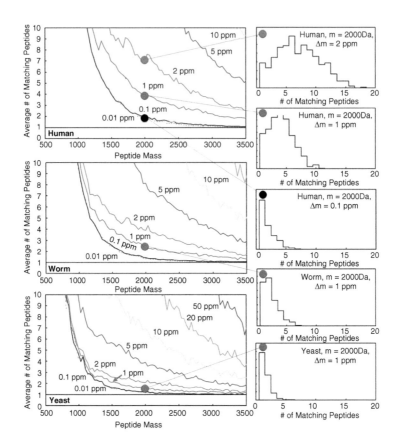

FIGURE 13.2 The average numbers of matching complete tryptic peptides from human, worm (*C. elegans*), and yeast (*S. cerevisiae*) proteins as a function of peptide mass for different accuracies of the mass measurements (left). The better the mass accuracy, the fewer tryptic peptides match a single peptide mass; e.g., a mass of 2000 Da correspond on the average to 7.0, 4.1, and 2.0 complete tryptic peptides from human proteins for mass accuracies of 2, 1, and 0.1 ppm, respectively. Below 0.1 ppm no further improvement is observed when improving the mass accuracy because the elemental composition of the peptide is uniquely defined by the mass. Organisms with fewer genes have fewer peptides that match a single tryptic peptide mass; e.g., a mass of 2000 Da with a mass accuracy of 1 ppm correspond on the average to 4.1, 2.3, and 1.6 complete tryptic peptides from human, worm, and yeast, respectively. The distributions of the number of matching peptides are shown for a few cases (right); e.g., the average value of 1.6 complete tryptic peptides from yeast matching a mass of 2000 Da with mass accuracy 1 ppm, correspond to matches to single peptides in 58.4% of the cases and to matches of 2, 3, 4, 5, and 6 peptides in 30.7, 8.4, 1.9, 0.4, and 0.2% of the cases, respectively.

FIGURE 14.1 3D interactive display of two classes of samples projected in UMSA component space.

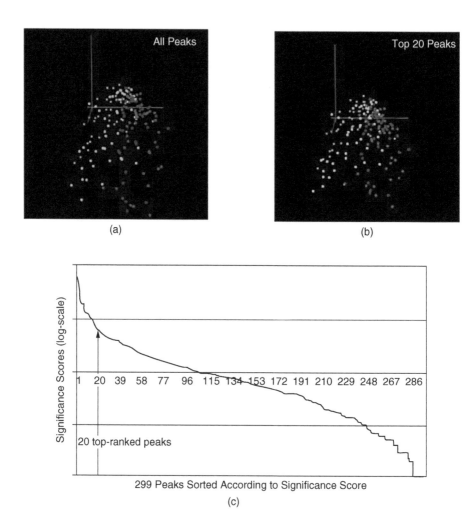

FIGURE 14.2 UMSA-based analysis of SELDI peak intensity data from 183 samples in two groups, each with 299 detected peaks. Group A: n = 92, 49 used for training (green) and 43 for test (olive); Group B: n = 91, 49 used for training (red) and 42 for test (blue). (a) UMSA component analysis of the training data using all 299 peaks; the fixed component projection was then applied to the test data. (b) Plot of significance scores of all 299 peaks in log-scale and descending order. Arrow indicates cutoff on significance scores where the score descending rates differ noticeably. With this cutoff, 20 top-scored peaks were selected. (c) UMSA component analysis of the sample training data using the 20 selected peaks; the fixed component projection was then again applied to the test data.

FIGURE 19.1 Example of antibody labeling in fixed tissue. The picture is a triple label confocal image of the cerebellum of the mouse mutant ataxia (axJ). Granule cells and other nuclei are labeled with DAPI (blue), Purkinje cells are labeled with antibody to calbindin (red), and synaptic contacts are identified by an antibody to syntaxin (green). (Courtesy of Dr. Rivka Rachel, NCI-Frederick).

Combined Use of DIC and Fluorescence Illumination

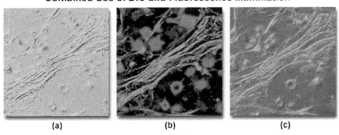

(a)   (b)   (c)

FIGURE 19.4 (a) A thin section of cat brain tissue infected with cryptococcus and imaged using DIC optics and a full-wave retardation plate. Note the pseudo three-dimensional appearance of the photomicrograph. (b) The same field of view, but imaged with fluorescence illumination and an Olympus WIB filter cube. The cells were stained with a combination of fluorescein-5-isothiocyanate (FITC) and Congo red (emission wavelength maxima of 520 and 614 nanometers, respectively). (c) The two techniques are used in combination, illustrating the infected cat brain tissue in both fluorescence and DIC illumination.

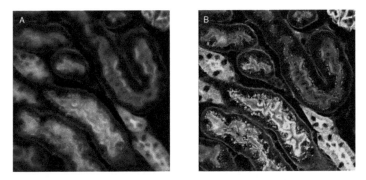

FIGURE 19.5 Tissue section of mouse kidney 15 μm thick, labeled with Alexa Fluor 488 wheat germ agglutinin (green) and Alexa Fluor 568 phalloidin (red). Images were acquired with an LSM 410 confocal microscope (Carl Zeiss Inc., Thornwood, NY). (A) Image acquired with the pinholes open to mimic conventional microscopy. (B) Image acquired with a small pinhole for confocal microscopy.

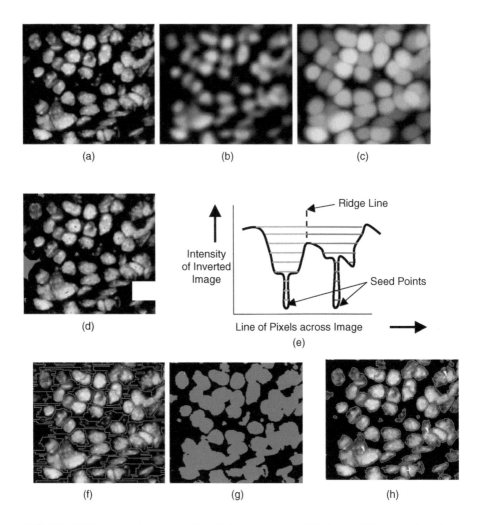

FIGURE 19.13 Segmentation of cell nuclei using matched filtering and the watershed algorithm. (a) Image of fluorescence-labeled cell nuclei. (b) The image in A after application of a matched filter. (c) The image in B after application of the dilation filter. (d) The seed image (red) generated by subtracting image C from image B and thresholding at intensity 0, overlaid on the image in A. (e) One-dimensional schematic illustration of the watershed algorithm. "Water" gradually fills up the valleys starting at the seed points. All pixels in the same valley as the seed are assigned to the same object (green and red). Locations of adjacent pixels from different objects form ridge lines, which water cannot cross. (f) Watershed ridges overlaid on the original image. (g) Original image after automatic thresholding to determine object and background regions of the image. (h) Final borders overlaid on the original image. "X" and "+" indicate errors. "X" marks undivided clusters of nuclei and "+" marks incorrectly split nuclei.

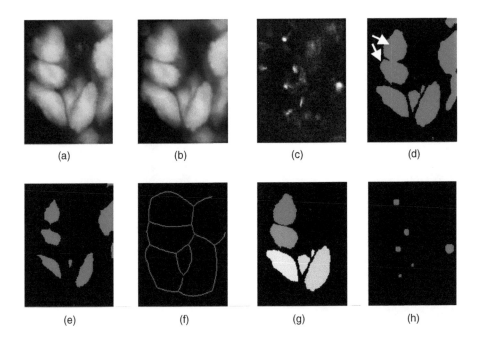

(a) (b) (c) (d)

(e) (f) (g) (h)

FIGURE 19.14 (a) Image of cell nuclei (green) and FISH signals (blue). (b) Grey image of the nuclei. (c) Grey image of the FISH signals. (d) Image B following setting of a threshold intensity to separate bright nuclei (red) with intensities > threshold intensity from background (black) with intensities < threshold intensity. Arrows indicate two touching nuclei, which the computer would consider as one object. (e) Image D after erosion to shrink the nuclei so that touching nuclei separate. (f) "Skeletonization" of the background in image E in order to determine the lines midway between the objects, which serve as divisions between touching objects. (g) The individually detected nuclei. (h) The detected FISH signals.

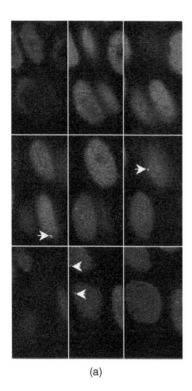

(a)

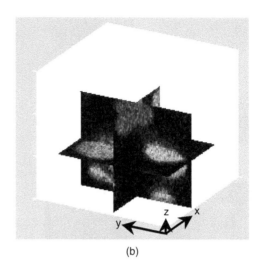

(b)

FIGURE 19.16 (a) "Gallery" of 2D slices from a 3D image. The red signal is from a fluorescent DNA dye that labels the entire volume of each nucleus. The yellow dots are FISH signals at the centromere of chromosome 1 (arrows). (b) Orthogonal views through a 3D image. (c) Generation of projection images in volume rendering. The intensity at coordinate (x',y') in the projection image is a function of the set of intensities along the line (x',y', z_1) to (x', y', z_n) in the 3D image, where n is the number of slices in the 3D image. (For maximum intensity projection, the function selects the maximum intensity.) The same transformation is applied at all coordinates in the x-y plane. (d) Example of maximum intensity projections after rotating the 3D image in 30° increments from 0 to 150°. In these projections, all FISH signals are shown in each projection, except those FISH signals that are directly behind another. (e) Three segmented nuclei displayed using surface rendering. The display looks flat without using texture (left rendering versus middle rendering). Wire-frame rendering (right rendering) enables the display of objects (segmented FISH signals) internal to outer objects (nuclei). (Lockett, S.J. Three-dimensional image visualization and analysis. *Current Protocols in Cytometry.* Copyright 2004 Wiley.)

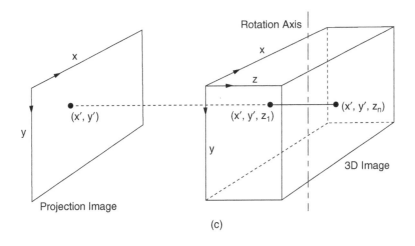

(c)

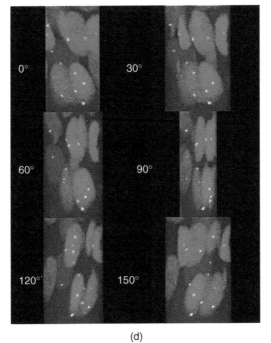

(d)

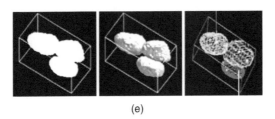

(e)

FIGURE 19.16 (Continued).

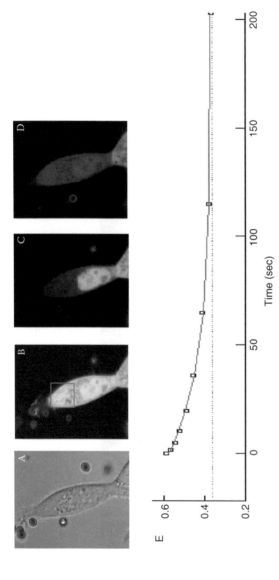

FIGURE 19.17 Fluorescence recovery after photobleaching (FRAP) measures nuclear-cytoplasmic transport rate. (A) Differential interference contrast image of the cell. (B) Image of the cell expressing GFP showing a higher concentration of fluorescence in the nucleus vs. the cytoplasm. After recording this image, a rectangular region of the cytoplasm was bleached. (C) Image immediately after bleaching. GFP diffuses very rapidly in the cytoplasm, therefore, GFP outside the red rectangle is exchanging with GFP inside the rectangle during the brief bleaching period. This results in GFP in the entire cytoplasm being bleached. However, translocation of GFP between the nucleus and cytoplasm is much slower, resulting in no alteration in the nuclear fluorescence immediately after bleaching. (D) Image 290 seconds after bleaching, showing less fluorescence in the nucleus and increased fluorescence in the cytoplasm compared to C, because unbleached GFP in the nucleus has exchanged with bleach GFP in the cytoplasm. Note that the adjacent, attached cell was unaffected during the study, suggesting a very slow exchange of GFP between the cells. (E) Ratio of fluorescence from the nucleus vs. the entire cell as a function of time after bleaching measured at nine time points (squares). The red curve is the best-fit exponential decay from which the exchange rate of GFP between the cytoplasm and nucleus can be determined.

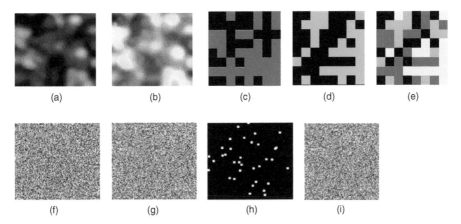

FIGURE 19.18 Visual interpretation of colocalization is very misleading. (a) and (b) Contrast enhancement leads to misleading conclusions. (a) Simulation of red and green fluorescence signals, where the low amount of yellow (red overlaid with green) implies very little colocalization of the two colors. (b) The same image as A after contrast enhancement now shows a significant amount of yellow implying significant colocalization. (c), (d), and (e) Apparent colocalization is not real. (c) and (d) Two 8 × 8 pixel binary images where approximately half the pixels were randomly assigned signals. (e) Overlay of images (c) and (d) shows approximately one quarter of the pixels contain both red and green (yellow), which is as expected to occur by chance. Quantitative colocalization analysis, Costes, et al.,[44] however, report a probability of 25% that the colocalization is not random; i.e., the presence of real colocalization is not significant. (f), (g), (h), and (i) Real colocalization is not visually apparent. (f) and (g) Two random images, where the probability of the colocalization between the two images is not random is 22%. (h) Mask image showing the regions of image (f) that replaced the same regions in image (g) in order to introduce real (nonrandom) colocalization between the images. Image (i) is the result, which, when compared to image (g), does not show any visual evidence of colocalization. However, quantitative analysis reports a highly significant probability of over 99% that real colocalization exists between the images.

10 Data Mining in Proteomics

R. Gangal

CONTENTS

Give me back the days of Yore,
BioScience is just not the same anymore.

Genomes, proteomes, metabolomes reign,
For days of Yore my heart does pine.

Each protein with its own story to tell,
Cajoling it to crystallize took efforts like hell.
To "understand" was a reward of its own,
Those were the days when "high throughput" was unknown.

Alas, days of Yore are long gone,
Data mining and statistics are today's norm.
Instrumentation, bioinformatics, data mining and all that jazz,
I feel like a shackled prisoner at Alcatraz.

10.1 INTRODUCTION

The poem above is reminiscent of a now distant past when high throughput was really unimaginable, genomics and proteomics were a long way away, and scientists built entire careers on researching just a couple of proteins and pathways.

While instrumentation has had a fundamental role in producing the required data, the role of informatics, and particularly data mining, has also been a significant one. Bioinformatics, as it is known today, was born out of the necessity of mining this large amount of data. Data mining plays a key role in the present context in informatics for molecular life science research, and especially in proteomics.

In relation to other chapters in this book, I will attempt to introduce the general discipline of data mining, relate it to insightful mining of proteomic data, and provide glimpses of emerging technologies in this area.

"Data mining" is the process of analyzing data (warehouses) to extract patterns, relationships, trends, or rules based on features (properties) derived from source data to obtain predictive models of the underlying biological process to reduce (not eliminate) the need for experimentation.

In the case of proteomics data mining, the task is of understanding the activation, abundance, dynamics, and interaction of the protein complement of genomes. This understanding is fundamental to many of the biological processes, but has particular relevance to disease diagnostics and drug discovery. The working definition here is

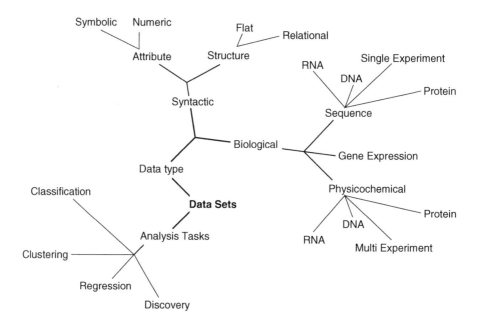

FIGURE 10.1 An unrooted tree diagram showing different types of data mining tasks classified by analysis task and data type.

the mining of data pertaining to the protein complement of any genome. The data to be mined might be information derived from symbolic (sequences), structural, physico-chemical (spectroscopic data, like molecular weight of peptides digested during MS), quantitative (expression), or interaction data.

Clearly, in proteomics, there are several data types and data mining tasks that a researcher might wish to perform. An excellent overview from the perspective of a bioinformatics scientist is available at http://bioinfo.cis.nctu.edu.tw/D_tree.html. It classifies data mining by the analysis task and data type. Figure 10.1 shows the same information in an unrooted tree.

10.2 THE GENERAL DATA MINING PROCESS

The Figure 10.2 work flow shows a generic approach to data mining. The "rate limiting" step in proteomic data mining is really the availability of clean, standardized data and the ability to derive the right properties/features to describe the underlying biology.

It is important to note that this work flow does not have any clear beginning or end. While it is customary to think about data mining as starting with data and ending with validated models used for prediction, in reality it has to be fed back for experimentation and is thus iterative. It is very important to build models on new data, better data, better features, and better data mining methods. The predictive models built have to be validated by experimental evidence whenever possible. This validated data can then be fed back into the data mining pipeline.

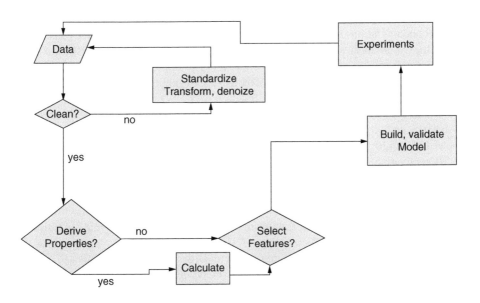

FIGURE 10.2 The general data mining work flow.

10.2.1 FILTER, CLEAN, AND STANDARDIZE

The inconsistency in nomenclature of proteins, their functions, the nonstandardization and variability of instrumentation, and the way data has been historically stored in online databases have plagued data mining attempts in life sciences in general, and data mining in particular.

In the last four to five years there has been an increasing realization that for meaningful data mining the underlying data has to be standardized. For example, until the pioneering efforts of ontology groups (http://www.geneontology.org/, http://obo.sourceforge.net/), there was no systematic way to define protein function. It is a sign of the urgency and importance of dealing with this issue that has led many pharmaceutical and biotech companies to lend their support to ontology groups.

Database querying is quite difficult even with improvements in web-based interfaces like the Sequence Retrieval System (SRS). The fundamental problem is that the underlying database design does not relate to unique biological entities, their properties and instances, and relationships between biological entities.

So a query for subcellular localization "cytoplasmic," not only returns cytoplasmic proteins but also those in possibly other locations. Proteins with a "putative" and "probable" cytoplasmic location will also turn up. The results of these queries have to be looked at in detail to be sure that one is getting the exact information needed. This is what "cleaning" or "scrubbing" the data in the context of this chapter means. Filtering is a step that should be performed before cleaning. It means refining queries to minimize the risk of obtaining wrong data. Thus, in our example, after several queries it will be noticed that putative, probable, cytoplasmic, OR nuclear, etc., keywords turn up in the detailed results. Negating these terms to filter them out can refine the earlier query.

Standardization of nomenclature can be achieved by using resources like gene ontology. Standardization, filtering, and cleaning of data is a prerequisite for successful data mining in every discipline, but more so in proteomics research. We will not go into further details about this crucial step of preprocessing since the same is covered elsewhere.

This phase of initial examination of data also includes calculating summary statistics and plotting data in various ways to get a feel for its structure. For example, the data might have outliers, missing values, and completely different ranges for different variables. Thus, during preprocessing one might need to normalize the data, fill in missing values, and filter outliers. Your friendly neighborhood statistician can be a big help in doing this.

Summary statistics can be calculated for every class individually or for the whole data set. It might give an idea about the discriminatory power of the variables or features used.

Since this becomes more and more difficult as the number of features and classes increases, principal components analysis (PCA) might be used at this stage to visualize the separation of the classes. Figure 10.3 shows a PCA plot for four protein fold types. Various physicochemical indices were first calculated for all four fold classes: serine protease, globins, virus capsid protein, and immunoglobin. The plot

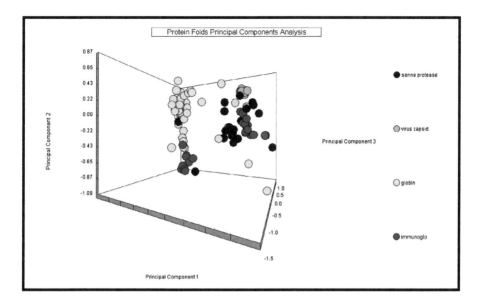

FIGURE 10.3 (Color insert follows page 204) A PCA plot of proteins with different folds.

has been generated from the PCA performed on these data points. Clear clusters are visible.

10.2.2 Feature Selection

As shown in the work flow diagram, the data miner has to decide whether or not to derive new features from input data. The next challenge is to determine which of the features are relevant to building the model. Features are also known as descriptors, dimensions, and properties.

If the data is, say, an image or DNA/protein sequences with no other information, then one has to calculate features from the raw data. In some cases the features will already be available; e.g., protein domain information, molecular function, secondary structure, etc. But in most cases, features will have to be calculated from input proteomic data. The question is once we do have features, which ones are really relevant? Are we taking any redundant ones? Are there features that do not change or change very little across different classes? This translates into a question of dimension reduction. The curse of dimension says that the cost of computation increases exponentially with the number of dimensions. Thus one might expect to model a process or problem better with more features, but not only does the complexity of calculations increase, the probability of finding a good model does not always increase. In fact, in many cases the accuracy might actually go down because of correlation between features and other statistical problems.

Various methods have been developed to deal with the above questions. These include two subtypes: (a) selecting features that contribute to the separation of classes termed as feature selection, and (b) transformation of features to a reduced subset of features, termed as feature extraction.

Various types of dimension reduction algorithms are used, e.g., principal components analysis, Karhunen-Loeve transformations, factor analysis, and multidimensional scaling. A detailed discussion of dimension reduction methods is out of the scope of this chapter, but the reader is referred to an excellent technical text on the same.[1]

10.3 APPROACHES TO DATA MINING

There can be several approaches to data mining based on different criteria. The best data mining method for a given problem needs to be chosen depending upon the nature of the data mining task:

1. Statistics or machine learning
2. Supervised or unsupervised
3. Pattern discovery or pattern recognition
4. The data type
5. High throughput/low throughput

Let us discuss each of the above criteria with examples.

10.3.1 STATISTICS OR MACHINE LEARNING

Statistics in the context of proteomics can be used for accepting or rejecting particular hypotheses. These hypotheses can pertain to

1. Examining the "goodness of fit" of experimental data to some known distribution or curve or fitting a curve to numerical data
2. Classifying an object in one or more classes with some measure of confidence

The core concepts in statistics are distributions, means, variances, and higher order moments. The concept of probability is central to "doing statistics." For example, bioinformatics scientists base their faith in a blast result on the p value rather than just the raw score of the hits obtained.

In case of classification of biological objects, simple statistical measures can sometimes suffice if the features that describe the biology clearly separate data into desirable discrete classes. For example, n-mer statistics have long been used as a measure to separate protein-coding genes from noncoding orfs. Thus if the frequencies of occurrence of 6 to 7 mers are used to separate coding and noncoding orfs, in at least simple cases for a given organism, the mean, variance, and some higher moments might be enough for highly accurate classification.

Consider an interesting aplication of data-mining, i.e., the prediction of essential bacterial proteins.[2] Those bacterial proteins that, if disrupted, can lead to the loss of viability or growth, are termed "essential proteins." Several experimental methods exist at the genomic and proteomic level to establish the essentiality of bacterial proteins. However, until recently there was no in-silico method for the same. This problem is of interest not only because it might give interesting bacterial drug targets, but also since it showcases the possibility of mining proteomic data before going for expensive experimentation.

In our work we took just about 50 known essential and nonessential proteins from *E. coli, H. influenzae,* and *S. pneumoniae.* The critical part of the research was to use features that represented the real biological differences between the two classes. This problem cannot be clearly dealt with by simple metrics like sequence similarity because the proteins were from very different functional classes. Some proteins were not universally conserved across different bacteria and were still essential. After substantial investigation the best features were found to be indices called Tsallis entropies (http://www.santafe.edu/sfi/publications/Bulletins/bulletinFall00/features/tsallis.html).

For years biologists have used shannon entropy as a measure of the information content or diversity or entropy of a system. Tsallis entropy is a generalized entropy measure that reduces to the shannon entropy under assumptions of a closed system at equilibrium. Since biological systems ranging from genes and proteins to cells, organisms, and ecosystems are open and far from equilibrium, Tsallis entropy might have an important role to play in chemical and biological dynamics in general.

Chi-squared analysis then clearly indicated that some protein sequences were essential and others were nonessential. There was no need to use machine learning methods to separate proteins into two classes. Thus, in the case of classification conventional statistics can suffice provided that the features for the biological problem are good. Techniques like machine learning should be used when the features used to model the biology are not sufficient or not well understood or cannot clearly separate the data into multiple classes clearly.

Any discussion regarding protein data mining is incomplete without a reference to the technique of Hidden Markov Modeling (HMM). HMMs stand at the interface between statistics and computing. The HMM is a dynamic kind of statistical profile. Like an ordinary profile, it is built by analyzing the distribution of amino acids in a training set of related proteins. However, an HMM has a more complex topology than a profile. It can be visualized as a finite state machine, familiar to students of computer science. Models for detecting remote homologs of proteins frequently rely on HMMs. Databases like Pfam and programs like HMMER (http://hmmer.wustl.edu/) and Meta-MEME (http://metameme.sdsc.edu/) are routinely used to build Markov models for protein families based on protein sequence alignments.

Statistics, however, is perhaps better used to test hypotheses about the confidence of classification/predictions rather than for classification/prediction itself.

10.3.1.1 The Role of Machine Learning

This brings us to the utility of machine learning for data that is not clearly separable into desirable classes. Most data mining problems in proteomics research might fall into this category since their underlying biology is not well understood or the features used to build predictive models do not sufficiently capture the underlying biology or chemistry. It has long been understood that the one-dimensional sequence of a protein dictates its 3D structure and its molecular function and interactions. However, none of the currently available features (e.g., secondary structure) for a protein sequence can help relate the chemistry and biology of proteins.

One possible definition of machine learning techniques is that it is a class of computational techniques that can build models or classifiers from raw or derived data that best separate multiple classes based on some known/unknown mathematical or statistical mapping between the input and the desired output (classes).

Machine learning itself has several distinct schools of thought. In layman terms the techniques can be classified into

1. Neural computing: mimic the biology of human brains and obtain some sort of mathematical mapping between the input/features and the output
2. Decision trees and association rules: derive interpretable rules relating the features and values directly with the output
3. Methods like support vector machines (SVM) that use a mathematical kernel to maximally separate classes
4. Expert system methods (Prolog, Lisp) and inductive logic programming methods (Progol)
5. Search/optimization methods like simulated annealing and genetic algorithms

Since the 1970s a substantial body of research has been devoted to such computational techniques and their application. A particular subtype—the feed-forward neural network—is the most widely used machine learning algorithm and consequently the most widely used methods in proteomics data mining (as in other disciplines). Thus most papers on secondary structure prediction, fold recognition, subcellular location prediction, and numerous other challenges have relied in such neural networks or some variation on the theme.

10.3.1.2 Basic Concept of Neural Networks

Figure 10.4 shows a typical feed-forward neural network. The network has three main parts or "layers"; viz. the input layer, the hidden layer(s), and the output layer. There are multitudes of computational units called "neurons" in each layer. Neurons in one layer are connected to the next layer to the right. There can be more than one hidden layer with variable number of neurons. There are very few and sometimes contradictory "thumb rules" for the number of neurons in hidden layers, their connections, and the overall topology. In certain cases, there is software available that can optimize the topology of a neural network; e.g., a utility called ENZO is available for the Stuttgart Neural Network Simulator (http://www-ra.informatik.uni-tuebingen.de/SNNS/).

The number of neurons in the input layer depends on the features used to represent a protein sequence or structure; the output is equal to the number of classes in which we wish to classify a protein. For example, in the above network, the output

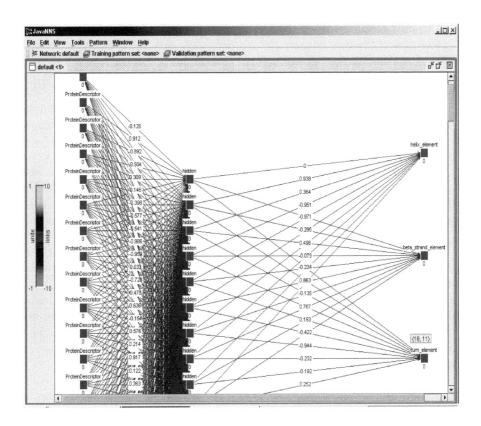

FIGURE 10.4 A typical feed-forward neural network.

layer has three neurons, one each representing a beta strand, helix, and turn classes. The number of inputs will depend on what features are used. In one possible scheme, a binary vector (array) of length 20 represents each amino acid. A particular amino acid is 1 and all others are 0. Thus if a "window" of a particular length (number of amino acid residues) is moved along the length of a sequence, a pattern of length 20 × 13 (length of binary representation × window size) is input for learning the secondary structural class at every iteration or epoch.

Other representations like n-grams and physicochemical profiles (hydrophobicity) can also be used to represent protein sequences. Patterns are input to the network from the input layer. Depending on the mathematical activation function used, the hidden layer neurons that get input from a multitude of input neurons "fire." The output is compared to the ideal output as present in the training examples and the error is propagated back to the input layer, passing through the hidden layers. Thus the weights connecting the neurons in different layers are updated to minimize the error between the desired output and current output. It is important to realize that all this is possible since numbers represent the output internally in neural networks. So the error that we are referring to is some function to measure the difference in numerical values as output by the network and as they are represented in the training file. For example, for our secondary structure problem the outputs might be internally represented as 0, 1, 2. Some previously decided threshold is needed to decide what class an example falls into; e.g., if the value is near 0 it is helix, near 1 it is a beta strand, and for 2 it is a turn.

The class of networks shown above is called a feed-forward network and the way the errors are propagated back from the output to the input (adjusting the weights in the process) defines the actual algorithm. The most common algorithm is back-propagation.

Although not discussed here, there are many other types of neural networks, notably the Kohonen and Hopfield networks. One area of caution while using neural networks in proteomic data mining is their tendency to overfit the data; i.e., over-learning. Although several methods like "pruning" nonchanging neuronal connections have been used, it still remains a significant drawback.

In addition to the problem of overlearning, another perceived problem has been the "black box" nature of such techniques. It perhaps reflects the chasm between practicing/bench scientists and bioinformatics scientists. Biologists and proteomics experts look for interpretable rules that link their concepts of sequences, structures, domains, conserved profiles, physicochemical properties, similarity/homology, and the final classification or result desired. While it is clearly desirable to understand a protein's function(s) or other classifications with reference to the above properties, it is not always possible in practice.

Algorithms like decision trees and rules (C4.5/C5.0, Association Rules & CART) should be used in cases where such an understanding of the biological rules is thought to be essential. An example of such techniques in proteomics data mining is for "Automatic rule generation for protein annotation with the C4.5 data-mining algorithm applied on peptides in Ensembl" (http://www.bioinfo.de/isb/gcb01/talks/kretschmann/main.html).[3]

Such approaches might be fruitful in certain cases where there is a clear statistical relationship between biological objects such as domains or sequence similarity and the end classification (say, membrane vs. nonmembrane proteins). But, in general, these biological features might not be enough to be able to predict/classify with a high accuracy. In these cases we might need to derive features from the sequence or structure of the protein and use these abstract features for machine learning.

The decision tree has been obtained by using the C4.5 algorithm to classify proteins into intracellular and extracellular locations (Figure 10.5). One can calculate some notional physicochemical indices[4] to represent protein sequences from the two subcellular locations. The algorithm then finds a tree-like rule based to classify the sequences into two subcellular locations. Decision tree algorithms use entropy or information content-based methodologies to decide which feature optimally divides the input data into n classes (here $n = 2$).

In such a case there really is no practical benefit of choosing decision tree–like methods over black box methods unless they outperform other algorithms in terms of classification accuracy. However, in case the features used are representative of the underlying biochemistry it does make sense to use decision tree–like

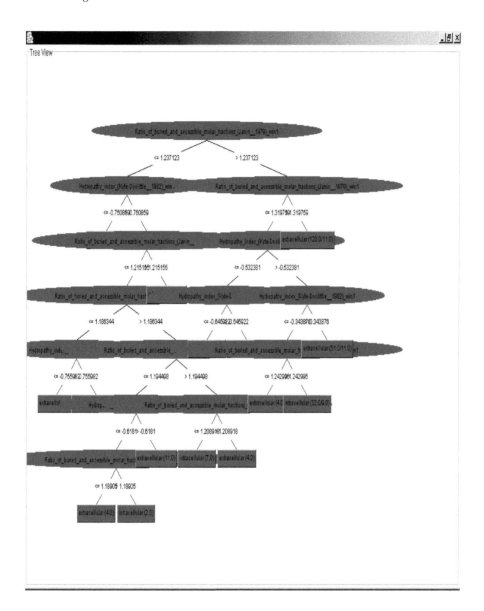

FIGURE 10.5 A decision tree.

methods even if the accuracy of the model obtained is a bit less. It is really a trade-off between the interpretability and accuracy of models that the investigator has to consider.

Due to the need to understand the logic of classification in case of black box techniques, attention is increasingly turning to research on methods to extract rules from neural networks and other approaches.

One method for classification that has taken the data mining community and bioinformatics scientists by storm is support vector machines (http://www.kernel-machines.org). As opposed to neural networks that tune coefficients or weight between neurons depending on error, SVMs use mathematical "kernels" or functions to separate classes in multivariate feature space.

Just as a line separates a two-dimensional space and a plane separates three dimensions, in algorithms like SVMs the idea is to build hyperplanes in more than three dimensions to separate the data optimally. SVMs use a particular formulation called structural risk minimization to achieve this separation.

From the point of view of proteomics data mining, it is important to note that in theory SVMs outperform methods like neural networks and decision trees. SVMs are less prone to overfitting or overlearning; thus they have the highly desirable ability to generalize, i.e., predict entirely unseen examples with high accuracy. SVMs also perform dimension reduction since the hyperplanes they construct or the number of support vectors that are finally used for classification are less than the number of properties or features input initially.

Until recently, SVMs were mainly used for two class problems. They have now been extended to multiclass classification. The approach for multiclass classification is to use two-class classification (binary) to make many building models for (1) each class in turn against all other classes taken together, OR (2) each class in turn against all other classes taken one at a time.

So if there are four super secondary classes into which we desire to classify proteins, i.e., alpha, beta, alpha+beta, alpha/beta, then the first approach will make 4 binary models internally, while the second approach will lead to 12 internal binary classification models. The output will then be a voted output of all these models.

Several kernels have been specifically devised for protein classification. For example, the Fisher kernel takes the sequence conservation for a given class into consideration. The Fisher kernel has been used in several data mining challenges in proteomics.[5,6] Polynomial kernels have been used in discriminating protein subcellular locations.[7–9]

However, in spite of all the progress made, there is still one area that has not been addressed until recently. This is the case of unbalanced data, a situation that occurs far too frequently in proteomics and genomics data mining.

Even in the case of well studies problems like subcellular location, the number of, say, experimentally known cytoplasmic proteins far outnumbers the number of secreted proteins. This might be a biological necessity, but it hampers most machine learning algorithms. The problem becomes more acute as we look at fold identification where some folds have many known examples and some have just a couple.

Another challenge occurs when a given protein might fall into several different classes, like in the case of subcellular location. There cannot always be a unique cellular location where a protein resides throughout its life cycle. In some ways this is also a problem of representation of the biology. If we had more knowledge about additional parameters, like the phase in the lifecycle, etc., that affect the localization of a protein at a given time, perhaps machine learning would be able to deal with such issues. But, in general, most machine learning models are built where there is expected to be just one Y ordinate, i.e., only one dimensional-dependent variable.

Most methods mentioned above can only deal with data of one type: feature-class (attribute-value pairs). It takes no account of relationships between features and the variable number of features. Relational data is very common in proteomics. One needs a different approach while dealing with relational data. One emerging method is that of inductive logic programming (ILP). ILP is a research area formed at the intersection of machine learning and logic programming. It uses a database of domain knowledge or facts, a rule base relating objects in the knowledgebase and propositional logic to induce new rules from data that extend the rule base.

This approach is perhaps more suitable for proteomics data mining in cases where it is clearly desirable to understand the underlying biology or chemistry of a problem. ILP has been recently used in protein fold recognition where the rules obtained are of clear interest to a proteomics scientist.[11]

10.3.2 Supervised vs. Unsupervised Learning

Generally an expert will have some data in hand about the relationship between proteins, their features, and the final desired classification. For example, there are at least some data on the subcellular location, tissue expression, fold, superfamily, molecular function, etc. This is also true for identification of proteins by mass spectrometry. In the last example cited, a database of peptides with masses for given proteins is available and methods like neural networks are used to identify the protein subjected to MS (http://us.expasy.org/tools/#proteome).

Clearly, supervised learning methods are used when sufficient examples per class are available for a given classification task. While the features used to represent the problem might not give the best possible classification, in general it is possible to achieve good accuracy since one tweaks the machine learning algorithm to force it to classify data into known classes.

In some cases, there might be no known or apparent way to classify proteins in the context of the problem under consideration. Questions that ask "How similar is my protein to this protein?" fall into this class. It might not be clearly understood what exactly is meant by "similar," since sequence similarity does not scale proportionally to 3D similarity.

One then attempts to calculate or derive features that one thinks relate to the underlying biology of the system and use some clustering algorithms like PCA, self-organized maps, or dendrograms to relate the proteins. It is basically searching for clusters of related information in large data sets to find associations that might reveal new insights between proteins, say, in relation to diseases and patient populations (biomarkers) or function (family members).

Many techniques that work for unsupervised classification can be modified to work as supervised classification techniques and vice versa (SOM, LVQ, neural networks).

10.3.3 PATTERN DISCOVERY OR PATTERN RECOGNITION

One of the most important ways to think about data mining in general, and proteomics in our case, is whether the problem under consideration is of pattern recognition or pattern discovery. Many times, the two approaches need to be used in a complementary fashion to really uncover new insights.

Pattern discovery with reference to proteomics refers to the discovery of patterns that are not pre-defined, e.g., prosite patterns, regular expressions derived from multiple alignments, motifs discovered by algorithms such as Pratt (http://www.ii.uib.no/~inge/Pratt.html), and teiresias (http://www.research.ibm.com/bioinformatics/).

The challenge of pattern discovery is far more difficult than the one posed by pattern recognition, where one uses features in an unsupervised or supervised fashion to search for other instances of the same class. It is assumed that the input to a pattern recognition algorithm is a known set of known patterns or features that define a class. The challenge in pattern discovery is to "discover" such unknown patterns from supplied raw data. There is an important difference between the meaning of "pattern" and "features."

By searching for patterns, one generally means searching for conserved signatures in protein sequences or structures that might not be a physically meaningful property like hydrophobicity. It might be a regular expression like AA*V[2].R that might be found to occur very frequently in a certain class.

But another interpretation of the term "pattern" is related to features. The trend of a particular feature like hydrophobicity over the length of a protein sequence or the profile can itself be treated as a pattern. Such patterns might be easier to identify than the pattern discovery in sequences, since here it is only a matter of using a mathematical function to identify the trend.

Patterns discovered by pattern discovery algorithms like mast, teiresias, etc., could then be used as features for data mining.

10.3.4 THE DATA TYPE (SPECTROSCOPIC DATA, SEQUENCE DATA, STRUCTURE DATA, IMAGE/BINARY OBJECT DATA)

As might be apparent from the preceeding discussion, not all data can be used directly for mining knowledge. Typically symbolic sequence data is normally analyzed by sequence alignment, pattern discovery, or feature calculation methods. The data thus derived is then used for further data mining.

As opposed to this, data from mass spectroscopy has to be used directly to infer the most probable protein. The identification of proteins by MS is an area of active research. A recent method for protein mass fingerprinting uses decision tree models.[12]

The quantification of proteins from gels (1D and 2D gel analysis) is a challenging task, since there is a lot of background noise in the form of spurious spots, hazy outlines of spots and different shapes of spots.

Another emerging area is the usage of image data for direct classification, e.g., crystal image classification for high throughput x-ray crystallography and protein subcellular location identification using 3D imaging.[13–15]

10.3.5   HIGH THROUGHPUT/LOW THROUGHPUT

In this era of high-throughput experimentation, a corresponding setup for high-performance computation is also necessary. In addition to high-performance computation, one also has to be able to use data mining algorithms that work fast, and more importantly, learn incrementally. In this form of data mining, algorithms incorporate or discard new data to refine the model. If one has to learn the models from scratch every time new data is added, it becomes a bottleneck. So another feature of incremental learning is just updating the model rather than rebuilding a model from scratch even if new data is incorporated. Most forms of machine learning algorithms, from decision trees to neural networks to SVMs, have been modified to incorporate incremental learning. While it might seem desirable in general to always use an incremental data mining system, there is always a trade-off in the decision to incorporate new data and to just refine the model. Sometimes building a model from scratch might really improve the classification accuracy.

Another way of thinking of the high- or low-throughput situation is whether a real-time or delayed response is needed. Real-time responses need more heuristics, better speed of learning, and fast response times.

10.4   MODEL PERFORMANCE EVALUATION AND VALIDATION

We are primarily talking about classification in this chapter as opposed to estimating a particular value by, say, regression. There are several ways to assess the performance of a model. Cross validation, receiver operator characteristics, and independent testing sets are some of the methods. Cross validation methods, although favored by many bioinformatics scientists, should not be used as the only measure of the model quality.

ROC curves are a powerful tool to find the optimal sample and model in which the ratio number of true positives/false positives is optimized (Figure 10.6).

In case of independent testing, the validation of the prediction accuracy is made only on an independent testing set not at all involved during model building.

With reference to building multiple models, e.g., in cross-validation, the Kappa statistic or Kendall's coefficient might be used to rate agreement between models. This point of using statistics was mentioned during the discussion of the role of statistics in proteomics data mining.

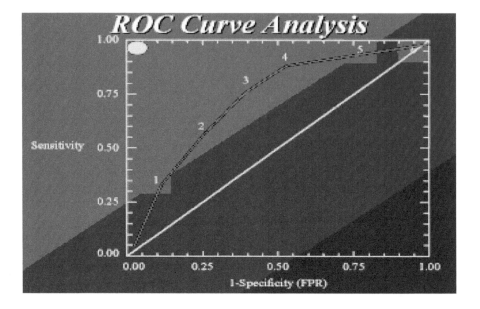

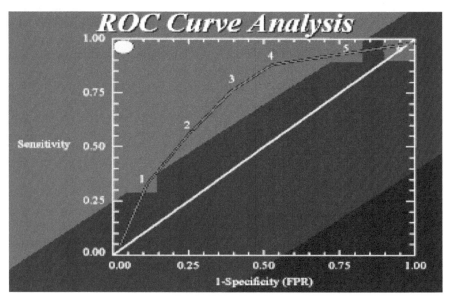

FIGURE 10.6 Receiver operator characteristic curve (ROC). The best region is the blob on the upper left.

10.5 EMERGING AREAS IN PROTEOMIC DATA MINING

At the beginning of this chapter we defined proteomics research as the study of dynamics, interaction, and activation of proteins. Data mining techniques are now being increasingly used in prediction of protein interactions, protein modifications, and their dynamics. What is significant is that one does not always need structural data to infer protein interactions. While identification of protein domains already known to be involved in interaction are clues to the possible interaction partners for a given protein, it is also possible to learn and predict possible interacting proteins based on features derived from protein sequences. In one such patented technology the authors have used SVMs to learn and then predict pairs of interacting proteins.[10] Since neither the features derived from the protein sequence nor the algorithm have any special significance in this analysis, one can easily extend this methodology to predict more specific types of interactions among proteins. Moreover, this technique can also be used for predicting interactions between any two ligands, and not only proteins. In addition to sequence-based methods, investigators have also derived protein interaction matrices based on structural data in the PDB. The propensity of amino acids to be at the interaction interfaces of proteins is taken into account and used to predict likely interaction partners.[16]

Investigators working in structural proteomics deal with the efficient and accurate prediction of 3D protein structures and their interactions with ligands. Examples of such efforts are (1) SPINE: Structural Proteomics in Europe (http://www.spineurope.org/), and (2) Protein Structure Initiative (PSI) (http://www.structuralgenomics.org/).

Although these projects aim to decipher the final quaternary or tertiary structure of proteins using experimental methods like NMR and x-ray crystallography, data mining has an important role to play in selecting the proteins of interest. The domains, the expectation of a novel fold, the solubility, and the importance of the protein in cellular processes are factors that influence the decision to choose particular proteins for structural proteomics. All of this information is not always known. Data mining techniques discussed earlier can address these issues. For example, fold prediction algorithms can be used to predict whether the fold expected for a protein is novel, abundant, or present in small numbers in the PDB.

Protein interaction mapping is the area evoking a great deal of interest since the dynamics of protein interactions with different ligands, their activation patterns, and abundances really drive biological processes at the cellular and tissue level.

Classical data mining, network and graph theory, and complexity theory are converging to provide new insights in life sciences. Efforts like Biospice are trying to model biological networks in terms of concepts from electrical and electronic components. This method does have some merits in that most biological processes and components can be described by analogous terms like transducers, activators, capacitors, and resistance. The journal *Omics* recently published a series of papers describing preliminary software tools developed as a part of the DARPA-funded program[17] (https://community.biospice.org/).

Mapping a significant part of any organism's protein interaction network is the key to this analysis. Only then can one try and elucidate missing pathways, model the dynamics, and use perturbations to analyze the network dynamics.

Linked to the above area is the question of activation of proteins due to post-translational modifications like glycosylation and phophorylation. Neural networks, SVMs, and decision trees have been used to predict such modifications (e.g., http://www.cbs.dtu.dk/services/NetPhos/).[18]

In terms of clinical applications of data mining in proteomics, the emerging area is of biomarker identification.[19] A recent review in *Bioinformatics* assesses the role of statistics and machine learning algorithms in this area of research.[20] A high sensitivity and selectivity is clearly of great importance in clinical biomarker data mining.

Another emerging area is one of inductive logic programming, the combination of logic programming and machine learning. Expert systems have been used for a long time to capture the "knowledge" of experts. Such systems also fall into the category of artificial intelligence. The challenge is in inducing new rules from existing data, facts, and rules. While decision trees or association rules, etc., are based on some frequency- and entropy-based methods to derive the rules, ILP induces new rules from existing data as well as existing rules. It has been applied to secondary structure and fold prediction with some success.[11] It is an exciting area because such a methods can easily incorporate biological rules and data, and new rules, if discovered, would probably make more sense to a life scientist.

In this chapter we have not dealt with heuristic search/optimization techniques like genetic or evolutionary programming. Topics like peptide identification using spectrometric data is also not dealt with in detail. It is very difficult to discuss the whole spectrum of data mining in proteomics in one chapter in a way that makes sense and is exciting to both the practicing proteomics researcher and data miners.

10.6 CROSS-POLLINATION OF CONCEPTS

Even with techniques like Hidden Markov Models and SVMs, and with all possible extensions to algorithms like BLAST, a significant proportion of data in sequence databases is still unannotated. One possible reason is that we might need to look at a sequence from a slightly different viewpoint.

There are many techniques like time-series analysis, chaos and complexity, or statistical thermodynamics that can potentially benefit proteomic data mining. By just transforming protein sequences to a time series by using any physicochemical parameter like, say, hydrophobicity or electron–ion interaction potential,[21] it is possible to open up the toolboxes of nonlinear dynamics and time-series analysis. These techniques have already been used successfully in medical informatics (ECG/EEG analysis) and there is no reason why they cannot be used in proteomics research. For example, Fourier and wavelet analysis has already been used to mine protein sequence data.[22,23]

SciNova Technologies has pioneered these cross-disciplinary approaches in its proteomics data mining tool, Prometheus (http://www.scinovaindia.com/prometheus.html). It calculates many physicochemical indices in addition to dynamical systems properties like Lyapunov exponent, Multifractals, Tsallis entropy, and

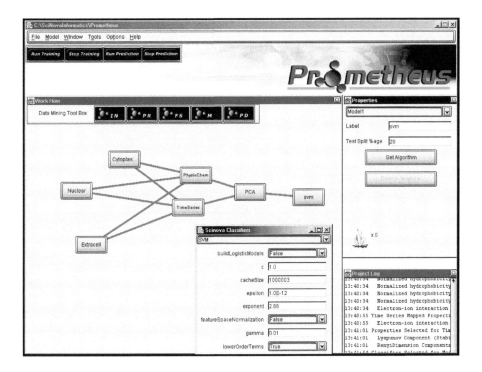

FIGURE 10.7 Prometheus, a work flow–based data mining software with time series and physicochemical indices and machine learning methods.

Fourier components (Figure 10.7).[2] The output of these calculations can then be the input for building predictive models using algorithms like SVMs.

It will be a matter of great satisfaction if this chapter goes some way toward contributing to widespread adoption of data mining and cross-talk between wet lab scientists and experienced data miners.

ACKNOWLEDGMENTS

My deep-felt thanks to Dr. Sudhir Shrivastava and the publishing group for giving us this chance to contribute our knowledge to an important area of informatics in proteomics.

REFERENCES

1. Webb, A. *Statistical Pattern Recognition*. Arnold Publishers, London, 213–273, 1999.
2. Human POL II Promoter Prediction: Time series descriptors and machine learning. *Nucleic Acids Res.*, 33.
3. Kretschmann, E., Fleischmann, W., Apweiler, R. Automatic rule generation for protein annotation with the C4.5 data mining algorithm applied on SWISS-PROT. *Bioinformatics*, 17, 920–6, 2001.

4. Kawashima, S. and Kanehisa, M. AAindex: amino acid index database. *Nucleic Acids Res.,* 28, 374, 2000.

5. Jaakkola, T., Diekhans, M., and Haussler, D. Using the Fisher kernel method to detect remote protein homologies. *Proc. Int. Conf. Intell. Syst. Mol. Biol.,* 4, 149–158, 1999.

6. Tsuda, K., Kawanabe, M., Ratsch, G., Sonnenburg, S., and Muller, K.R. A new discriminative kernel from probabilistic models. *Neural Comput.,* 14, 2397–2414, 2002.

7. Park, K.-J. and Kanehisa, M. Prediction of protein subcellular locations by support vector machines using compositions of amino acids and amino acid pairs. *Bioinformatics,* 19, 1656–1663, 2003.

8. Chou, K.C. and Elrod, D.W. Prediction of membrane protein types and subcellular locations. *Proteins,* 34, 137–153, 1999.

9. Reinhardt, A. and Hubbard, T. Using neural networks for prediction of the subcellular location of proteins. *Nucleic Acids Res.,* 26, 2230–2236, 1998.

10. Bock, J.R. and Gough, D.A. Predicting protein–protein interactions from primary structure. *Bioinformatics,* 17(5), 455–460, 2001.

11. Turcotte, M., Muggleton, S.H., and Sternberg, M.J.E. Application of Inductive Logic Programming to Discover Rules Governing the Three-Dimensional Topology of Protein Structure. In Proceedings of the 8th International Conference on Inductive Logic Programming, 1446, 53–64, 1998.

12. Gay, S., Binz, P.A., Hochstrasser, D.F., and Appel, R.D. Peptide mass fingerprinting peak intensity prediction: Extracting knowledge from spectra. *Proteomics,* 2, 1374–1391, 2002.

13. Oh, J.M., Hanash, S.M. and Teichroew, D. Mining protein data from two-dimensional gels: tools for systematic post-planned analyses. *Electrophoresis,* 20, 766–774, 1999.

14. Berntson, A., Stojanoff, V., and Takai, H. Application of a neural network in high-throughput protein crystallography. *J. Synchrotron. Radiat.,* 10, 445–449, 2003.

15. Boland, M.V. and Murphy, R.F. A neural network classifier capable of recognizing the patterns of all major sub-cellular structures in fluorescence microscope images of HeLa cells. *Bioinformatics,* 17, 1213–23, 2001.

16. Lu, H., Lu, L., and Skolnick, J. Development of unified statistical potentials describing protein-protein interactions. *Biophys. J.,* 84, 1895–1901, 2003.

17. Kumar, S.P. and Feidler, J.C. BioSPICE: A computational infrastructure for integrative biology. *OMICS.* 7, 225, 2003.

18. Blom, N., Gammeltoft, S., and Brunak, S. Sequence and structure-based prediction of eukaryotic protein phosphorylation sites. *J. Mol. Biol.,* 294, 1351–1362, 1999.

19. Creighton, C. and Hanash, S. Mining gene expression databases for association rules. *Bioinformatics,* 19, 79–86, 2003.

20. Wu, B., Abbott, T., Fishman, D., McMurray, W., Mor, G., Stone, K., Ward, D., Williams, K., and Zhao, H. Comparison of statistical methods for classification of ovarian cancer using mass spectrometry data. *Bioinformatics,* 19, 1636–1643, 2003.

21. Lazovic, J. Selection of amino acid parameters for Fourier transform-based analysis of proteins. *Comput. Appl. Biosci.,* 12, 553–562, 1996.

22. Liò, P. Wavelets in bioinformatics and computational biology: State of art and perspectives, *Bioinformatics,* 19, 2–9, 2003.

23. Kauer, G. and Blöcker, H. Applying signal theory to the analysis of biomolecules. *Bioinformatics,* 19, 2016–2021, 2003.

APPENDIX

Web sites of interest for data mining in proteomics:

DESCRIPTION	URL
Swiss Institute of Bioinformatics	http://www.expasy.ch
SciNova Informatics	http://www.scinovaindia.com/prometheus.html
EBI Proteomics	http://www.ebi.ac.uk/proteome/
Biospice Protein Network Modeling	https://community.biospice.org/
NCTU Data Mining Archive	http://bioinfo.cis.nctu.edu.tw/d_tree.html
SPINE: Structural Proteomics in Europe	http://www.spineurope.org
Protein Structure Initiative (PSI)	http://www.structuralgenomics.org
Inductive Logic Programming Network	http://www.cs.bris.ac.uk/%7EILPnet2/index.html
Machine Learning Archive	http://www.mlnet.org
Kernel Machines (SVM resources)	http://www.kernel-machines.org
Gene Ontology	http://www.geneontology.org
SNNS Neural Network	http://www-ra.informatik.uni-tuebingen.de/SNNS/
Introduction to HMMs	http://www.cse.ucsc.edu/research/compbio/ismb99.handouts/KK185FP.html
Pfam HMM Database	http://www.sanger.ac.uk/Software/Pfam/
HMMER HMM Software	http://hmmer.wustl.edu/
Meta Meme Software	http://metameme.sdsc.edu/
Pratt Pattern Discovery Software	http://www.ii.uib.no/~inge/Pratt.html
IBM Teireias Pattern Discovery	http://www.research.ibm.com/bioinformatics/
Protein Phosphorylation Prediction	http://www.cbs.dtu.dk/services/NetPhos/

11 Protein Expression Analysis

Guoan Chen and David G. Beer

CONTENTS

11.1   INTRODUCTION

The technologies currently available that allow the quantification of protein expression in clinical samples include two-dimensional polyacrylamide gel electrophoresis (2D PAGE), multidimensional chromatography, protein arrays, matrix-assisted laser desorption ionization time-of-flight mass spectrometry (MALDI-TOF MS), liquid chromatography coupled with tandem mass spectrometry (LC-MS/MS), surface-enhanced laser desorption/ionization (SELDI), and tissue arrays.[1,2] 2D PAGE has been the mainstay of electrophoretic technologies and is the most widely used tool for protein separation and quantification. In this chapter, as a practical example of the use of expression proteomics, we will describe the methods utilized for the analysis of the quantitative expression of a large number of proteins in a relatively large number of individual lung tissue samples using 2D PAGE.

11.2   SPECIMEN PREPARATION

Proteins can be extracted from blood, fluid, cell lines, or fresh tissue by the use of various cell lysis buffers. Sample preparation is critically important because it may affect the reproducibility and thus comparability of a given set of proteins. Variability in protein expression between samples may result from the heterogeneity of cell populations in a sample.[3] Efforts should be made to obtain comparable tissue or cell preparations and process all samples in a similar manner. At one extreme is the homogenization of a tissue or tumor without ensuring that the percentage of a given cell population is similar. At the other extreme one may utilize laser capture microdissection (LCM),[4] which enables the one-step procurement of highly selected cell populations from sections of a complex or heterogeneous tissue sample. Although the latter method is superior to the former, in our experience it is difficult to obtain sufficient protein from LCM-captured material to run 2D gels.

In our analyses of normal lung tissues and lung adenocarcinomas[5,6] we have utilized both primary tissue and lung tumor cell lines. The method we have used to obtain fairly comparable samples for 2D PAGE is as follows: All tumors and adjacent nonneoplastic lung tissue are collected immediately at the time of surgery and transported to the laboratory in Dulbecco's modified Eagle's medium (Life Technologies, Gaithersburg, MD) on ice. A portion of each tumor and/or lung tissue is embedded in OCT (Miles Scientific, Naperville, IL), or a comparable embedding medium for cryostat sectioning and frozen in isopentane cooled to the temperature of liquid nitrogen. The samples are then stored at −80°C. Hematoxylin-stained cryostat sections (5 μm), are prepared from tumor pieces to be utilized for protein and/or mRNA isolation. The sample to be used for protein isolation is then evaluated by a pathologist and compared to standard H&E sections, made from paraffin blocks of the same tumors, as necessary. This method provides an assessment of the quality of the material to be used for 2D analysis as well as a determination of whether it is representative of the tumor as a whole. Special care is required to obtain comparable regions of each tumor with a similar percentage of tumor cells in the specimen used for analysis and to avoid necrotic areas, as this will result in degraded proteins for analysis. We exclude tumors for our analysis of lung adenocarcinomas if they show any mixed histology (e.g., adenosquamous), tumor cellularity less than 70%, potential metastatic origin as indicated by previous tumor history, extensive lymphocytic infiltration, fibrosis, or history of prior therapy. To compare early and advanced disease, we examined stage I and stage III tumors.

11.3 TWO-DIMENSIONAL POLYACRYLAMIDE GEL ELECTROPHORESIS (2D PAGE)

11.3.1 PROTEIN EXTRACTION

Both normal lung and tumor tissues (approximately 50 to 70 mg wet weight) were homogenized and solubilized in a lysis buffer containing 9.5 M urea, 2% Nonidet P-40, 2% ampholines (pI 3.5 to 10; Pharmacia/LKB, Piscataway, NJ), 2% β-mercaptoethanol, and 10 mM phenylmethylsulfonyl fluoride. Insoluble material was pelleted by centrifugation at $16,000 \times g$ for 5 min at room temperature. Protein concentrations of the soluble extract for each sample were determined using a colorimetric protein assay (Bio-Rad, Hercules, CA). Aliquots of the solubilized extract of both tumor and normal lung tissue as well as cultured cells were stored at −80°C.

11.3.2 2D PAGE

Proteins derived from the tissue extracts were separated in two dimensions as previously described,[7,8] with the following modifications: Approximately 30 micrograms of protein in a volume of 35 µl were applied to isofocusing gels. Isoelectric focusing, using pH 4 to 8 carrier ampholytes, was conducted at 700 V for 16 hours, followed by 1000 V for an additional 2 hours. The first-dimension tube gel was loaded onto a cassette containing the second-dimension gel, after equilibration in second-dimension sample buffer (125 mM Tris (pH 6.8), containing 10% glycerol, 2% SDS, 1% dithiothreitol, and bromophenol blue). For the second-dimension separation, an acrylamide gel gradient of 11 to 14% was used. The samples were electrophoresed until the dye front reached the opposite end of the gel. The gels were then removed and either stained, as described below, or used for Western blot analyses in which the separated proteins were transferred to Immobilon-P PVDF membranes (Millipore Corp., Bedford, MA). Protein patterns in some gels were visualized by either direct silver staining or using Coomassie Brilliant Blue following protein transfer to Immobilon-P membranes with subsequent Western blot analysis.

11.3.3 SILVER STAINING

After separation, the protein spots are visualized utilizing a silver-based staining technique described below.

11.3.3.1 Procedure

All steps are performed with the gels placed on an orbital shaker set to rotate
at 36 rpm.

Following the second-dimension separation, the gels are carefully removed
from the glass plates and fixed in 50% methanol with 10% acetic acid for
2 hours at room temperature.

The gels are rinsed in deionized water for 1 hour.

The gels are impregnated with a silver nitrate solution (2 g silver nitrate/L
deionized water) for 25 min.

The gels were washed in deionized water twice, 1 min. each.

The gels are developed in a solution containing sodium carbonate (30 g/L)
and formaldehyde (10 ml of a 37% solution) for up to 10 min.

Protein staining is then stopped with an acetic acid solution (1% in distilled
water).

11.4 PROTEIN SPOT DETECTION, QUANTIFICATION, AND MATCHING

11.4.1 2D GEL DIGITIZATION (SCANNING)

Following silver staining, each gel is scanned using a Kodak CCD camera. We used
a 1024×1024 pixel format, yielding pixel widths of 163 µm where each pixel had
256 possible gray-scale values (optical density).

11.4.2 PROTEIN SPOT DETECTION AND QUANTIFICATION

Spot detection was accomplished utilizing a 2D gel analysis software (Bioimage Corp.,
Ann Arbor, MI) (Figure 11.1). The company that makes this software is now called
Genomic Solutions, and the newest version of their software is called Investigator
ProImage 2D Analysis System. Each gel generated 1600 to 2200 detectable spots. The
background-subtracted integrated intensity of each spot is obtained in units of optical
density units multiplied by mm^2. The software actually runs on a central machine
(running a UNIX operating systems), but the interface can be displayed on any computer running X-server software, which includes any computer running UNIX or
LINUX operating systems, as well as any computer that has X-server software
installed. We typically use Microsoft operating systems with the "Exceed" X-server
installed (Hummingbird Communications, Ltd., Burlington, MA).

11.4.2.1 Procedure

Open Bio Image 2D Analyzer.

Select the "Analyze" icon, then select "Visage Image" under the "Load" icon.

Select the gel number to be analyzed and open it. Select a project name (for
example, Lung) that will be used for the gels analyzed in the project.

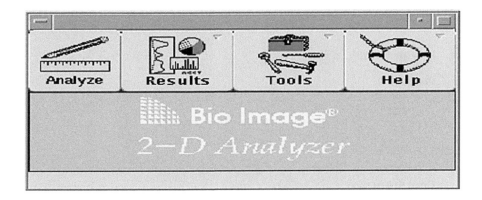

FIGURE 11.1. Bio Image 2D Analyzer software utilized for protein expression analysis. This software was used for the scanning, spot detection, quantification (open "Analyze"), and also to correct or edit incorrectly identified spots (open "Results"). Another software called "Geled" was used for the matching and neighboring reference spot adjustment (normalization).

Select spot detection parameters. We utilized a minimum size of 17, minimum intensity of 20, filter width of 4 and minimum filter value of 5. Then click "Find spots" (Figure 11.2).

Select "Continue Analysis." The program will automatically finish the spot detection and quantification and send this gel to the next program, called "Results."

Select the "Result" icon and open the gel to view each spot. Correct incorrectly detected spots manually (Figure 11.3). For example, the program may incorrectly identify two spots as one. This step requires at least 1 to 2 hours for each gel with approximately 800 to 900 spots.

11.4.3 Matching the Protein Spots

The "Geled" program is utilized to perform the automatic protein spot match work.[9] We usually first create some initial protein spot matches to reduce the time needed for later editing of poorly matched spots. The selected protein spots are designated on one gel termed the "master" gel.

11.4.3.1 Procedure

Open Geled program for spot matching.

Determine which 2D gel will be used as the "master gel." We chose a gel within one batch that was particularly rich in spots and without defects of any kind as our master gel.

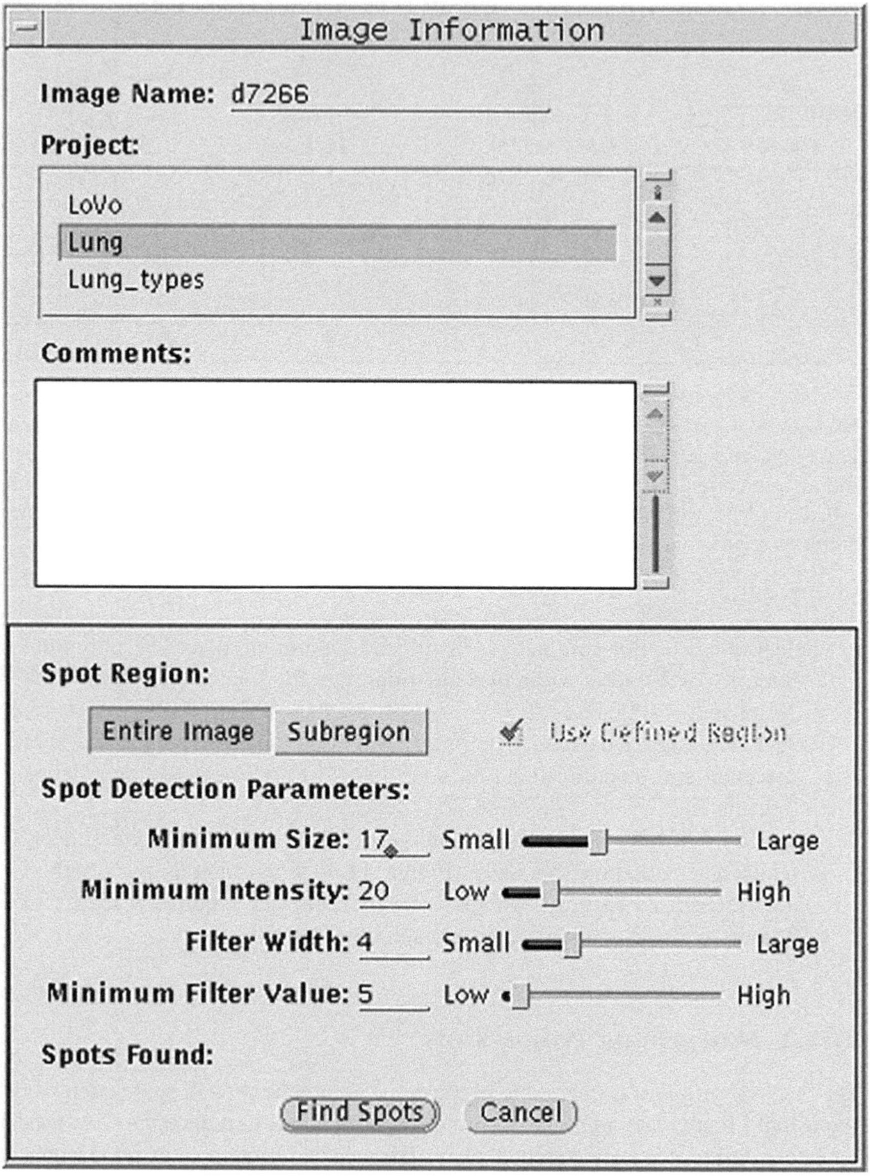

Image Information

Image Name: d7266

Project:

LoVo
Lung
Lung_types

Comments:

Spot Region:

| Entire Image | Subregion |

✓ Use Defined Region

Spot Detection Parameters:

Minimum Size: 17   Small ▬▬▬█───────── Large

Minimum Intensity: 20   Low ▬█──────────── High

Filter Width: 4   Small ▬▬█─────────── Large

Minimum Filter Value: 5   Low ◄├──────────── High

Spots Found:

(Find Spots)   (Cancel)

FIGURE 11.2 Image information and spot detection parameters using Bio Image 2D Analyzer software. These include the minimum size, minimum intensity, filter width, and minimum filter value.

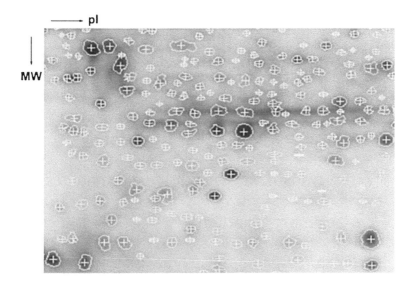

FIGURE 11.3 Partial region of a 2D gel showing spot boundaries detected by the Bio Image software based on the chosen parameters in Figure 11.2.

Determine which spots are to be matched. In our initial analysis we chose only 600 spots in the first batch of gels but later added another 220 spots.

Manually match 30 to 60 spots of the other sample gels, termed the "children" gels, to the corresponding protein spots on the master gel.

Match all other spots using the selected parameters (Figure 11.4). In our analyses, we matched a total of 820 spots in 103 samples that included 93 lung adenocarcinomas and 10 normal lung. We usually open five child gels and one master gel during this procedure (Figure 11.5).

Correct mismatched spots manually.

Return to Bioimage software to correct incorrect density spots.

11.5 QUANTITATIVE ADJUSTMENT OF PROTEIN SPOTS

11.5.1 Neighboring Reference Spot Adjustments

Slight variations in protein loading or silver staining from gel to gel can be mathematically adjusted to potentially remove this source of variation. The spot size integrated intensities that are present on the "child" gel's spot-list are adjusted to the "master" gel; that is, the spot sizes on the master gel are treated as references to which to adjust the spot sizes on the child gel. This procedure is briefly described below. The algorithm utilized for this purpose has been previous

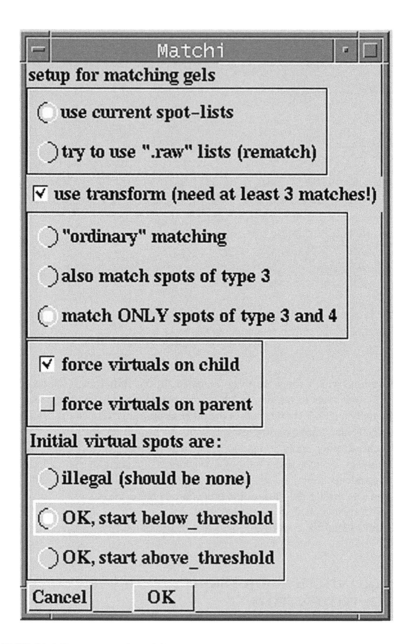

FIGURE 11.4 Parameters selected for matching all spots of children gels to a master gel. Initial reference spots can be matched manually prior to the automatically matching procedure to reduce time required for editing matches.

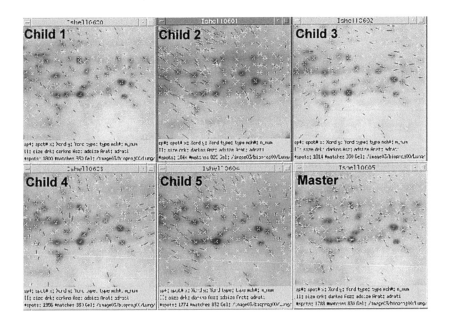

FIGURE 11.5 Spots on children gels are matched to a master gel automatically based on the parameters in Figure 11.4, and all mismatched spots are corrected manually. We usually open five child gels and one master gel during this procedure.

described.[10] Various versions of this method differ according to how the reference set of spots are selected and how missing data in the reference set are treated. The ratios of the spot size on the child image to the master image are computed. The largest and smallest of the ratios are then disregarded, and the remaining ratios are averaged. (The average is actually computed as the antilogarithm of the average of the logarithm of the ratios.) This average is called the "darkness measure" or "adjustment factor" for the spot on the child. The adjusted size of the spot is the raw size divided by the darkness measure. For example, if the child's spots are all about 10% larger than the master's spots, then the darkness measure will be approximately 1.10, and the raw spot sizes will be divided by 1.10 to obtain the adjusted spot size values. In our study we selected 250 spots on the master gel that were ubiquitously expressed to serve as reference spots for the adjustment. The ratios of the reference spot sizes on the child image to the master image are computed. For each spot on the child image, the 10 closest neighboring reference spot ratios are considered. The two largest and two smallest ratios were discarded and the remaining six were averaged (antilogarithm of the average log ratio). The spot on the child image is then divided by this local darkness measure (Figure 11.6 and Figure 11.7). The resulting data can be exported into Excel office software for further analysis.

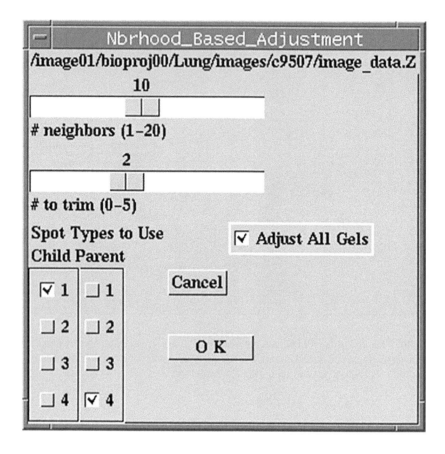

FIGURE 11.6 Parameters selected for neighboring based reference spot adjustment. Here "Parent" = "Master." We chose ten spots and trimmed two neighboring reference spots on master (parent) gel to adjust all "child" gels.

11.5.2 BATCH ADJUSTMENT

11.5.2.1 Why Do We Need Batch Adjustment?

Many commercial 2D PAGE systems are available that allow multiple gels to be run concurrently and that can therefore reduce the variation from run to run. The system we have utilized allows 20 gels to be run together at one time (one batch). This means that 100 samples will require five separate batches. The "Geled match software" also includes a function that could adjust for variations between gels based on neighboring reference spots, but there are also small differences that may occur between each batch (Figure 11.8). It is therefore necessary to perform a batch adjustment and normalize gel data prior to analysis for potential correlation between protein expression and other variables such as relationship to mRNA expression values or clinical outcome. We have performed this adjustment in our studies involving protein expression and variables such as lung adenocarcinoma stage, histological

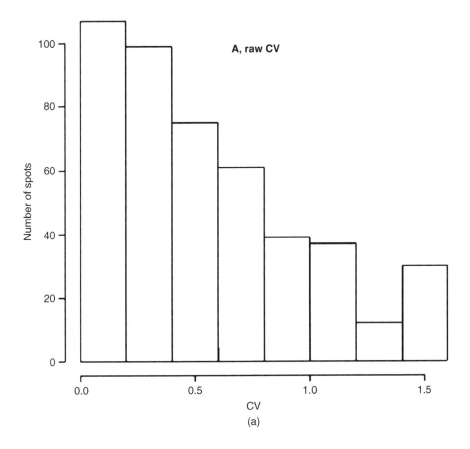

FIGURE 11.7 CV (coefficient of variation = SD/X) distribution of the raw and adjusted data from two gels (in two batches). (a) The values of all spots in one gel are always higher than another gel. (b) After neighboring based reference spot adjustment (Figure 11.6), the CV of most spots is still around 0 (between −1 and 1), but the values of all spots in one gel are not always higher than another gel.

subtype, and other clinical-pathological parameters using an ANOVA and F test rather than a simple t-test.

11.5.2.2 Procedure

First, missing data values were filled in with the spot means, and a log-transformation $Y = \log(1 + X)$ applied. For any spot let Y_{ijk} denote the log quantity of the protein that was present in sample k, where sample k was a member of batch i and belonged to tumor class j. We fit the additive linear models $Y_{ijk} = \mu + \beta_i + \gamma_k + \varepsilon_{ijk}$ to the data, where μ is the mean expression level for the protein, β_i is the influence of batch i, and Y_k is the influence of tumor class k. Standard linear model F tests were used to test for significance of the Y_k's. Cox proportional hazards models were fit to the values $Y_{ijk} - b_i$ where the b_i were the estimates of β_i from models including batch and stage (I or III) effects.

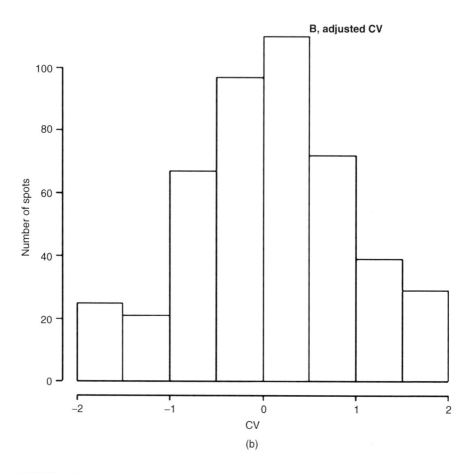

FIGURE 11.7 (Continued).

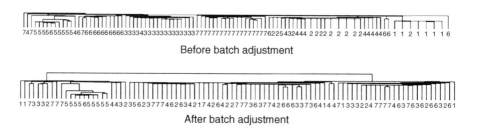

FIGURE 11.8 Cluster analysis showing the batch effect. The same number indicates the gels that are run together in the same batch. (*Top*) The same batch of gels are clustered together prior to batch adjustment. (*Bottom*) The same batch of gels are observed to be separated into different clusters after batch adjustment.

11.6 IDENTIFICATION, CONFIRMATION, AND VALIDATION OF THE PROTEINS OF INTEREST

11.6.1 MASS SPECTROMETRY

Protein spots that are of interest due to potential relationships to specific tumor-related or clinical variables are cut from preparative 2D gels using extracts from either lung cell lines or primary lung tissues for potential identification by mass spectrometry. The conditions utilized for preparative gels are identical to the analytical 2D gels except there is 30% greater protein loading. Following the run, these gels are stained with a modified silver-staining procedure incorporating successive incubations in 0.02% sodium thiosulfate for 2 min., 0.1% silver nitrate for 40 min., and 0.014% formaldehyde plus 2% sodium carbonate for 10 min. We have utilized two types of mass spectrometry for protein identification. The first method is matrix-assisted laser desorption ionization time-of-flight mass spectrometry (MALDI-TOF MS) using a Perspective Voyager Biospectrometry Workstation (Perseptive Biosystem, Framingham, MA), which provides a "fingerprint" for each spot based on the molecular weight of trypsin-digested products. The resulting masses of the products are then compared to known trypsin digests of a large number of proteins using the MSFit database (http://prospector.ucsf.edu/ucsfhtml3.2/msfit.htm).[11] Another method we utilized is nanoflow capillary liquid chromatography of the tryptic protein digests coupled with electrospray tandem mass spectrometry (ESI MS/MS) using a Q-TOF micro (Micromass, Manchester, U.K.). MS/MS spectra produced by ESI MS/MS are automatically processed and searched against a nonredundant database using ProteinLynx Global SERVER (http://www.micromass.co.uk/).[12] This latter technique is a more sensitive approach and can provide precise protein identification as the protein sequence can be determined (Figure 11.9).

11.6.2 VERIFICATION BY 2D WESTERN BLOT

Although mass spectrometry is used to identify the proteins of interest, we have also attempted when available, to utilize specific protein-directed antibodies to provide validation and to potentially identify other isoforms of the protein not previously appreciated in the given tissue or cell line. Protein extracts are run on 2D gels using the identical conditions as for analytical 2D gel preparations. All of the gel proteins are then transferred onto polyvinylidene fluoride membranes using standard electroblotting procedures. The blots are then incubated for 2 hours at room temperature with a blocking buffer consisting of TBST (Tris-buffered saline, 0.01% Tween 20) and 5% nonfat dry milk. Individual membranes are washed and incubated with primary antibody for 1 hour at room temperature. After additional washes with TBST, the membranes are incubated for 1 hour with a species-specific secondary antibody conjugated with horseradish peroxidase (HRP), usually at a 1:5000 dilution depending on the antibody. The membranes are carefully washed and incubated for 1 min. with ECL (enhanced chemiluminescence; Amersham, Piscataway, NJ) and exposed to x-ray film (Amersham) for variable times (10 seconds to 2 min) to obtain an ideal exposure of the protein of interest (Figure 11.10).

MALDI-TOF MS

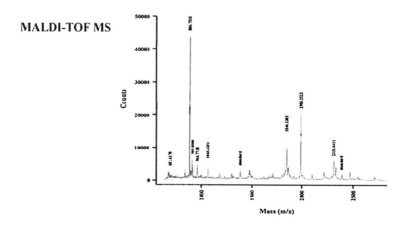

Mass (m/z)

ESI MS/MS

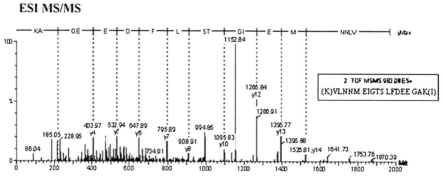

2 TOF MSMS 980.00 ES+
(K)VLNNM EIGTS LFDEE GAK(I)

FIGURE 11.9 Representative results of matrix-assisted laser desorption ionization time-of-flight mass spectrometry (MALDI-TOF MS) and tandem mass spectrometry (ESI MS/MS).

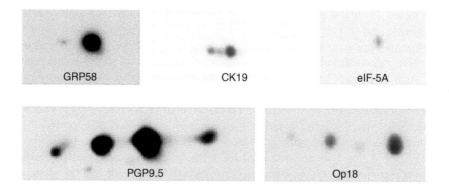

GRP58   CK19   eIF-5A

PGP9.5   Op18

FIGURE 11.10 2D Western blot of candidate proteins. Multiple isoforms of proteins are detected for most proteins using this technique.

11.6.3 TISSUE ARRAYS

2D PAGE and MS-based approaches can identify large numbers of proteins in a given tissue of interest. However, it is a significant challenge to then validate many of the best candidate proteins. An important aspect of the validation of protein candidates in addition to the 2D Western blot procedure described above is the use of tissue arrays containing representative tissue cores from hundreds of individual tumors. This method can provide significant clinical information that may aid in the evaluation of candidate proteins for potential use in the diagnosis, staging, and monitoring of the response to chemo or other therapies.[13] The most critical advantage is the ability to determine the protein's localization at the cell or tissue level. We have utilized tissue microarrays (TMA) constructed using triplicate tumor cores from representative formalin-fixed paraffin blocks of normal lung and lung tumors. After utilizing microwave antigen retrieval methods, the tissue sections made from these arrays are treated to remove endogenous peroxidase by incubation with 1% hydrogen peroxide for 60 min. at room temperature. Following blocking steps to reduce nonspecific binding, sections are incubated with primary antibodies for 1 hour at room temperature, washed, and incubated with the appropriate peroxidase-labeled secondary antibody. Following the final washes, the protein of interest is visualized with an avidin-biotin-based (ABC-kit, Vector Laboratories, Burlingame, CA) amplification method. The sections are lightly counterstained with hematoxylin and after permanent mounting they are independently scored by two collaborating study pathologists for the presence and level of staining (Figure 11.11). We are investigating the use of image-based methods for quantification of immunoreactive staining; however, use of a level-based method (0 = no staining, 1 = low, 2 = moderate, and 3 = high level) and the determination of the proportion of cells showing staining can provide useful data.

11.7 PRACTICAL APPLICATION OF PROTEIN EXPRESSION ANALYSIS OF 2D PAGE IN LUNG CANCER

11.7.1 REPRODUCIBILITY ANALYSIS

The reproducibility of the 2D separation process can be affected by a number of factors, including differences in sample preparation and loading, staining, and image acquisition. To accurately compare the quantity of any spot across a large number of sample gels, it is essential to compensate for these variations. It is widely accepted that multiple gels must be analyzed to reduce experimental and systematic errors.[14,15] To examine the reproducibility of 2D PAGE analyzed in our study, the same protein extracts from ten lung tumors and nine normal lung samples were run in two different batches and quantitatively analyzed. A comparison of the spot values for the same 820 spots in these samples revealed a correlation coefficient of 0.80 for the lung tumors (paired individual tumor range: 0.54 to 0.78), and 0.92 for the normal lung (paired individual normal lung range: 0.72 to 0.93). When all samples were compared a greater correlation was observed in the normal lung samples compared to the lung tumors, suggesting that the tumor samples have much greater variance between

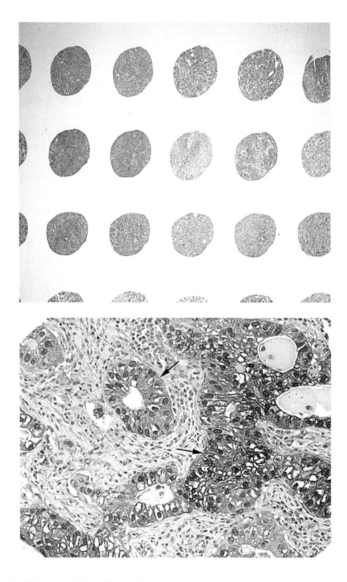

FIGURE 11.11 Immunohistochemical analyses using tissue arrays. Top panel shows an H&E stained region of the array containing sections of individual tumor samples, ×10; bottom panel shows abundant expression of a tumor marker in a lung adenocarcinoma (arrow) using immunohistochemistry, ×400.

individual protein spot values (Figure 11.12). We also analyzed one pair of gels (same tumor tissues but the protein extraction was performed separately and the gels run in two batches). The correlation coefficient was 0.75 for all 820 spots with CV (coefficient of variation) = −2 to 2. A strong correlation (r = 0.94) was found, however, among the 208 protein spots (no 0 value spots) whose CV was −0.4 to 0.4

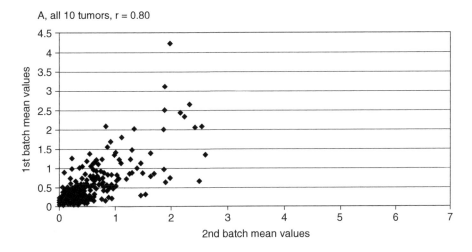

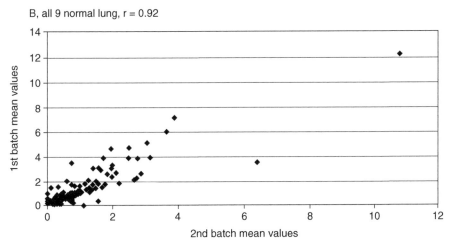

FIGURE 11.12 Correlation analysis examining the reproducibility of 2D gels for the analyses of the same 820 spots in lung tumors and normal lung samples. Mean values of each spot were used. (A) The correlation coefficient (r, Pearson) of ten lung adenocarcinomas is 0.8; (B) the correlation coefficient of nine normal lung tissues is 0.92.

(Figure 11.13). These results indicate that reasonable levels of reproducibility of protein expression values are obtained using this 2D PAGE–based approach.

11.7.2 IDENTIFICATION OF POTENTIAL LUNG TUMOR BIOMARKERS

Cancer remains a major public health challenge despite progress in detection and therapeutic intervention. Among the important tools critical to detection, diagnosis, treatment monitoring, and prognosis are biomarkers that have the potential to be utilized in patient tissues or body fluids. 2D PAGE technology allows the simultaneous

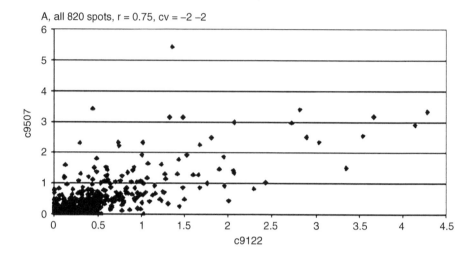

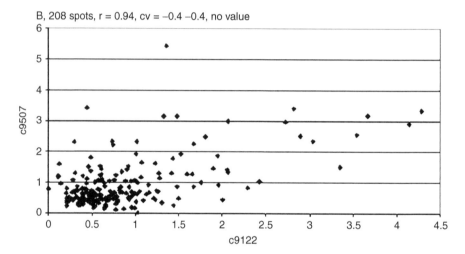

FIGURE 11.13 Correlation analysis of the reproducibility of 2D gels using one tumor sample. Two separate protein extractions were performed and the gels (c9122, c9507) run in two batches. (A) The correlation coefficient is 0.75 based on all 820 spots with a CV −2 to 2; (B) the correlation coefficient is 0.94 if 208 spots are chosen based on CV = −0.4 to 0.4 and containing none with a 0 value.

examination of hundreds of polypeptides in a given tissue sample. It has been widely used for the detection and identification of potential tumor markers from normal and malignant tissue.[5]

A total of 725 protein spots were analyzed in a series of 93 lung adenocarcinomas (64 stage I and 29 stage III) and 10 uninvolved lung samples (Table 11.1) for quantitative differences in protein expression between lung adenocarcinomas and normal lung using 2D PAGE. Of the 725 protein spots 392 were significantly

TABLE 11.1
Clinicopathologic Variables
in Lung Adenocarcinoma

Variables	n
Age:	
<65	49
>65	44
Gender:	
Female	53
Male	40
Smoking:	
Smoker	79
Nonsmoker	10
Stage:	
Stage I	64
Stage III	29
T status:	
T1	49
T2–T4	44
N status:	
N0	68
N1, N2	25
Classification:	
Bronchioalveolar	14
Bronchial-derived	76
Differentiation:	
Poor	23
Moderate	47
Well	22
Lymphocytic response:	
Yes	41
No	52
P53 nuclear accumulation:	
Positive	28
Negative	54
K-ras12th/13th codon mutation:	
Positive	36
Negative	40

different between these two types of samples (p value < .05) (Figure 11.14). Candidate proteins were identified using MALDI-TOF mass spectrometry or with tandem MS/MS. For those candidate proteins for which antibodies were available, confirmation using 2D Western blot was performed, and determination of the cellular localization of the proteins was examined with tissue array. The frequency of expression for each protein in the lung tumors was determined using a cutoff value defined

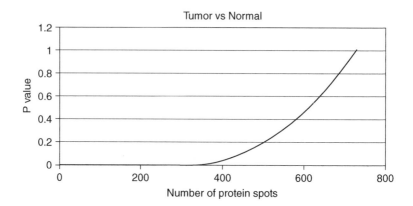

FIGURE 11.14 The distribution of *p* values of all 725 protein spots analyzed based on a comparison between tumor and normal lung using a *t*-test. In this comparison, 392 of 725 spots are observed with *p* value < 0.05.

as the mean value in the normal lung samples plus two standard deviations. Table 11.2 lists some of the candidate proteins demonstrating either high or low expression levels as well as the frequency of expression relative to normal lung. The examination of the association of specific isoforms of these proteins and clinical variables is one approach to help determine the potential of these proteins as biomarkers for lung cancer.

11.7.3 DISCOVERY OF SURVIVAL-RELATED PROTEINS

The identification of survival-related genes or their protein products is an important area of investigation as it allows either individual candidates or pathways potentially important to cancer development or progression to be determined. We have reported that gene-expression profiles can predict survival of patients with lung adenocarcinoma.[16] The use of a proteomic-based analytical approach also provides a powerful tool to discover survival-related proteins. For example, using 2D PAGE analysis, we reported that individual isoforms of cytokeratins (CK) were correlated with patient outcome.[8] We have also performed a detailed analysis of potential survival-related proteins in lung adenocarcinomas.[17] In this study, a total of 682 individual protein spots were quantified in 90 lung adenocarcinomas using quantitative 2D PAGE. Cox proportional hazards regression methods were used to investigate the relationship of protein expression values and patient survival or separate clinical-pathological variables. Using protein expression profiles, a risk index based on the top 20 survival-associated proteins identified using leave-one-out cross-validation from each set of 89 samples was used to categorize the remaining left-out sample as either low or high risk. The leave-one-out approach has been recently discussed[18] and is an appropriate method for these types of analyses. Significant differences in survival between stage I tumor patients categorized as low and high risk were observed (Figure 11.15). No differences were detected between survival of low- and high-risk stage III patients. Thirty-three

TABLE 11.2
Protein Expression Level in Lung Adenocarcinomas and Normal Lung Tissues

Spot#	Protein Name	p Value (T vs. N T test)	Fold Change Tumor/Normal	Increased Protein		Decreased Protein	
				Frequency[a] in Tumor (%)	Frequency[a] in Normal (%)	Frequency[b] in Tumor (%)	Frequency[b] in Normal (%)
1252	26S proteasome p28	<0.0001	NA	67.7	0.0		
1479	Amyloid B4A	<0.0001	40.5	60.2	10.0		
1193	Antioxidant enzyme AOE37-2	<0.0001	10.6	60.2	10.0		
1737	Calumenin precursor	<0.0001	4.3	81.7	10.0		
529	Cytokeratin 18	<0.0001	3.2	60.2	0.0		
446	Cytokeratin 8	<0.0001	5.5	50.5	0.0		
902	Cytosolic inorganic pyrophosphatase	<0.0001	7.6	54.8	0.0		
1354	F1F0-type ATP synthase subunit d	<0.0001	1.7	54.8	0.0		
1138	Glutathione-s-transferase M4 (GST m4)	<0.0001	4.0	96.8	0.0		
1229	Glyoxalase-1	<0.0001	NA	53.8	0.0		
1250	Hsp27	<0.0001	4.1	52.7	10.0		
1445	Huntingtin interacting protein 2 (HIP2)	<0.0001	22.1	76.3	10.0		
1060	MHC class II antigen	<0.0001	2.3	59.1	0.0		
1456	Nm23 (NDPKA)	<0.0001	58.4	88.2	10.0		
1064	Novel protein similar to zinc finger protein (CAC15900)	<0.0001	7.2	65.6	10.0		

(*Continued*)

TABLE 11.2
Protein Expression Level in Lung Adenocarcinomas and Normal Lung Tissues (Continued)

Spot#	Protein Name	p Value (T vs. NT test)	Fold Change Tumor/Normal	Increased Protein		Decreased Protein	
				Frequency[a] in Tumor (%)	Frequency[a] in Normal (%)	Frequency[b] in Tumor (%)	Frequency[b] in Normal (%)
1054	Nuclear chloride channel (RNCC protein)	<0.0001	NA	78.5	0.0		
1494	OP18 (Stathmin)	<0.0001	5.3	89.2	0.0		
350	Phospholipase C	<0.0001	2.9	52.7	10.0		
353	Protease disulfide isomerase	<0.0001	2.2	52.7	0.0		
1280	Pulmonary surfactant-associated protein (apoprotein)	<0.0001	5.2	78.5	0.0		
1161	Triose phosphate isomerase (TPI)	<0.0001	2.2	58.1	0.0		
1246	UCH-L1/PGP9.5	<0.0001	2.0	61.3	0.0		
594	Vimentin-derived protein (vid4)	<0.0001	13.4	61.3	10.0		
1181	Annexin varient I	<0.0001	0.2			77.4	0.0
780	Apolipoprotein J (ApoJ)	<0.0001	0.3			79.6	0.0
615	B-haptoglobin	<0.0001	0.4			51.6	0.0
1039	Cytoskeletal tropomyosin	<0.0001	0.5			88.2	0.0
1405	Ferritin light chain	<0.0001	0.4			61.3	0.0
866	Troponin T	<0.0001	0.3			78.5	0.0

[a] Cutoff value is mean value of normal lung + 2SD.

[b] Cutoff value is mean value of normal lung − 2SD.

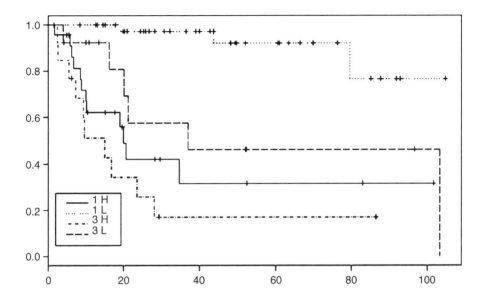

FIGURE 11.15 Kaplan-Meier survival plots showing the relationship between patient survival and the risk index based on the leave-one-out cross-validation procedure using the top 20 survival-associated proteins among the 62 stage I tumors and 28 stage III tumors. 1H, stage I high risk; 1L, stage I low risk; 3H, stage III high risk; 3L, stage III low risk.

survival-related proteins have been identified using matrix-assisted laser desorption/ionization mass spectrometry and tandem mass spectrometry. The expression of individual candidate proteins was confirmed using 2D Western blot and as tumor-derived using immunohistochemical analysis and tissue arrays.

11.7.4 PROTEIN–mRNA CORRELATION ANALYSIS

Gygi, et al. were the first study to perform a quantitative comparison of mRNA transcript and protein expression levels for a relatively large number of genes (128 genes) in yeast.[19] Relatively little is known, however, of the regulatory mechanisms controlling the complex patterns of protein abundance and posttranslational modification of proteins in human tumors. Most reports examining the regulation of protein translation have focused on one or several protein products.[20] By combining proteomic and genomic analyses of the same samples, however, a greater understanding may be possible of the complex mechanisms influencing protein expression in human cancer. It would be very useful to the gene array community to determine whether the extent of mRNA abundance is indeed predictive of the corresponding protein's abundance.

In a recent study performed by our group,[6] a quantitative comparison was made for mRNA and protein expression levels for a large number of genes (165 proteins representing 98 gene products) expressed in a large number of lung tissue samples (76 lung adenocarcinomas and 9 normal lung). We focused on two aspects: the number

of genes and the specific protein isoforms that showed a statistically significant correlation with their mRNA levels. These studies included many important quality
controls previously discussed in this chapter to assure the reliability of the data. We
also used the Spearman correlation coefficient analysis to identify potentially significant correlations between gene and protein expression. We used an analytical strategy
similar to SAM (Significance Analysis of Microarrays),[21] a permutation technique to
determine the significance of changes in gene expression between different biological
states. Correlation plots are necessary because it shows how many spots contribute the
correlation coefficient value (r). Twenty-eight of the 165 protein spots (17%) or 21 of
98 genes (21.4%) had a statistically significant correlation between protein and mRNA
expression ($r > 0.2445$; $p < .05$); however, among all 165 proteins, the correlation
coefficient values ranged from -0.467 to 0.442, indicating that some proteins are very
negatively to very positively correlated to each other.

We also tested the global relationship between mRNA and the corresponding
protein abundance across all 165 individual protein spots in the lung samples. Similar
analyses were examined by Gygi, et. al. using 106 genes in yeast.[19] Anderson, et al.
also examined these relationships but used only 19 genes in liver cell lines.[22] Because
protein abundance may affect these types of correlation analyses, we examined the
correlations among both low and high abundance proteins but found that the correlation coefficient values were not related to protein abundance. Further, no significant
correlation between mRNA and protein expression was found ($r = -0.025$) if the
average levels of mRNA or protein among all samples was applied across the 165
protein spots (98 genes). The analyses of mRNA/protein correlation coefficients also
indicated that significant variation is observed among proteins with multiple isoforms, suggesting separate isoform-specific mechanisms for the regulation of protein
abundance. The analysis of the mechanisms of protein expression is complex and
will require other techniques in addition to combining 2D PAGE and microarray data.

11.7.5 PROTEINS RELATED TO TUMOR DIFFERENTIATION

In our analyses of proteins expressed in lung adenocarcinomas, we have also obtained
information regarding how the expression of different proteins are related to other
clinical-pathological variables such as tumor stage, lymphocytic response, smoking
status, and tumor differentiation (Table 11.1). The relationship between alteration in
protein expression and tumor differentiation can provide insight into changes in important cellular pathways and are of significant interest. In a comparison of proteins that
were significantly different according to tumor differentiation status, we observed 102
out of 820 protein spots that showed a $p < .05$ (F test with batch adjustment) in 93 lung
adenocarcinomas. Of these, 40 spots were identified by MS (Table 11.3), including
stathmin, or oncoprotein 18 (Op18), which is believed to act as a relay for a variety of
cell signaling pathways. The phosphorylated and unphosphorylated Op18 protein isoforms were significantly increased in poorly differentiated tumors compared to moderately or well-differentiated lung adenocarcinomas.[23] This suggests that upregulation of
the expression of Op18 protein may reflect a poorly differentiated and a higher cell
proliferative status. This was further verified in A549 and SKLU1 lung adenocarcinoma
cell lines by examining overall Op18 protein levels and isoform phosphorylation status

TABLE 11.3
Protein Expression Correlated to Differentiation in Lung Adenocarcinomas

Spot No.	Protein Name	p Value[a]	Up/Down in Poor[a]
1874	Aldo keto reductase	0.0038	Up
1479	Amyloid B4A	0.0176	Up
1193	Antioxidant enzyme (AOE372)	0.0193	Up
325	Beta + alpha tubulin	0.0077	Up
1338	Clathrin light chain A	0.0015	Up
707	Crk	0.0309	Up
1124	Cytokeratin 1	0.0294	Up
514	Cytokcratin 18	0.0355	Up
352	Cytokeratin 8	0.0275	Up
470	Cytosol aminopeptidase	0.0081	Up
1527	eIF-5A	0.04	Up
2336	Glial fibrillary acidic protein, astrocyte	0.0031	Up
450	Glucose-6-phosphate 1-dehydrogenase variant A	0.015	Up
1036	HSPC089	0.0194	Up
1547	HSPC321	0.0059	Up
1445	Huntingtin interacting protein 2 (HIP2)	0.0047	Up
974	IGFBP3	0.031	Up
1728	L-FABP	0.0006	Up
1064	Novel protein similar to zinc finger protein (CAC15900)	0.0401	Up
1492	OP18 (Stathmin)	0.0364	Up
459	Pyruvate kinase M1/M2	0.0009	Up
1155	VAMP-associated 33 kDa protein	0.0028	Up
511	Vimentin	0.0234	Up
2503	Aldehyde dehydrogenase	0.0027	Down
1034	Alternative splicing factor-associated 32 kDa chain	0.0014	Down
760	Annexin I	0.0166	Down
994	Annexin IV	0.0167	Down
963	Annexin V	0.0107	Down
1083	Cathepsin L	0.0074	Down
609	Cytokeratin 19	0.0034	Down
1405	Ferritin light chain	0.0178	Down
989	HLA-Cw5	0.0139	Down
1253	Hypothetical protein KIAA0053	0.0379	Down
891	Microsomal epoxide hydrolase	0.0311	Down
867	PCNA	0.0171	Down
1583	R33729 1	0.0203	Down
460	Selenium-binding protein 1	0.001	Down
1190	Serum amyloid p-component precursor [precursor]	0.0236	Down
278	T-complex protein I, alpha subunit	0.0134	Down

[a] p Value of F test in tumor differentiation. Poor, poorly-differentiated tumor.

following treatments that alter either cell proliferation or differentiation. The observed overexpression of Op18 protein in poorly differentiated lung adenocarcinomas and increase of the phosphorylated forms of Op18 may have potential utility as a tumor marker or as a new target for drug or gene-directed therapy.

An additional protein of interest identified in these analyses is the eukaryotic initiation factor 5A (eIF-5A), the only cellular protein known to contain the unusual amino acid hypusine, a modification that appears to be required for cell proliferation. Higher levels of eIF-5A protein expression are present in tumors showing poor differentiation, 12/13th codon K-*ras* mutations, p53 nuclear accumulation, or a positive lymphocytic response. Patients having a higher level of eIF-5A protein showed a relatively poor survival, suggesting targeted inhibition of eIF-5A expression may be potentially beneficial for improving patient survival with lung adenocarcinoma.[24] We also found that CRK protein was increased in poorly differentiated tumors.[25]

Proteins showing a decreased expression in poorly differentiated tumors relative to well-differentiated tumors were also observed (Table 11.3), including selenium-binding protein 1 (SBP1). It has been proposed that the effects of selenium in preventing cancer and neurologic diseases may be mediated by this protein, thus the loss of expression of proteins like SBP1 may reflect loss of normal regulatory pathways in lung adenocarcinomas.

11.8 CONCLUSIONS

The use of 2D PAGE and mass spectrometry is a powerful tool to identify proteins of interest in a cell or tissue type of interest.[26] Analyses of a large number of individual proteins in a large set of tissue samples raise a number of important issues that must be understood to obtain accurate measures of protein expression. By applying appropriate methods that allow for corrections for slight gel variations, the 2D PAGE approach described in this chapter can be successfully applied to clinical samples for the analysis of protein expression in human lung adenocarcinomas and associated normal lung tissue.

ACKNOWLEDGMENT

The authors acknowledge our many collaborators who have provided invaluable contributions to the studies we describe, and especially Rork Kuick for his comments on portions of this chapter.

REFERENCES

1. Wu, W., Hu, W., and Kavanagh, J.J. Proteomics in cancer research. *Int. J. Gynecol. Canc.,* 12, 409–423, 2002.
2. Srinivas, P.R., Verma, M., Zhao, Y., and Srivastava, S. Proteomics for cancer biomarker discovery. *Clin. Chem.,* 48, 1160–1169, 2002.
3. Herbert, B. Advances in protein solubilisation for two-dimensional electrophoresis. *Electrophoresis,* 20, 660–663, 1999.

4. Banks, R.E., Dunn, M.J., Forbes, M.A., Stanley, A., Pappin, D., Naven, T., Gough, M., Harnden, P., and Selby, P.J. The potential use of laser capture microdissection to selectively obtain distinct populations of cells for proteomic analysis— Preliminary findings. *Electrophoresis,* 20, 689–700, 1999.

5. Chen, G., Gharib, T.G., Huang, C.C., Thomas, D.G., Shedden, K.A., Taylor, J.M., Kardia, S.L., Misek, D.E., Giordano, T.J., Iannettoni, M.D., Orringer, M.B., Hanash, S.M., and Beer, D.G. Proteomic analysis of lung adenocarcinoma: Identification of a highly expressed set of proteins in tumors. *Clin. Cancer Res.,* 8, 2298–2305, 2002.

6. Chen, G., Gharib, T.G., Huang, C.C., Taylor, J.M., Misek, D.E., Kardia, S.L., Giordano, T.J., Iannettoni, M.D., Orringer, M.B., Hanash, S.M., and Beer, D.G. Discordant protein and mRNA expression in lung adenocarcinomas. *Mol. Cell. Proteomics,* 1, 304–313, 2002.

7. Strahler, J.R., Kuick, R., and Hanash, S.M. Two-dimensional gelelectrophoresis of proteins. In *Protein Structure: A Practical Approach,* Creighton, T., Ed. IRL, Oxford, pp. 65–92, 1989.

8. Gharib, T.G., Chen, G., Wang, H., Huang, C.C., Prescott, M.S., Shedden, K., Misek, D.E., Thomas, D.G., Giordano, T.J., Taylor, J.M., Kardia, S., Yee, J., Orringer, M.B., Hanash, S., and Beer, D.G. Proteomic analysis of cytokeratin isoforms uncovers association with survival in lung adenocarcinoma. *Neoplasia,* 4, 440–448, 2002.

9. Kuick, R.D., Skolnick, M.M., Hanash, S.M., and Neel, J.V. A two-dimensional electrophoresis-related laboratory information processing system: Spot matching. *Electrophoresis,* 12, 736–746, 1991.

10. Asakawa, J., Kuick, R., Neel, J.V., Kodaira, M., Satoh, C., and Hanash, S.M. Genetic variation detected by quantitative analysis of end-labeled genomic DNA fragments. *Proc. Natl. Acad. Sci. USA,* 91, 9052–9056, 1994.

11. Bonk, T. and Humeny, A. MALDI-TOF-MS analysis of protein and DNA. *Neuroscientist,* 7, 6–12, 2001.

12. Reid, G.E. and McLuckey, S.A. "Top down" protein characterization via tandem mass spectrometry. *J. Mass Spectrom.,* 37, 663–675, 2002.

13. Skacel, M., Skilton, B., Pettay, J.D., and Tubbs, R.R. Tissue microarrays: A powerful tool for high-throughput analysis of clinical specimens: A review of the method with validation data. *Appl. Immunohistochem. Mol. Morphol.,* 10, 1–6, 2002.

14. Mahon, P. and Dupree, P. Quantitative and reproducible two-dimensional gel analysis using Phoretix 2D Full. *Electrophoresis,* 22, 2075–2085, 2001.

15. Blomberg, A., Blomberg, L., Norbeck, J., Fey, S.J., Larsen, P.M., Larsen, M., Roepstorff, P., Degand, H., Boutry, M., Posch, A., et al. Interlaboratory reproducibility of yeast protein patterns analyzed by immobilized pH gradient two-dimensional gel electrophoresis. *Electrophoresis,* 16, 1935–1945, 1995.

16. Beer, D.G., Kardia, S.L., Huang, C.C., Giordano, T.J., Levin, A.M., Misek, D.E., Lin, L., Chen, G., Gharib, T.G., Thomas, D.G., Lizyness, M.L., Kuick, R., Hayasaka, S., Taylor, J.M., Iannettoni, M.D., Orringer, M.B., and Hanash, S. Gene-expression profiles predict survival of patients with lung adenocarcinoma. *Nat. Med.,* 8, 816–824, 2002.

17. Chen, G. Gharib, T.G., Wang, H., Huang, C.C., Thomas, D.G., Shedden, K.A., Misek, D.E., Taylor, I.M.G. Giordano, T.J., Iannettoni, M.D., Orringer, M.B., Hanash, S., Beer, D.G. Protein expression profiles predict patient survival in stage I lung adenocarcinoma. *Proc. Natl. Acad. Sci* (USA) 100:13537–13542, 2003.

18. Simon, R., Radmacher, Dobbin, K., and McShane, L.M. Pitfalls in the use of DNA microarray data for diagnostic and prognostic classification. *J. Natl. Canc. Inst.*, 95, 14–18, 2003.

19. Gygi, S.P., Rochon, Y., Franza, B.R., and Aebersold, R. Correlation between protein and mRNA abundance in yeast. *Mol. Cell Biol.*, 19, 1720–1730, 1999.

20. Tew, K.D., Monks, A., Barone, L., Rosser, D., Akerman, G., Montali, J.A., Wheatley, J.B., and Schmidt, D.E., Jr. Glutathione-associated enzymes in the human cell lines of the National Cancer Institute drug screening program. *Mol. Pharmacol.*, 50, 149–159, 1996.

21. Tusher, V.G., Tibshirani, R., and Chu, G. Significance analysis of microarrays applied to the ionizing radiation response. *Proc. Natl. Acad. Sci. USA*, 98, 5116–5121, 2001.

22. Anderson, L. and Seilhamer, J.A. comparison of selected mRNA and protein abundances in human liver. *Electrophoresis*, 18, 533–537, 1997.

23. Chen, G., Wang, H., Gharib, T.G., Huang, C.C., Thomas, D.G., Shedden, K.A., Kuick, R., Taylor, J.M., Kardia, S.L., Misek, D.E., Giordano, T.J., Iannettoni, M.D., Orringer, M.B., Hanash, S.M., and Beer, D.G. Overexpression of oncoprotein 18 correlates with poor differentiation in lung adenocarcinomas. *Mol. Cell Proteomics*, 2, 107–116, 2003.

24. Chen, G., Wang, H., Gharib, T.G., Thomas, D.G., Huang, C.C., Misek, D.E., Kuick, R., Giordano, T.J., Iannettoni, M.D., Orringer, M.B., Hanash, S.M., and Beer, D.G. Proteomic analysis of eIF-5A in lung adenocarcinomas. *Proteomics*, 2, 107–116, 2003.

25. Miller, C.T., Chen, G., Gharib, T.G., Wang, H., Thomas, D.G., Misek, D.E., Giordano, T.J., Yee, J., Orringer, M.B., Hanash, S.M., and Beer, D.G. Increased C-CRK proto-oncogene expression is associated with an aggressive phenotype in lung adenocarcinomas. *Oncogene*, 3, 11–97, 2003.

26. Westermeier, R. and Naven, T. *Proteomics in Practice.* Wiley-vch, Germany, 2002.

12 Nonparametric, Distance-Based, Supervised Protein Array Analysis

Mei-Fen Yeh, Jeanne Kowalski, Nicole White, and Zhen Zhang

CONTENTS

12.1 INTRODUCTION

With proteomic and other high-throughput genomic profiling technologies, such as microarrays that allow for the simultaneous analysis of expression levels of thousands or even tens of thousands of biomarkers, we are able to expand our ability to characterize and understand disease processes at the molecular level and the heterogeneity surrounding them. As advances in biotechnology continue to support this remarkable expansion, however, the need for extracting and synthesizing information from the volumes of expression data has created an equally challenging research area in the development of corresponding statistical methods for their analysis. Within the context of biomedical and clinical research, a common objective of proteomic data analysis is the selection of biomarkers, from among the thousands profiled, that characterize groups of similar phenotypes based on a small number of samples.

Cluster analysis (CA)[1] is a technique used to identify samples with similar genetic patterns. The application of CA to gene expression data is based on the biologic premise that biomarkers displaying similar expression patterns may be coregulated and share a common function or contribute to a common pathway. Because CA is a technique used to identify samples (or biomarkers) with similar intensities based on a metric that is irrespective of phenotype, its use requires secondary analyses to describe samples (biomarkers) that characterize the clusters formed and the differences among them. A related approach, recursive partitioning (RP)[2] is a technique used to identify subsets of biomarkers that explain most of the variability in some (continuous) phenotypic response, but the actual phenotypic differences between subsets is not apparent and, similar to CA, requires further exploratory analyses for this purpose. A pertinent issue related to both CA and RP is the subjective defining of the number of sample clusters, with no reference to defined population clusters. As primarily a descriptive rather than inferential tool, CA has enticed the development of many ad hoc approaches toward statistical inference. By instead conditioning upon samples of similar phenotypes, linear discriminant analysis (LDA)[3] is a tool used to identify (linear) combinations of variables (peak heights) that best predict phenotypic group membership. Unlike the dependent observations that may be formed by the use of distance pairs, such as in CA, LDA is similar to RP in that it requires the use of independent observations. With LDA, the observations within each group are typically assumed to belong to a multivariate normal distribution that is characterized by different means and a similar covariance matrix that can be estimated by pooled samples. These assumptions, when they are met, guarantee that the LDA will be the optimal predictive model with an error rate approaching the lowest possible rate (Bayes' error rate) as the number of training samples increases. However, restrictions such as independence, distributional assumptions and equal second (and higher) moments, altogether, preclude the usefulness of LDA and its derivatives as data analytic tools for studying differential molecular heterogeneity among groups, since it is quite plausible that the two groups differ in terms of higher moments, such as skewness and kurtosis.

This latter consideration is especially important when working with genomic data of complex, high-dimensional structure. Support vector machines (SVM) is a supervised data modeling and classification tool.[4] With its structural risk-minimization learning algorithm, SVM circumvents the need for explicit estimation of data distributions and has become a popular tool for array (microarray) analysis for its ability to handle high-dimensional data.[5,6] As a supervised classification model, SVM can be used to identify linear or nonlinear (with the use of kernel functions) combinations of the variables (biomarkers) to best predict phenotypic group memberships and provide the basis for gene discovery.[7]

It is often clinically desirable to be able to simultaneously query changes in expression patterns of tens of thousands of biomarkers. However, when the expression levels of these biomarkers are regarded as variables, the individual array experiments, or observations, become data points in an extremely high dimensional space. The convergence of any estimator to the true value of a smooth function defined in such a space where $p \gg n$, will be very slow, as is often reflected in the expression of "the curse of dimensionality." The availability of clinical samples, and the cost and effort associated with proteomics and microarray experiments, dictate that, for the time being, the majority of large-scale genomic expression analysis studies will not have a statistically sufficient number of observations to allow for a "good" estimate of a function of the biomarkers that identifies, for example, an altered expression pattern associated with a specific tumor type. All in all, this is a pretty dim scenario from all perspectives.

Fortunately, however, we may reasonably be able to hope that in many cases, there are really "a few things that matter," and thus the function of interest is expected to be constant along most dimensions of the space. This opens up the possibility of conducting statistical analyses in a meaningful and novel way, and, in particular, motivates the potential for an inference framework.

Despite the numerous clustering and discriminating algorithms available, there remains a lack of a formal framework in which to conduct analysis of data from proteomic or microarray experiments that in particular is flexible in terms of analytical assumptions. In some cases, array analysis involves the combining of CA and LDA features into a unified framework. For example, the high dimensionality of the problem may be approached by the construction of a single composite measure that summarizes information among the dimensions into a single statistic, or perhaps through a few dimensions, such as in principal components analysis, followed by the characterization of genomic differences based upon this composite measure, such as in LDA. We focus upon a recent development in the areas of dimension reduction and discrimination that utilizes a general distance measure to characterize genomic and proteomic heterogeneity. In particular, we describe a novel approach to nonparametric inference for high-dimensional comparisons involving two or more groups, based on a few samples and very few replicates from within each group. In this way, we extend traditional cluster and linear discriminant methods to address current as well as future challenging analyses settings presented by the continued and more creative use of microarray and proteomic technology, apart from the initial (two-sample) cancerous versus non-cancerous comparisons.

12.2 NONPARAMETRIC METHOD FOR SUPERVISED PROTEOMIC ANALYSIS

By comparing and characterizing genomic heterogeneity among defined groups, within a nonparametric framework, we introduce in this section some recently developed methods for analyses of high-dimensional data, within the context of a supervised proteomic analysis. In particular, we discuss a nonparametric inference approach for comparisons of two or more groups, based on a few, or as little as a single, sample from within each group. To illustrate the method, we focus upon the simple setting of two groups, each with a few samples from within each group. In this context, we discuss a distance-based approach to analysis that requires the construction of a composite measure of proteomic heterogeneity within and between each group to formulate hypotheses.

12.2.1 CONSTRUCTION OF REPLICATE-COMPOSITE MEASURE

We first construct a measure that collapses available replicate information on peak heights from within each individual by implementing an algorithm that applies singular value decomposition (SVD). This method in particular, accommodates data with varying dimension, due to presence of differing amount of available replicate information among samples. To this end, let z_i denote a $P \times R_i$ matrix of replicate peak heights on P peaks for an ith subject, and R_i replicates for $1 \leq i \leq n$ and $1 \leq k \leq R_i$. By applying SVD to each z_i matrix as a way of combining replicate information into a single, composite statistic, we reduce the information contained in z_i into a single, P-dimensional column vector. In particular, we define $z_i = P_i \Lambda_i Q_i^T$, where

$$P_i = [\mathbf{p}_{i1}, \dots, \mathbf{p}_{iR_i}], \quad \mathbf{p}_{ik} = (\mathbf{p}_{i,1k}, \dots, \mathbf{p}_{i,Pk})^T, \quad Q_i^T = (\mathbf{q}_{i1}, \dots, \mathbf{q}_{iR_i}); \ \mathbf{q}_{ik}^T = (q_{i,k1}, \dots, q_{i,kRi})$$

such that $\mathbf{p}_{ik} \in \mathbf{R}^P$ and $\mathbf{q}_{ik} \in \mathbf{R}^R$ are orthonormal vectors and Λ is a diagonal matrix of ordered decreasing eigenvalues of dimension R_i. We summarize in a nonparametric framework the information obtained in z_i into P-dimensional column vectors through the statistic, $\tilde{\mathbf{z}}_i = (\lambda_1 \bar{q}_{1.})\mathbf{p}_1$, where $\bar{q}_{1.} = R_i^{-1} \sum_r q_{i,1r}$. Hereafter, unless otherwise stated, denote by z_i a replicate-composite measure of peak heights for an ith subject.

12.2.2 RELATING PROTEOME HETEROGENEITY TO PATHOLOGICAL DISEASE STATES

In this section, we highlight recent work,[8,9] for relating genetic heterogeneity to phenotypes, within a proteomic context. This approach may be viewed as nonparametric inference for discriminant analysis in the sense that groups are conditioned upon and the differences between them, in terms of the degree of heterogeneity, are formally compared through hypothesis testing. Similar to principal components analysis, the dimension of the problem is reduced by collapsing information contained in all peaks into a single composite measure to facilitate hypothesis testing.

Let G (indexed by g) denote the number of mass spectrometry experimental proteins arrayed, i.e., the number of peaks. Suppose that the n subjects are classified into 2 (distinct) pathologic conditions (normal versus cancerous), h and k, with sample sizes n_h and n_k, respectively. Denote by $\mathbf{z}_j = (z_{1j}, \dots, z_{Gj})^T$ the G-dimensional vector of (replicate-composite) intensities obtained from G peaks. Let $\mathbf{z}_h$ refer to peak heights obtained from the reference (control) tissue that is to be compared against peak heights obtained from the target tissue, $\mathbf{z}_k$. We propose the following algorithm, applied to individual, replicate-composite peak heights for hypothesis testing of equal degrees of proteome heterogeneity within and between the two distinct pathological disease states:

1. Construct a composite measure of peak height heterogeneity within the reference, within the comparison condition, and between the reference and

comparison conditions, respectively, among all peaks, by

$$d^2_{ii';hh}(\mathbf{z}_{hi}, \mathbf{z}_{hi'}; \alpha) = (\mathbf{z}_{hi} - \mathbf{z}_{hi'})^{\mathsf{T}} W(\alpha)(\mathbf{z}_{hi} - \mathbf{z}_{hi'}), \quad 1 \le i < i' \le n \quad (12.1)$$

$$d^2_{jj';kk}(\mathbf{z}_{kj}, \mathbf{z}_{kj'}; \alpha) = (\mathbf{z}_{kj} - \mathbf{z}_{kj'})^{\mathsf{T}} W(\alpha)(\mathbf{z}_{kj} - \mathbf{z}_{kj'}), \quad 1 \le j < j' \le m$$

$$d^2_{ij;hk}(\mathbf{z}_{hi}, \mathbf{z}_{kj}; \alpha) = (\mathbf{z}_{hi} - \mathbf{z}_{kj})^{\mathsf{T}} W(\alpha)(\mathbf{z}_{hi} - \mathbf{z}_{kj}), \quad 1 \le i \le n, \quad 1 \le j \le m$$

where W denotes a symmetric, nonnegative definite $G \times G$ weighting matrix, characterized by an $n_\alpha \times 1$ parameter vector α.

2. Using the above distances, we form matrices and their corresponding distributions to formulate hypotheses of equal degrees of heterogeneity within and between conditions. To this end, consider the following distance matrices, conditional on each pathologic state:

Pathologic States	Reference (h)	Comparison (k)
Reference (h)	$D_{hh} = [d_{ii';hh} = d_{ii'}(\mathbf{z}_{hi}, \mathbf{z}_{hi'})]$	$D_{hk} = [d_{ij;hk} = d_{ij}(\mathbf{z}_{hi}, \mathbf{z}_{kj})]$
Comparison (k)	$D_{kh} = [d_{ji;kh} = d_{ji}(\mathbf{z}_{kj}, \mathbf{z}_{hi})]$	$D_{kk} = [d_{jj';kk} = d_{jj'}(\mathbf{z}_{kj}, \mathbf{z}_{kj'})]$

The within-condition matrices, D_{hh} and D_{kk}, are symmetric with zero on the diagonal, while the between-condition distance matrices, D_{hk} for $h \ne k$ is nonsymmetric with a nonzero diagonal. The distribution of each of these stochastic distance matrices, D_{hk}, provides information for testing peak heterogeneity within and between pathologic states.

3. Let $F_{hk}(d) = \Pr(d_{ij;hk'} \le d)$ denote the cumulative distribution function (cdf) based on D_{hk}, where d is a mass point among the elements of D_{hk}. Let $\{d_{\min}, \ldots, d_{\max}\}$ denote the ordered distinct mass points, such that $F_{hk}(d_{\min}) = 0$ and $F_{hk}(d_{\max}) = 1$, respectively. Since $d_{ij;hk}$ is a discrete random variable, the set, $\{d_{\min}, \ldots, d_{\max}\}$ is finite. For a given distance, d, $F_{hk}(d)$ refers to the within-group ($h = k$) and between-group ($h \ne k$) distance distributions.

4. For a given sample, a nonparametric estimator of $F_{hk}(d)$ is defined through an empirical cdf (ecdf), as given below

$$\hat{F}_{hk}(d) = \begin{cases} \binom{n_h}{2}^{-1} \sum_{(i,j) \in C_h} I_{\{d_{ij \cdot hh} \le d\}} & \text{for} \quad h = k \\[2ex] \prod_{b=h,k} \binom{n_b}{1}^{-1} \sum_{i,j} I_{\{d_{ij \cdot hk} \le d\}} & \text{for} \quad h \ne k \end{cases} \quad (12.2)$$

where $\binom{n}{k}$ is the binomial coefficient, C_h denotes the set of all distinct (i, j) pairs with $1 \le i, j \le n_h$, and $I_{\{x \le d\}}$ is a binary indicator with the value 1 if $x \le d$, and 0 otherwise. For notational convenience, we have suppressed the dependency of $\hat{F}_{hk}$ on α. By applying the theory of U-statistics, $\hat{F}_{hk}(d)$ has been shown to be consistent and asymptotically normal.[9] Regardless of whether α is treated as known values (e.g., W is an identity matrix) or estimates (e.g., $W_2 = \hat{\Sigma}_d^-$), an estimate of the asymptotic variance of $\hat{F}_{hk}(d)$, which may be readily calculated. To eliminate the dependency of these statistics on d, let $\hat{\mathbf{F}}_{hk} = (\hat{F}_{hk}(d_{min}),\ldots,\hat{F}_{hk}(d_{max}))^{\mathsf{T}}$ denote an estimator of $\mathbf{F}_{hk} = (F_{hk}(d_{min}),\ldots,F_{hk}(d_{max}))^{\mathsf{T}}$. The vector statistic, $\mathbf{F}_{hk}$, is well defined, since $d_{ij;hk}$ is a discrete random variable and in addition, it may also be used to characterize within ($h = k$) and between ($h \neq k$) group differences.

5. A probability "density" function (pdf) may be defined based on the first-order differences, $f_{hk}(d) = F_{hk}(d+1) - F_{hk}(d)$. By substitution of the respective ecdfs into this expression, i.e., $\hat{f}_{hk}(d) = \hat{F}_{hk}(d+1) - \hat{F}_{hk}(d)$, we obtain the vector statistic based on the density function, $\hat{\mathbf{f}}_{hk}$. Although equivalent in terms of statistical inference, the density statistics are preferable for visually displaying within and between group heterogeneity.

6. Construct the following hypotheses to formally compare heterogeneity within and between groups:

$$H_{01} : \mathbf{F}_{hh} = \mathbf{F}_{kk}, \quad H_{02} : \mathbf{F}_{hk} = \mathbf{F}_{hh}, \quad H_{03} : \mathbf{F}_{hk} = \mathbf{F}_{kk} \qquad (12.3)$$

The first hypothesis, H_{01}, tests equality in the degree of peak heterogeneity between the two groups, while H_{02} (H_{03}) examines such equality in the combined group versus each of the other groups. In comparison to the other two hypotheses, H_{01} tests for differential within-group heterogeneity between the reference and comparison tissue groups. For example, if the kth group is more homogenous than the hth group, then $d_{ij;kk}$ will be smaller, on average, relative to $d_{ij;hh}$ and thus, constitute a difference in their distributions. However, since the distance measure $d_{ij;kk}$ only discriminates between different types of heterogeneity, but not heterogeneity of the same type, the hypothesis H_{01} does not test for a group difference when the same degree of within-group heterogeneity is characterized differently between the two groups. In such cases, H_{02} and/or H_{03} may be used to test for homogeneity within groups that is characterized differently between them, since the between-group distances will be larger, on average, relative to the within-group distances.

7. To test each hypothesis in Eq. (2), define the following difference statistics:

$$\hat{\delta}_j^2 = \hat{\Delta}_j^{\mathsf{T}} W_{3j} \hat{\Delta}_j, \quad \hat{\Delta}_1 = \hat{\mathbf{F}}_{hh} - \hat{\mathbf{F}}_{kk}, \quad \hat{\Delta}_2 = \hat{\mathbf{F}}_{hk} - \hat{\mathbf{F}}_{hh},$$

$$\hat{\Delta}_3 = \hat{\mathbf{F}}_{hh} - \hat{\mathbf{F}}_{kk}, \quad j = 1, 2, 3, \qquad (12.4)$$

where W_{3j} denotes a symmetric, nonnegative $n_d \times n_d$ weighting matrix. In the simplest case, for $W_{3j} = I_{n_d}$, $\hat{\delta}_j^2$ is the squared Euclidean distance. Each of the difference statistics, $\hat{\Delta}_j$, have been shown to be asymptotically normally distributed[9] (in which case, $\hat{\delta}_j^2$ is asymptotically χ^2 under H_{0j}, provided that a consistent estimate of the inverse [or generalized inverse] $\Sigma_{\hat{\Delta}_j}$ of the asymptotic variance is used as the weighting matrix for each W_{3j}). Inferences may also be based on a Monte Carlo approximation to the permutation distribution of $\hat{\delta}_j^2$ under each null hypothesis.[10,11] Note that the hypotheses in Eq. (12.3) may be equivalently expressed in terms of pdfs, with the epdf statistics used for testing them. Letting $W_{3j} = \Sigma_{\hat{\Delta}_j}$ enables both asymptotic and exact inference methods to be used in obtaining p values for each respective hypothesis, where for the exact methods, a Monte Carlo approximation may be implemented.

12.2.3 Translating Observed Proteome Heterogeneity to Peak Selection

The distance measure and associated hypotheses based upon it enabled a reduction in the high dimension typically associated with proteomic analyses. In addition to formal hypothesis testing of differential heterogeneity between groups, it is typically of interest to determine the peaks that attribute to such observed differences in heterogeneity. The goal here is similar to an analysis of variance, in which statistical significance from an all-group comparison is typically followed by various subgroup, pairwise analyses.

Our approach to this problem is based upon partitioning the multivariate test statistic, $\hat{\delta}_j^2$ ($j = 1, 2, 3$), into an additive sum of L terms, with the following properties:

1. Each term is positive and thus its magnitude estimates the relative contribution to the underlying test statistic.
2. The relative magnitude of each term estimates the degree of heterogeneity among locations. For ease of exposition, we break down the process into the following general steps. In each step, a re-expression of a statistic is carefully constructed with emphasis on the properties contained in each re-expression for latter-use interpretation of results.
   a. For a given pair of peaks, we re-express the distance measure, d_{ij}^2, into a (nonnegative) linear sum of L terms involving the peak distances by:

$$d_{ij;hk}^2 = \mathbf{d}_{ij,hk} W_2 \mathbf{d}_{ij,hk} = \left(W_2^{\frac{1}{2}} \mathbf{d}_{ij,hk} \right) \left(W_2^{\frac{1}{2}} \mathbf{d}_{ij,hk} \right) = \sum_{l=1}^{L} (\mathbf{w}_{2l} \mathbf{d}_{ij,hk})^2 = \sum_{l=1}^{L} \xi_{ijl,hk}^2,$$

$$(12.5)$$

where $\mathbf{w}_{2l}$ is the lth row of $W_2^{1/2}$.

For $W_2 = I_L$, d_{ijl}^2 (or $\rho_l = d_{ijl}^2/d_{ij}^2$) is interpreted as the contribution (proportionate contribution) from the lth location to d_{ij}^2. For $W_2 \neq I_L$, let $\xi_{ij} = W_2^{1/2} \mathbf{d}_{ij}$, where $W_2^{1/2}$ is the square-root matrix.[12] In this case, $\xi_{ij,l}^2$ is associated with the contribution from groups of individual locations by interpreting the rows of $W_2^{1/2}$ as "factor-type loadings" in the sense that their (absolute) values indicate the contribution from each of several locations to $\xi_{ij,l}^2$, depending upon the structure of W_2.
   b. Each egrcdf is expressed as a function of $\xi_{ijl,hk}$. For simplicity, we focus on the within-group case ($h = k$), with the results readily applied to the between-group case. It follows from Eq. (12.2) that

$$\hat{F}_{hh}\left(\xi_{ijl,hh}^2; d_m\right) = \sum_{l=1}^{L} \binom{n_h}{2}^{-1} \sum_{i<j} \frac{\xi_{ijl,hh}^2}{d_{ij\cdot,hh}^2} I_{\{d_{ij\cdot,hh} \leq d_m\}} = \sum_{l=1}^{L} \rho_{l,hh}(d_m) \quad (12.6)$$

By partitioning each $I_{\{d_{ij;hh} \leq d\}}$ into the sum of $\frac{\xi^2_{ijl,hh}}{d^2_{ij;hh}} I_{\{d_{ij;hh} \leq d_m\}}$, $\rho_{l,hh}(d_m)$ not only indicates the relative contribution from L terms, but also expresses them in terms of proportionate contributions.

c. In this step, we readily express the peaks difference statistic as a linear combination of L terms. Without loss of generality, consider the statistic Δ_1. It follows from Eq. (12.4) that

$$\Delta_1(d_m) = \sum_{l=1}^{L} [\rho_{l,SS}(d_m) - \rho_{l,RR}(d_m)] = \sum_{l=1}^{L} \Delta_{1l}(d_m), \quad d_m = d_{min}, ..., d_{max}$$

$$(12.7)$$

If the $\Delta_{1l}(d)$ are all positive, then their magnitudes indicate the relative contributions to $\Delta_{1l}(d)$. In general, the $\Delta_{1l}(d)$ may involve different (mixed) signs; thus, this statistic can be used to identify the over- or underexpressed peaks under different pathologic states.

12.3 DATA APPLICATION

A protein array experiment was conducted for the purpose of characterizing two common pathologic states of a disease, i.e., malignant or normal, in terms of peaks. To this end, sera were absorbed onto ProteinChip® arrays and read on a surface enhanced laser desorption/ionization time-of-flight mass spectrometry (SELDI TOF-MS) instrument, from which peaks were manually selected and evaluated. For further details on the experiment, see Zhang et al.[13,14]

Displayed in Figure 12.1 are the empirical CDF and PDF plots for the protein distance, based on the training data set. The distinct distances were grouped into 12 equally sized intervals for a better depiction of overall differences among groups. The plots indicated some difference in the protein distances between the cancer

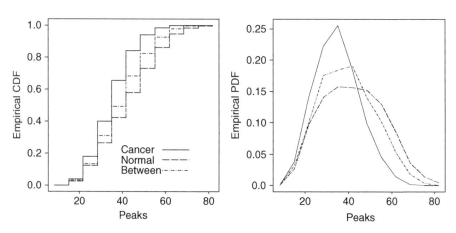

FIGURE 12.1 Distribution of protein (peak) distances within the cancer group ($n = 50$) and normal group ($n = 44$) and between the cancer and normal subjects (training data set).

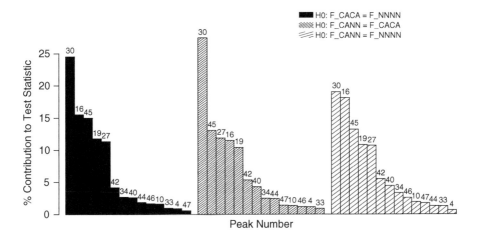

FIGURE 12.2 Peaks with estimated contribution of greater than 2% within each test statistic [NNNN = normal group (hh); CACA = cancer group (kk); CANN = between cancer and normal groups (hk)].

group and normal groups. In particular, the cancer group appeared to be associated with smaller protein distances, on average, relative to the normal group. The between-group distribution of peak distances appeared to be different from the cancer group but similar to the normal group.

To assess the statistical significance of the observed differences described above, we tested the three hypotheses in Eq. (12.4) based on the empirical CDF defined in Eq. (12.3).

By implementing the asymptotic inference methods, we rejected the first two null hypotheses ($p_1 = 0.0448$, $p_2 = 0.0315$, $p_3 = 0.1881$), reflecting our previous observations. Our analysis suggested that the heterogeneity among peak measurements between the cancer and normal group appeared to be different as a whole at a statistically significant level.

In proteomic research, it is of interest to identify peaks that are differentially expressed under different conditions. For this task, we decomposed the test statistics relating to each hypothesis to estimate a contribution from each peak to such differences, following Eq. (12.5) and Eq. (12.6). Figure 12.2 displays peaks with high contributions to each test statistic.

Among these peaks, 30, 16, 45, 19, and 27 were of special interest, as each were estimated to contribute more than 10% to the test statistics. It is worth noting that these peaks were expressed as quadratic forms, so the magnitudes indicate their relative contributions in general. To identify "condition-specific" (i.e., cancer or normal) peaks that can only be detected in the cancer or normal groups, we re-expressed the test statistics, which involves different signs to identify the direction of differences in peak heterogeneity. The re-expressed difference statistics are shown in Figure 12.3. It was indicated that candidate peaks for the positives expressed were 16, 19, 27, and 30. The direction of peak 45 was ambiguous.

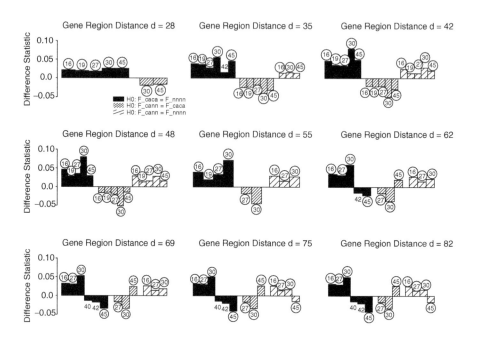

FIGURE 12.3 Plots of the re-expressed difference statistics [Δ_1 = top plot; Δ_2 = middle plot; Δ_3 = bottom plot, as defined in Eq. (12.5) for peaks with an observed location difference by distance mass point].

12.4 DISCUSSION

In this chapter we present a formal approach for companing and characterizing contributions from combinations of peaks at biomarkers that are associated with different pathologic sites. The most flexible feature of the proposed approach is the ability to identify potentially important combinations of biomarkers to observed differences in the total variation between two (empirical) distributions, without imposing a specific distribution assumption. By implementing an SVD approach to summarize replicate measurements, this method is robust to outliers and retains dependencies within observations.

Another consideration of this approach is that we model the variation. Protein array data are characterized by its high biologic variation.[15] We caution, however, that observed variation may spawn from biologic variation other than pathologic status. We also note that it is possible that while there may be no notable, overall difference, in terms of peak heterogeneity, there may exist differentially expressed peaks under various pathologic states.

While we focus our discussion on a distance-based approach, we note that there are several statistical methodologies proposed to analyze mass spectra. A data analysis tool for biomarker pattern discovery, Proteome Quest, beta version 1.0 (http://www. correlogic.com/; Correlogic Systems, Inc. Bethesda, MD), was developed for the analysis of biomarker discovery. The software implements a pattern-discovery algorithm and integrates elements from genetic algorithms proposed by Holland[16] and self-organizing adaptive pattern recognition systems proposed by Kohonen.[17,18] Genetic algorithms organize and analyze complex data sets as if the information were comprised of individual elements that can be manipulated through a computer-driven analog of the natural selection process. Self-organizing systems cluster data patterns into similar groups.

12.5 SUMMARY

For a data set of n samples and p variables, the standard inference theory developed for multivariate data analysis requires that $n \gg p$. Proteomic technology, similar to microarrays, is characterized by the reverse inequality, i.e., $n \ll p$, and has thus introduced many challenges for data analysis. The high dimensionality of the data introduces statistical problems for inference, while the notably large variability reflected in protein data compounds the complexity of the problem. To address analysis for the setting of $n \ll p$ within the context of protein arrays, we applied a method to select peaks that may be characteristic of distinct pathological disease states through the construction of a composite distance measure and formulate hypothesis tests based on this measure. The method is robust to outliers, retains dependencies within observations, and assumes no specific analytical distribution upon the data. In addition, the inherent variability in protein array data is also considered as part of the method. Peaks are selected based on their relative contributions to observed heterogeneity differences reflected in the distance measure among the various pathological states. Our results are comparable with those obtained using other analysis methods for similar investigations.

REFERENCES

1. Everitt, B. *The Analysis of Contingency Tables*. Chapman & Hall, London, 1992.
2. Breiman, L. *Classification and Regression Trees*. Wadsworth International Group, Belmont, CA, 1984.
3. Fisher, A., Dickson, C., et al. (1922). *The Mathematical Theory of Probabilities and Its Application to Frequency Curves and Statistical Methods*. The Macmillan Co., New York, 1922.
4. Vapnik, V.N. *Statistical Learning Theory, Adaptive and Learning Systems for Signal Processing, Communications, and Control*. New York: Wiley, xxiv, p. 736, 1998.
5. Brown, M.P., Grundy, W.N., et al. Knowledge-based analysis of microarray gene expression data by using support vector machines. *Proc. Natl. Acad. Sci. USA*, 97, 262–267, 2000.
6. Furey, T.S., Cristianini, N., et al. Support vector machine classification and validation of cancer tissue samples using microarray expression data. *Bioinformatics*, 16, 906–914, 2000.
7. Weston, J., Mukherjee, S., Chapelle, O., Pontil, M., Poggio, T., and Vapnik, V. Feature selection for SVMs. In *Advances in Neural Information Processing Systems 13*, Solla, S.A., Leen, T.K., and Muller, K.-R., Eds. MIT Press, Cambridge, pp. 668–674, 2001.

8. Kowalski, J. A non-parametric approach to translating gene region heterogeneity associated with phenotype into location heterogeneity. *Bioinformatics,* 17, 775–790, 2001.

9. Kowalski, J., Pagano, M., et al. A nonparametric test of gene region heterogeneity associated with phenotype. *J. Am. Stat. Assoc.,* 97, 398–408, 2002.

10. Fisher, R.A. The use of multiple measurements in taxonomic problems. *Annals Eugenics,* 7, 179–188, 1936.

11. Efron, B. and Tibshirani, R. *An Introduction to the Bootstrap.* Chapman & Hall, New York, 1993.

12. Johns, R.A. and Wichern, D.W. *Applied Multivariate Statistical Analysis.* Prentice Hall, Englewood Cliffs, NJ, 1998.

13. Zhang, Z., Page, G., et al. Applying classification separability analysis to microarray data. In *Methods of Microarray Data Analysis: Papers from CAMDA '00,* Lin, S.M. and Johnson, K.F., Eds. Kluwer Academic Publishers, Boston, pp. 125–136, 2001.

14. Zhang, Z., Page, G. Fishing expedition—A supervised approach to extract patterns from a compendium of expression profiles. In *Microaway Data Analysis II: Papers from CAMBLA '01,* Lin, S.M. and Johnson, K.F., Eds. Kluwer Academic Publishers, Boston, 2002.

15. Molloy, M., Brzezinski, E., et al. Overcoming technical variation and biological variation in quantitative proteomics. *Proteomics,* 3, 1912–1919, 2003.

16. Holland, J.H. *Adaptation in Nature and Artificial Systems: An Introductory Analysis with Applications to Biology, Control, and Artificial Intelligence,* 3rd ed. MIT Press, Cambridge, MA, 1994.

17. Kohonen, T. Self-organized formation of topologically correct feature maps. *Biol. Cybern.,* 43, 59–69, 1982.

18. Kohonen, T. The self-organizing map. *Proc. IEEE,* 78, 1464–1480, 1990.

13 Protein Identification by Searching Collections of Sequences with Mass Spectrometric Data

D. Fenyö, J. Eriksson, and R.C. Beavis

CONTENTS

13.1 INTRODUCTION

Protein identification is a process that involves matching mass spectrometric data with collections of protein sequences. This procedure has become an essential part of proteomics-based biological research.[1] The mass spectrometric data is represented by a set of observed signal intensity mass-to-charge ratio pairs. These pairs are compared directly to the set of similar pairs that should be representative of a subset of proteins in the sequence collection. The scores that result from the comparison are then analyzed and clustered to find the best model set of protein sequences that fits the experimental data.

Protein identification experiments commonly involve a similar set of sample preparation protocols. The specific protocol to be used is selected based on the type of mass spectrometer (single or tandem) that will be used to generate the information for comparison. The work flows associated with the most common protocols are represented in Figure 13.1. The proteins in the sample are first fractionated, using an experimental design that will enrich a mixture of the proteins that are relevant to the biological hypothesis being answered by the overall experiment. Often, this initial mixture of proteins will also undergo additional steps of separation by means of multidimensional chromatography and/or gel electrophoresis. These further steps of

267

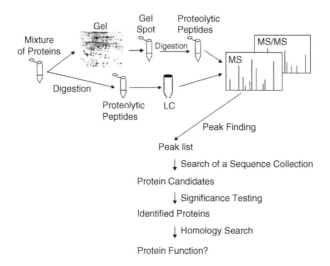

FIGURE 13.1 Common protein identification work flows: In the gel-based work flow: the proteins of interest are separated by 2D gel electrophoresis, the spots are visualized and excised, the proteins are digested, and the proteolytic peptides are analyzed by mass spectrometry. In the LC-based work flow a complex mixture of proteins is digested and the resulting peptide are separated by liquid chromatography (LC) followed by mass spectrometric analysis. Data analysis for both work flows consist of first processing the mass spectra to find the masses of the peptides and peptide fragments, followed by searching a sequence collection to obtain a list of protein candidates. The significance of the candidates is subsequently assessed to obtain a list of identified proteins. The function of these proteins can in some cases be assigned by homology searching.

purification are carried out so that additional information about the component proteins can be determined (e.g., pI, amount), as well as to simplify the analysis of the mass spectrometry data set. Following the separation steps, fractions (or bands) of interest are digested with a proteolytic enzyme to generate a set of peptides in a form compatible with further analysis. A final step of chromatography may be required, if the resulting mixture of proteolytic peptides is expected to be too complex for direct mass analysis.

   Once the original sample has been processed and appropriate samples for mass spectrometry have been obtained, the masses of the resulting peptides and their fragments can be measured by mass spectrometry and compared with calculated peptide masses from a protein sequence collection.[2–8] The scores are calculated for comparison and the protein sequences in the collection are ranked according to the scores.

   The significance of the protein candidates can then be assessed by calculating the probabilities that they are false positives.[9–13] The experimental design will dictate some acceptable cutoff for the probability of having a false-positive result in the final list of sequences. The candidate proteins in the list that pass this quality test are considered to be "identified" and can be incorporated into the model for the system.

Subsequent to identification, there are a number of possible informatics alternatives that can be used to add value to the model. For example, in cases where proteins of unknown function are identified, it may be possible to assign a function to the sequence by homology searching. It may also be possible to create additional mapping of the sequences, such as gene linkage maps or protein–protein interaction maps.

13.2 THE INFORMATION VALUE OF MASS MEASUREMENTS

Protein identification using a single stage mass analyzer is commonly referred to as "peptide mass fingerprinting" or simply "fingerprint analysis." This type of simple instrumentation can be performed because the accurately determined molecular mass of a single peptide generated by a sequence-specific proteinase contains a significant amount of information about that peptide's amino acid composition[14] (Figure 13.2). If the uncertainty in the peptide ion mass measurement is low, then there are only a few proteolytic peptides that can match that single mass measurement. In a typical experiment that deals with proteins that must be matched with a relatively large collection of potential protein sequences, e.g., a eukaryotic proteome, it will always be necessary to obtain the mass of more than one proteolytic peptide to obtain a confident protein identification.

The actual number of peptides required to obtain a confident identification will depend on the size of the proteome, the enzyme used in the experiment, the measured peptide mass, the accuracy of the measurement, and the sequence of the protein. When the allowed uncertainty in the mass measurement is very low (0.1 ppm), the number of matching peptides does not decrease with improved mass accuracy, because the elemental composition of the peptide is uniquely defined by the mass. The elemental composition does not supply a unique amino acid composition, as there are isobaric combinations of amino acid residues. This type of high accuracy mass measurement is not often available: an allowed mass accuracy of 5 to 10 ppm is much more common. At this level, there are significantly more potentially isobaric compositions and the number of peptides required is normally greater than 10. The addition of any further hints about the potential residue composition, such as a chemical tag indicating the presence of cysteine, can also be strong evidence in the statistical analysis of the data. The incorporation of these hints into the scoring system for sequence comparison will result in improved assignment confidence, even with a relatively small number of matching peptides.

The use of a tandem mass spectrometer, which allows the isolation of a particular parent ion and subsequent fragmentation, can give enough sequence-dependent information to compensate for low accuracy peptide ion mass measurements. Mass accuracies of 1000 to 5000 ppm can be compensated for quite effectively using this additional information. A particular protein sequence can often be confidently identified from one or two tandem peptide spectra, depending on the protein's sequence and the potential for alternative exon splicing.

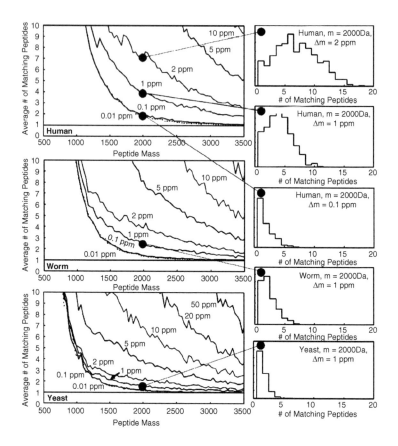

FIGURE 13.2 (Color insert follows page 204) The average numbers of matching complete tryptic peptides from human, worm (*C. elegans*), and yeast (*S. cerevisiae*) proteins as a function of peptide mass for different accuracies of the mass measurements (left). The better the mass accuracy, the fewer tryptic peptides match a single peptide mass; e.g., a mass of 2000 Da correspond on the average to 7.0, 4.1, and 2.0 complete tryptic peptides from human proteins for mass accuracies of 2, 1, and 0.1 ppm, respectively. Below 0.1 ppm no further improvement is observed when improving the mass accuracy because the elemental composition of the peptide is uniquely defined by the mass. Organisms with fewer genes have fewer peptides that match a single tryptic peptide mass; e.g., a mass of 2000 Da with a mass accuracy of 1 ppm correspond on the average to 4.1, 2.3, and 1.6 complete tryptic peptides from human, worm, and yeast, respectively. The distributions of the number of matching peptides are shown for a few cases (right); e.g., the average value of 1.6 complete tryptic peptides from yeast matching a mass of 2000 Da with mass accuracy 1 ppm, correspond to matches to single peptides in 58.4% of the cases and to matches of 2, 3, 4, 5, and 6 peptides in 30.7, 8.4, 1.9, 0.4, and 0.2% of the cases, respectively.

13.3 ALGORITHMS FOR PROTEIN IDENTIFICATION

The searching of a collection of sequences is performed by mimicking the experiment in silico. The basic steps shown in Figure 13.3 are common to all algorithms. The difference between algorithms is how the score is calculated from the comparison of

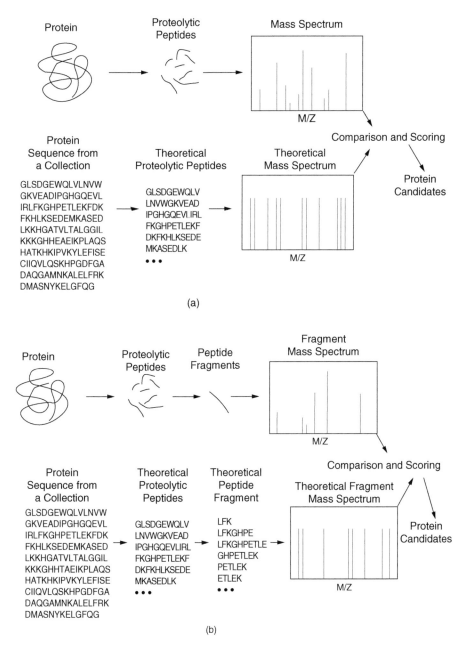

FIGURE 13.3 Searching a sequence collection with (a) peptide masses and (b) peptide fragment masses.

experimental and theoretical mass spectra. It is critical that the algorithm makes optimal use of the experimental information to allow for identification of low-abundance proteins from noisy data.

For a peptide mass fingerprinting experiment (Figure 13.3a), each protein sequence in a collection is theoretically digested using the same cleavage rules as the enzyme used in the experiment. The masses of the resulting peptides are calculated and a theoretical mass spectrum is constructed. The measured mass spectrum is then compared to the theoretical mass spectrum and a score is calculated. This procedure is repeated for each protein in the sequence collection. Finally, the proteins in the sequence collection are ranked according to the calculated scores. The simplest scoring algorithm simply counts the number of matching masses between the experimental and theoretical mass spectra. This works well only when the data quality is high. The main problem is that the probability of matching a protein with a long sequence by chance is high; therefore, the highest ranked protein is often a random large protein. Several, much more sophisticated algorithms have been developed to overcome this problem, e.g., MOWSE,[4] ProFound,[15] Mascot,[16] PeptIdent,[17] and Probity.[12]

The basic steps for protein identification using data from tandem mass spectrometry experiments are similar to peptide mass fingerprinting, except for the additional step of calculating the peptide fragment masses. In contrast to peptide mass fingerprinting where several peptides are necessary, a tandem mass spectrum of a single peptide can be sufficient to uniquely identify the peptide. The most common search engines for tandem mass spectrometry data are SEQUEST,[7] Mascot,[16] Sonar,[13] ProbID,[18] Popitam,[19] and Tandem.[20]

Prior to searching with peptide mass fingerprinting or tandem mass spectrometry data, the user selects a set of parameters, including the sequence collection to search, properties of the protein like mass and pI, and experimental information like specificity of the enzyme and modifications to the amino acids.

In the case of peptide mass fingerprinting, the sequences searched are protein sequences using an enzyme with high specificity (e.g., trypsin). The information content in the data is usually not sufficient to search genomic sequences or for using a nonspecific enzyme. Additional information like protein mass and pI from a 2D gel can be very useful to restrict the number of sequences searched. The ranges of mass and pI for allowed proteins should not, however, be set too narrow because the sequences used in the search do not often correspond to the mature protein and the proteins analyzed might be degradation products. Most search engines allow the user to specify amino acid modifications of two types: complete (e.g., alkylation of cysteines) and partial (e.g., phosphorylation and methionine oxidation). The assumption of partial modifications should be avoided when searching with peptide mass fingerprinting data because: (1) for most proteins only a few of the proteolytic peptides are modified, and (2) the increased number of theoretical masses for each protein increases the random matching and the risk for false positive results.

Tandem mass spectra can successfully be used to search genomic and expressed sequence tag (EST) sequences in addition to protein sequences. Genomic and EST sequences are translated in six reading frames prior to the steps in Figure 3b. Most protein sequences have been predicted from the genome and therefore contain errors due to the difficulty in predicting intron/exon boundaries and alternative splicing. The use of the genomic sequences in the searches has the advantage of overcoming many of the potential frame shifts introduced by these predictions.

13.4 SIGNIFICANCE TESTING

It is critical to test the significance of protein identification results because false identifications are possible due to random matching between the measured and calculated masses. In the result of a search of a sequence collection with both peptide mass fingerprinting and tandem mass data, there will always be a highest ranked protein sequence. This protein sequence might correspond to a protein in the sample analyzed (true positive) or simply get the highest score because of random matching between the calculated proteolytic peptide masses and the measurement (false positive). The probability that a protein candidate is a false positive can be estimated by comparing its scores to the distribution of scores for random and false identifications (Figure 13.4). The distribution of scores for random and false identifications can be obtained from computer simulations,[10,11] calculations based on a model,[12] or by collecting statistics during the search.[9] Figure 13.5 illustrates how the statistics collected during the search can be used to estimate the significance of the results. During the search, a score is calculated for each protein sequence in the collection. For the majority of sequences the matching with the experimental data is random. An example of the distribution of the scores for proteins in a sequence collection matching a peptide mass fingerprint is shown in Figure 13.5. Typically, a distribution of scores from randomly matching protein sequences is observed at low scores (Figure 13.5a). This distribution is an extreme value distribution, having a linear tail when plotted on a log-log scale (Figure 13.5b). The significance of high-scoring protein sequences is estimated by linear extrapolation.

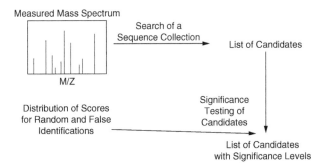

FIGURE 13.4 Significance testing is performed by comparing the score of protein candidates to the distribution of scores for random and false identifications.

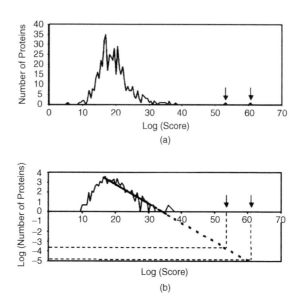

FIGURE 13.5 Expectation values are calculated by extrapolation of the tail of the distribution of scores for random and false identifications. The expectation value for a protein candidate is the number of protein sequences in the collection searched expected to get the same or higher score by random matching. (a) An example of the distribution of scores for proteins in a sequence collection matching a peptide mass fingerprint. The score of the two high-scoring sequences are indicated with arrows. (b) The same distribution shown on a log-log plot. The expectation value of high-scoring protein sequences is calculated by linear extrapolation.

13.5 SUMMARY

There are several critical steps that are necessary for being successful with protein identification and obtaining meaningful results:

1. The masses of peaks in the mass spectrum corresponding to the monoiso-topic peptide masses have to be assigned
2. A collection of sequences have to search using a sensitive and selective algorithm
3. The significance of the results have to be tested
4. The functions of the identified proteins have to be assigned

REFERENCES

1. Aebersold, R. and Mann, M. Mass spectrometry-based proteomics. *Nature,* 422, 198–207, 2003.
2. Henzel, W.J., Billeci, T.M., Stutts, J.T., Wong, S.C., Orimley, C., and Watanabe, C. Identifying proteins from two-dimensional gels by molecular mass searching of peptide fragments in protein sequence databases. *Proc. Natl. Acad. Sci. USA,* 90, 5011–5015, 1993.

3. Mann, M., Hojrup, P., and Roepstorff, P. Use of mass spectrometric molecular weight information to identify proteins in sequence databases. *Biol. Mass Spectrom.*, 22, 338–345, 1993.

4. Pappin, D.D.J., Hojrup, P., and Bleasby, A.J. Rapid identification of proteins by peptide-mass fingerprinting. *Curr. Biol.*, 3, 327–332, 1993.

5. Yates, J.R., III, Speicher, S., Griffin, P.R., and Hunkapillar, T. Peptide mass maps: A highly informative approach to protein identification. *Anal. Biochem.*, 214, 397–408, 1993.

6. James, P., Quadroni, M., Carafoli, E., and Gonnet, G. Protein identification by mass profile fingerprinting. *Biochem. Biophys. Res. Commun.*, 195, 58–64, 1993.

7. Eng, J.K., McCormack, A.L., and Yates, J.R. *J. Am. Chem. Soc.*, 5, 976, 1994.

8. Mann, M. and Wilm, M. Error-tolerant identification of peptides in sequence databases by peptide sequence tags. *Anal. Chem.*, 66, 4390–4399, 1994.

9. Fenyo, D. and Beavis, R.C. A method for assessing the statistical significance of mass spectrometry-based protein identifications using general scoring schemes. *Anal. Chem.*, 75, 768–774, 2003.

10. Eriksson, J., Chait, B.T., and Fenyo, D. A statistical basis for testing the significance of mass spectrometric protein identification results. *Anal. Chem.*, 72, 999–1005, 2000.

11. Eriksson, J. and Fenyo, D. A model of random mass-matching and its use for automated significance testing in mass spectrometric proteome analysis. *Proteomics*, 2, 262–270, 2002.

12. Eriksson, J. and Fenyo, D. The statistical significance of protein identification results as a function of the number of protein sequences searched. *J. Proteome Res.*, 3, 979–982, 2003.

13. Field, H.I., Fenyo, D., and Beavis, R.C. RADARS, a bioinformatics solution that automates proteome mass spectral analysis, optimizes protein identification, and archives data in a relational database. *Proteomics*, 2, 36–47, 2002.

14. Fenyo, D., Qin, J., and Chait, B.T. Protein identification using mass spectrometric information. *Electrophoresis*, 19, 998–1005, 1998.

15. Zhang, W. and Chait, B.T. ProFound: An expert system for protein identification using mass spectrometric peptide mapping information. *Anal. Chem.*, 72, 2482–2489, 2000.

16. Perkins, D.N., Pappin, D.J., Creasy, D.M., and Cottrell, J.S. Probability-based protein identification by searching sequence databases using mass spectrometry data. *Electrophoresis*, 20, 3551–3567, 1999.

17. Wilkins, M.R, Gastieger, E., Bairoch, A., Sanchez, J.C., Williams, K.L., Appel, R.D., and Hochstrasser, D.F. Protein identification and analysis tools in the ExPASy server. *Methods Mol. Biol.*, 112, 531–552, 1999.

18. Zhang, N., Aebersold, R., and Schwikowski, B. ProbID: A probabilistic algorithm to identify peptides through sequence database searching using tandem mass spectral data. *Proteomics*, 2, 1406–1412, 2002.

19. Hernandez, P., Gras, R., Frey, J., and Appel, R.D. Popitam: Towards new heuristic strategies to improve protein identification from tandem mass spectrometry data. *Proteomics*, 3, 870–878, 2003.

20. Craig, R. and Beavis, R.C. A method for reducing the time required to match protein sequences with tandem mass spectra. *Rapid Commun. Mass. Spectrom.*, 17, 2310–2316, 2003.

14 Bioinformatics Tools for Differential Analysis of Proteomic Expression Profiling Data from Clinical Samples

Zhen Zhang

CONTENTS

14.1 CHALLENGE OF EXPRESSION PROFILING USING CLINICAL SAMPLES

With recent advances in technologies for high-throughput genomic and proteomic expression analysis, the simultaneous measurement of expression levels of a large number of molecular entities or so-called profiling has become an important screening tool for the discovery of new biomarkers that are associated with a particular disease process. In addition to the potential for direct clinical applications, such as for early detection and diagnosis of diseases and use as therapeutic targets, results from such "target-hunting" activities also facilitate the generation of hypotheses that may lead to new discoveries that help us to better understand the disease process itself.

Recently reported work in using profiling methods to identify individual biomarkers or genomic/proteomic expression patterns for disease detection, staging, and classification has greatly raised the expectation for the clinical utility of such approaches.[1-12] However, the unique characteristics of clinical samples and the conditions under which the samples are collected and processed impose a special set of challenges to the analysis of expression data from clinical samples. Bioinformatics tools for such data analysis tasks will have to address these special issues and be aware of their implications on the interpretation of the results.

The purpose of proteomic expression profiling of clinical samples typically involves the differential analysis of the expression levels of a large subset of the proteome of a particular type of clinical specimens to identify those proteins whose change in expression levels might be associated with a given disease process. This effort is often hindered by the fact that the observed differences in expression levels are influenced by multiple factors, of which many are irrelevant to the particular disease of interest. Examples of such nondisease-associated factors include

1. Within-class biological variability which may include unknown sub-phenotypes among study populations
2. Pre-analytical variables such as systematic differences in study populations and/or in sample collection, handling, and processing procedures
3. Analytical variables such as inconsistency in instrument conditions that result in poor reproducibility
4. Measurement imprecision

Among them, factors (1) and (2) pose a much more significant problem for the analysis of expression data from clinical samples than those from cells or animal models under well-controlled experimental conditions. This problem is further worsened by the fact that the dynamic range of protein expression levels could be much greater than what is typically seen in genomic expression data. Finally, another important issue for clinical samples is the mislabeling of samples, which happens frequently due to imperfection in the current gold standard diagnostic methods (e.g., false negatives in biopsy results) or ethical constraints (e.g., inability to biopsy controls for a low prevalent disease).

In this chapter, we first briefly compare the current two approaches used by practitioners in genomic and proteomic expression data analysis, one grounded on statistical theories and the other on machine learning and other computational data mining techniques. This will be followed by discussions on how the special characteristics of clinical samples might affect the observed expression profile data and their impact on the analytical results by these two different approaches. We will then present the bioinformatics approach that we used for biomarker discovery, which involves the analysis of SELDI TOF-MS (surface-enhanced laser desorption/ionization time-of-flight mass spectrometry)-generated proteomic profiling data from clinical samples. In particular, we will introduce the unified maximum separability analysis (UMSA) algorithm that forms the foundation of many of our bioinformatics tools and its relative advantages for the analysis of data from clinical samples. Finally, we discuss the importance of study design in expression profiling studies of clinical samples.

14.2 TWO SCHOOLS OF BIOINFORMATICS APPROACHES TO EXPRESSION DATA ANALYSIS

One of the common characteristics of expression profile data is the high dimensionality (number of measured variables) in comparison to a relatively small sample size. An immediate consequence of this is the lack of a stable estimate of the covariance matrix, a prerequisite in many traditional statistical multiple variable analysis methods. Because of this difficulty, much of the development in statistical and computational methods for differential expression data analysis has been focused on univariate analysis methods using a modified form of the t-statistic[13] and a cutoff value on the adjusted p values or on the test statistic directly based on estimation of false-discovery rates (FDR).[14,15] A shortcoming of such univariate methods is that the analyses of the "informativeness" of individual variables are independent of one another, an assumption that in many situations clearly does not reflect the reality of biology. In order to use a multivariate approach and at the same time to circumvent the issue of covariance matrix estimation, a compromising approach would be to first apply an unsupervised (e.g., singular value decomposition) or supervised (e.g., partial least squares) dimension-reduction step to project the large number of genes or proteins onto a smaller and more manageable number of "clusters",[16,17] or some types of latent variables or "supervariables."[4] More traditional multivariate analysis methods may then be applied to ascertain the significance of such supervariables.

While more robust and statistically sound multivariate approaches are being carefully derived, in the clinical genomics and clinical proteomics fields, however, driven by the need to identify biomarkers that can be associated with a particular aspect of the disease process and carried out by computational scientists specialized in machine learning and "data-mining" techniques, algorithms have been developed to directly fit parametric or nonparametric models to the data and from which to assess the significance of individual variables.[2,6,10,18–20] Some even went a step further to forgo the analysis of contributions by the selected individual variables altogether and decided to combine variable selection and multivariate predictive model derivation into a single iterative process.[8] An advantage of the model-fitting approaches is that the significance of individual variables are determined collectively with all the other variables. The real issue, however, is that for any nonlinear multivariate models with a modest level of complexity it is, in general, difficult or even impossible to quantify the contribution of individual variables. In addition, overfitting can be a serious problem for complex nonlinear multivariate models, which in turn greatly amplifies the impact of nondisease-associated data variability on the analysis results.

For the analysis of expression profile data from clinical samples, the permutation-based methods for estimation of FDR and selection of cutoff on significance values can be readily applied to help to assess FDR caused by within-class biological variability. However, many of the recently developed statistical and computational approaches do not address the issues related to the above-mentioned specific characteristics (1) and (2) of clinical samples. The issue of labeling error among clinical samples has also been generally ignored.

14.3 CONSIDERATIONS OF BIOINFORMATICS TOOLS FOR DIFFERENTIAL ANALYSIS OF PROTEOMIC EXPRESSION PROFILE DATA

A generic supervised procedure using multivariate models to rank and select informative variables in an expression dataset involves (a) constructing a multivariate classification model that best separates the classes of samples in the training data; and (b) ranking the variables according to their relative contribution in the constructed model and selecting a subset of variables that contributes the most toward the separation of the classes of samples.

To use this procedure, however, one still has to decide what constitutes the best model and how to estimate the contributions of individual variables in a meaningful and consistent way. From a statistical point of view, if the underlying conditional distributions of the different classes of samples are known, the Bayes' decision rule[21] provides the optimal separation of the classes with the available information from the variables (the Bayes' error rate is a reflection of the imperfect information carried by the variables). The construction of the best predictive model, therefore, depends on (1) the model's capacity (flexibility) of fitting the underlying distributions of the data; (2) the availability of training data representative of the true distributions; and (3) the ability of the associated-learning algorithm to extract information from the training data. For most expression profile data from studies using clinical specimens, the available data may not fully represent the true data distributions, mostly due to insufficient sample sizes, and many times due to the biases and labeling errors in the data. This reality will not likely change much for the time being. The selection of the best predictive model, therefore, will have to take into consideration the efficiency of the model's learning algorithm in information utilization with a small, noisy, and possibly partially mislabeled sample set. The choice of model complexity will also have to be balanced between its ability to model the underlying data distribution and the availability of informative training samples to support the learning process. The latter, or the lack of it, is one of the major reasons why most reported models for differential expression data analysis have been the relatively simple ones.

The attribution of significance to individual variables in a nonlinear multivariate model is a difficult problem and there are no generally applicable analytical solutions. Computationally, a simple approach is to correlate the output of a single trained model with perturbations in its input variables one at a time, which essentially calculates the "partial derivatives" of the model. A computationally more demanding approach is to repeatedly construct and evaluate models, each time with one variable left out, using some type of performance criteria (e.g., validation error). However, other than the training data and input variables used, the construction of a nonlinear model also depends on factors such as the complexity of the models, initial values of model parameters, and the associated problem of nonuniqueness in solutions from multiple local optima. Consequently, the ranking and selection results may also depend on the particular settings of the model construction processing that do not bear any biological meanings. One also has to note that the above procedure selects a subset of top-performing variables using only a single rank order. Due to the potentially complicated interaction among the variables, to truly search for the "optimal" subset of variables one would have to repeat the procedure over all possible subsets of the total variables, which could be impractical even for a modest number of variables. There has been effort to reduce the search space by converting the variable selection problem into that of minimizing a vector of continuous objective functions using a gradient-based search algorithm.[18] However, it is not clear from a theoretical point of view how the terrain of the error surface would look like and whether the gradient-based search will lead to a global optimal solution.

There are two basic approaches for the derivation of classification models. The first, which is the traditional statistical approach, is based on the parametric or nonparametric estimation of the conditional distributions of the classes of samples. The classification model is then constructed based on the Bayes' decision rule. The second, a machine-learning approach, is based on the direct determination of a classification model by minimizing an empirical risk function (e.g., the mean square classification error). An example for the first approach is the Fisher's Linear Discriminant Analysis (LDA) function[22]; for the second approach is the support vector machine (SVM).[23] These two approaches differ significantly in their utilization of training samples. In the traditional statistical approach, the training samples, regardless of their relative locations to the underlying class boundaries in the variable space, contribute equally to the estimation of the distributions and the construction of the classification function. For instance, the LDA is completely determined by the estimated means and the pooled covariance matrix for which all data are used in the exactly the same way. On the other hand, in the empirical risk minimization approach, the samples that are close to the class boundaries are weighted much more heavily than the interior samples. As an extreme example, the solution of an SVM model is solely determined by the so-called support vectors, which consist of only the boundary data points. The removal of any interior samples does not affect the solution at all. When the training sample size is sufficiently large, both approaches will asymptotically reach their own expected solutions. However, for many clinical proteomic profiling studies, at least at the variable (proteins) selection stage, the total number of available samples is far below what is required for the asymptotic behavior of the algorithms in these approaches to actually take effect. For situations where each clinical sample represents a considerable amount of effort and cost and the purpose is for data classification instead of representation, treating all samples equally or using only the support vectors might not necessarily be the most efficient use of information from a very small number of samples. In addition, for small sample problems, models that rely solely on the support vectors could be very sensitive to labeling errors in the training samples.

14.4 THE UNIFIED MAXIMUM SEPARABILITY ANALYSIS ALGORITHM

With the above described shortcomings of the existing approaches in mind, we developed the unified maximum separability analysis (UMSA) algorithm for the analysis of genomic and proteomic expression data.[2,6,10,24,25] The conceptual framework of UMSA is very straightforward. In the original SVM learning algorithm,[23,26] there is a constant C that limits the maximum influence of *any* sample point on the final SVM model solution. In UMSA, this constant becomes an individualized parameter for each data point to incorporate additional statistical information about the data point's position relative to the distribution of all the classes of samples. The rationale behind UMSA is that information about the overall data distribution (even though the estimation itself might not be perfect) can be used to prequantify the "trustworthiness" of any training samples to be a support vector. The final solution, therefore, will rely on the weighted contributions of the support vectors and be less sensitive to labeling errors of a small percentage of the samples.

As a concrete example, a linear UMSA classifier for a set of n training samples $x_1, x_2, ..., x_n$ drawn from distributions D^+ and D^- with the corresponding class membership labels $l_1, l_2, ..., l_n \in \{-1, 1\}$ may be obtained by solving the following constrained optimization problem:

$$\text{Minimize } \frac{1}{2} v \cdot v + \sum_{i=1}^{n} p_i \xi_i$$

$$\text{Subject to } l_i(v \cdot x_i + b) \geq 1 - \xi_i, \ i = 1, 2, ..., n,$$

where the nonnegative variables $\xi_1, \xi_2, ..., \xi_n$ represent errors in the constraints that are penalized in the object function, and the coefficients $p_1, p_2, ..., p_n$ are the "individualized" positive constants reflecting the relative "importance" of the n individual data points. In UMSA, $p_i = \phi(x_i, D^+, D^-) > 0$ is typically related to the level of disagreement of a sample x_i to a statistical classifier derived based on estimates of distributions D^+ and D^- from the n training samples (e.g., an LDA or quadratic classifier). Let this level of disagreement be δ_i, the following positive decreasing function is used to compute p_i:

$$p_i = \phi(\delta) = C \cdot e^{-\delta_i^2/\sigma^2}, \text{ where } C > 0.$$

The two parameters, σ and C modulate the amount of influence an individual sample may have upon the solution of v in the optimization problem above. One may notice that for a very large σ relative to the range of δ_i, p_i would essentially turn into a constant close to C. The UMSA algorithm then becomes equivalent to the optimal soft-margin classifier in SVM. On the other hand, a very small σ relative to the spread of the data would make the p_is for those samples that have a high level of disagreement with the statistical classier so small that they would essentially be rendered useless in the final solution.

The solution of the linear UMSA learning algorithm υ is an n-element vector. Same as in SVM, the quantity $2/\|v\|$ defines the margin between the two classes and may be viewed as a measure of class separability. The unit projection vector $v/\|v\|$ represents the direction along which the two classes of samples are best separated in the n-dimensional variable space by a linear UMSA model for the given training data and learning parameters used. The magnitude of the elements in $\upsilon/\|\upsilon\|$ (weights for linear combination) may be viewed as a measure of the relative contributions (significance) of individual variables toward the separation of data.

It should be noted that, similar to SVM, using nonlinear kernel functions to map the original data to a high-dimensional space, UMSA can also be used to derive nonlinear classifiers. Other than its potential of being less sensitive to labeling errors, the previously discussed problems in variable selection using nonlinear models remain to be true for nonlinear UMSA models.

14.5 UMSA-BASED PROCEDURES FOR EXPRESSION PROFILE DATA ANALYSIS

The construction of a linear UMSA classifier provides a supervised multivariate method to rank a large number of variables. For it to be useful in the differential analysis of genomic or proteomic expression data, we have incorporated it into a number of analytical procedures.

The first UMSA-based procedure is a supervised component analysis method for reduction of data dimension. Similar to unsupervised component analysis methods such as principle component analysis or singular value decomposition (PCA/SVD), the UMSA-based procedure is also a linear projection of data. However, in PCA/SVD, the axes in the new space represent directions along which the data demonstrate maximum variations (indicated by the eigenvectors of the covariance matrix). In UMSA component analysis, the new axes represent directions along which the two classes of data are best separated by a linear UMSA classifier. When it is used for dimension reduction, the smaller number of new axes may be viewed as composite features that retain most of the information relevant to the separation of data classes. This is done by iteratively computing a projection vector along which two classes of data are best separated by a linear UMSA classifier. The data are then projected onto a subspace (one dimension lower) that is perpendicular to this vector. In the next iteration, UMSA is applied to compute a new projection vector within this subspace. The iteration continues until a desired number of components have been reached.

For most practical problems, the top few UMSA components are sufficient to extract most of the separation information between the classes of data. In our software implementation, the data are projected onto a space spanned by the top three UMSA components in interactive 3D display (see Figure 14.1) for easy assessment of data distribution and separation. The 3D view also helps the user to interactively try to optimize the two UMSA parameters σ and C, which, in general, depend on the sample size, distributions, and quality of the data.

The second procedure (see Box 1) is a stepwise backward variable elimination/selection method similar to that in the stepwise backward multivariate logistic regression. In each step, it constructs a linear UMSA classifier and computes a significance score for the variable that carries the smallest absolute weight in the UMSA classifier. The variable is then eliminated from further consideration. The iteration continues until there is only one variable left. The computation of

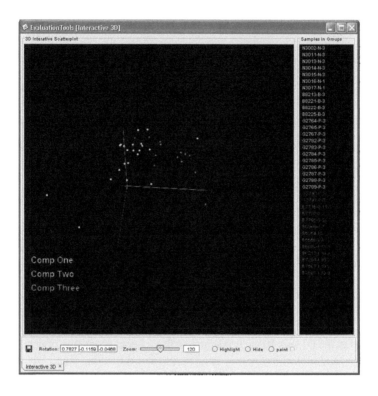

FIGURE 14.1 (Color insert follows page 204) 3D interactive display of two classes of samples projected in UMSA component space.

significance scores needs to be adjusted to take into account the decreasing dimensionality of the variable space. The algorithm in Box 2 ensures that these significance scores are monotonically increasing over the variable elimination/selection process (i.e., the significance score w^{k-1} for the variable eliminated at iteration $k-1$ w^k, for all k. Proof omitted).

With the number of simultaneously measured proteins typically much greater than the number of samples and the presence of considerable biological variability in expression levels, there is a high probability that a multivariate model, even a linear one, could separate the classes of samples well by picking up expression patterns that are particular to the given sample set yet have no relevancy to the disease of interest. In fact, for such data sets, even after randomly reassigning class labels to the samples, one may still be able to derive a classifier to separate the "classes." In order to reduce false discovery, we implemented a third UMSA-based procedure that uses bootstrap to rank and select variables that offer consistent performance across multiple subpopulations. The working assumption is that truly disease-associated changes in proteomic expression profiles should persist over multiple resampled populations. In this procedure,

---

Box 1   Procedure: UMSA component analysis for a two-class dataset with m variables and n samples

inputs:

UMSA parameters C and σ,
number of components $q \leq \min(m, n)$;
data $X = (x_1, x_2, \ldots, x_n)$; and
class labels $L = (l_1, l_2, \ldots, l_n)$, $l_i \in \{-1, +1\}$.

initialization:

component set $D \leftarrow \{\}$;
$k \leftarrow 1$.

while $k \leq q$

1. applying UMSA (σ, C) on $X = (x_1, x_2, \ldots, x_n)$ and L;
2. $d_k \leftarrow v/\|v\|$; $D \leftarrow D \cup \{d_k\}$;
3. $x_i \leftarrow x_i - (x_i^T d_k) d_k$, $i = 1, 2, \ldots, n$;
4. $k \leftarrow k + 1$.

return D.

---

in each bootstrap run, a subpopulation is randomly selected (based on a fixed resampling scheme such as the $e0$ bootstrap, which evaluates n cases sampled with replacements from a size n data set) to obtain a rank order for all variables. The results are then used to calculate the mean, median, and standard deviation of a variable's ranks from multiple bootstrap runs. In general, we are interested in variables with top mean and/or median ranks and a small rank standard deviation. We also apply this procedure on the same data set with all the sample labels randomly permutated. The minimum of rank standard deviations from such mislabeled data sets provides the basis of a selection threshold for variables for which both their performance and the consistency in their performance are less likely by random chance.

The three UMSA-based analytical procedures are implemented in JAVA as independent modules for the NetBeans platform.[27] The software user interface and functionality are designed to facilitate the above discussed generic supervised procedure using multivariate models to rank and select informative variables. The system allows a user to apply iteratively and recursively the analytical procedures to select subsets of the original variables that retain sufficient discriminatory information.

**Box 2   Procedure: Stepwise backward UMSA variable selection for a
two-class dataset with *m* variables and *n* samples**

inputs:

UMSA parameters C and σ,
data $e = \{e_{ji} | j = 1,2,\ldots, m; i = 1,2,\ldots, n\}$; and
class labels $L = (l_1, l_2, \ldots, l_n), l_i \in \{-1,+1\}$.

initialization:

$G_k \leftarrow G_m = \{g_j = (e_{j1}, e_{j2}, \ldots, e_{jn})^T, j = 1,2,\ldots, m\}$;
score vector $w = (w^1, w^2, \ldots, w^m)^T \leftarrow (0,0,\ldots,0)^T$.

while $|G_k| > 1$

1. forming $X = (x_1, x_2, \ldots, x_n) \leftarrow (g_1, g_2, \ldots, g_k)^T$.
2. applying UMSA (C, σ) on X and L;

   $S_k \leftarrow 2/\|v\|$ and $d_k \leftarrow v/\|v\|$.

3. for all $g_j \in G_k$, if $s_k|d_k^j| > w^j$, $w^j \leftarrow s_k|d_k^j|$.
4. $G_{k-1} \leftarrow G_k - \{g_r\}$, where r is determined from $w^r = \min_{g_j \in G_k}\{w^j\}$.

return w.

14.6 APPLICATION OF UMSA-BASED TOOLS FOR BIOMARKER DISCOVERY USING CLINICAL PROTEOMICS DATA FROM MASS SPECTROMETRY

The UMSA-based software system has been used for genomic expression data analysis[2,24,25] and, more recently, for biomarker discovery using clinical proteomic profiling data[5,6,10,28] generated by surface-enhanced laser desorption/ionization time-of-flight mass spectrometry (SELDI TOF-MS). SELDI is an affinity-based MS method in which proteins are selectively adsorbed to a chemically modified surface and impurities are removed by washing with buffer. By combining different surfaces and wash conditions, SELDI allows on-chip protein capture and micropurification, which facilitates high-throughput protein expression analysis of a large number of clinical samples.[29,30] After preprocessing steps such as mass calibration, baseline subtraction, and peak detection, the mass spectra from *n* individual samples are converted into peak intensity data typically organized as an *m* (peaks) × *n* (samples) matrix, where the intensity of a peak corresponds to the relative abundance of proteins at a particular molecular mass (as mass-to-charge ratio, or m/z). The goal of bioinformatics analysis is to select a subset of the total detected peaks that are most informative in separating the different classes of clinical samples. As an

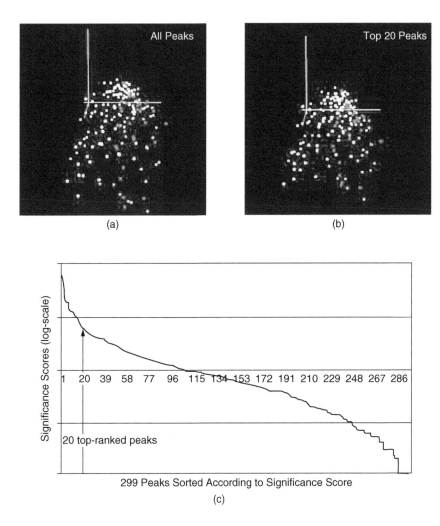

(a)   (b)

(c)

FIGURE 14.2 (Color insert follows page 204) UMSA-based analysis of SELDI peak intensity data from 183 samples in two groups, each with 299 detected peaks. Group A: n = 92, 49 used for training (green) and 43 for test (olive); Group B: n = 91, 49 used for training (red) and 42 for test (blue). (a) UMSA component analysis of the training data using all 299 peaks; the fixed component projection was then applied to the test data. (b) Plot of significance scores of all 299 peaks in log-scale and descending order. Arrow indicates cutoff on significance scores where the score descending rates differ noticeably. With this cutoff, 20 top-scored peaks were selected. (c) UMSA component analysis of the sample training data using the 20 selected peaks; the fixed component projection was then again applied to the test data.

example, we analyzed a two-class data set of 183 samples, each with 299 detected peaks. The samples were randomly divided into a training set (49 from group A and 49 from group B) and a test set (43 from group A and 42 from group B). Figure 14.2a shows in 3D view the separation of the samples by the UMSA component analysis using all of the 299 peaks. The component projection vectors were determined using the training data and then applied to the test data. The training and test data demonstrated very similar distribution patterns and degree of separation in the UMSA component space. Figure 14.2b plots in log-scale and descending order the significance scores of the 299 peaks obtained from the stepwise backward UMSA variable selection procedure. The plotted curve has roughly three segments with different descending rates, very steep for the left-most 20 peaks, slower in the middle, and becoming fast again toward the end*. It provided a natural cutoff for us to select the top 20 peaks. The UMSA component analysis procedure was applied again to the data using only the selected peaks. The result in Figure 14.2c proves that these 20 peaks indeed retained almost all the discriminatory power of the original data and the training and test results were again in good agreement.

14.7 THE IMPORTANCE OF STUDY DESIGN

Among the issues associated with expression profiling using clinical samples, systematic biases from preanalytical variables could be the most damaging. While careful statistical examination of analysis results and their correlation with possible nondisease-related variables may reveal the existence of such biases, no amount of statistical or computational processing will be able to correct such problems within a single set of samples collected under the same conditions. Since such biases are often specific to institutions (sites), the use of specimens from multiple institutions combined with sound study design might be the only way for us to alleviate the impact of such biases in our effort to discover biomarkers that are truly associated with the disease process. The typical way of using multiple data sets is to pool them together and then randomly divide them into a discovery/training set and a test/validation set. The advantage of such an approach is that the discovery set will be more representative of the actual target population. Statistically, the discovery set and the validation set are guaranteed, albeit artificially, to satisfy the independently and identically distributed (i.i.d.) condition, a prerequisite for most statistical inference and learning algorithms. However, unless the number of sites is large and diverse enough to form a true representative sample of the target population, depending on the type of multivariate models used for analysis, this "mix-and-split" use of multicenter samples may still turn out to be overly optimistic, with results unsustainable in actual field use. With the large number of simultaneously measured variables, it is possible for a complex multivariate model to pick up from a pooled data set the different types of systematic biases that existed in the original individual data sets in its training process. The artificially created i.i.d. condition ensures that the constructed model will perform well in the validation set even though the model's performance may rely on information unrelated to the disease. An alternative and more conservative approach, which we used in our biomarker discovery study for early detection of ovarian caner, is to conduct independent discovery sessions using the data sets separately. The top-ranked targets from these sessions are then

---

* We have observed similar multisegment descending patterns of significance scores in a number of genomic and proteomic profile analyses by stepwise backward UMSA variable selection.

crosscompared/validated to select a common target set (one would often be surprised to see how small this common set could be). These targets will then be further validated using data from additional sites that have not been involved in the discovery phase. This approach will likely miss some useful information for biomarker discovery. However, it mimics during discovery the multicenter validation process that any clinical biomarkers will eventually have to pass before they can be used in practice. Considering the effort and cost required for post-discovery validation, we believe that it is important to incorporate a sound and sometimes even conservative study design into the discovery phase of biomarker research.

Another important issue in study design involves the selection of samples. With the large number of variables available in expression profile analysis, one could easily forget that the selection of informative variables is done in the space spanned and characterized by the samples used in the experiment. The quality of the discovered targets can be only as good as the quality of the samples. Many of the false discovery and overfitting problems are often the results of "trying too hard" when there is actually insufficient information from the variables due to under- or poorly represented samples of the intended population. It is critically important for an expression profiling study of clinical samples to clearly define the endpoint (markers for screening, early detection, or monitoring, etc.) and the target population (high-risk or general population, age groups, etc.) of the potential biomarkers using available clinical and epidemiological knowledge about the disease process. The study design, population inclusion/exclusion criteria, and sample size requirement all have to be based on these choices.

14.8 FINAL NOTES

In this chapter, we discussed a number of issues involved in expression profiling studies of clinical samples. We also presented several expression data analysis procedures based on the unified maximum separability analysis algorithm. With the advances in genomic and proteomic expression profiling technologies, more powerful and complex bioinformatics tools are being developed to screen the large volumes of expression data for potential targets. The commonly observed problem of high dimensionality and small sample sizes poses severe challenges to many traditional multivariate analysis approaches. Methods based on direct empirical risk minimization such as support vector machine may be able to bypass computationally the "curse of high dimensionality"; however, they do not necessarily eliminate the consequence of the problem; i.e., statistically unstable solutions and the possibility of multiple solutions that lead to very different biological interpretations. Expression data analysis to a large degree remains an ill-posed problem for which there is not enough information to reach a definitive solution by informatics tools alone. We have to incorporate biological and clinical knowledge into the analysis process as much as possible to help to identify those solutions that are biologically plausible and meaningful. We need to develop true bioinformatics tools instead of informatics tools that just happen to be used for biological problems.

REFERENCES

1. Golub, T., Slonim, D., Tamayo, P., Huard, C., Gaasenbeek, M., Mesirov, J., Coller, H., Loh, M., Downing, J., and Caligiuri, M. Molecular classification of cancer: Class discovery and Class prediction by gene expression monitoring. *Science,* 286, 531–537, 1999.
2. Buckhaults, P., Zhang, Z., Chen, Y.C., Wang, T.L., St. Croix, B., Saha, S., Bardelli, A., Morin, P.J., Polyak, K., Hruban, R.H., Velculescu, V.E., and Shih Ie, M., Identifying tumor origin using a gene expression-based classification map. *Canc. Res.,* 63, 4144–4149, 2003.
3. Cazares, L.H., Adam, B.L., Ward, M.D., Nasim, S., Schellhammer, P.F., Semmes, O.J., and Wright, G.L., Jr. Normal, benign, preneoplastic, and malignant prostate cells have distinct protein expression profiles resolved by surface enhanced laser desorption/ionization mass spectrometry. *Clin. Canc. Res.,* 8, 2541–2552, 2002.
4. Huang, E., Cheng, S.H., Dressman, H., Pittman, J., Tsou, M.H., Horng, C.F., Bild, A., Iversen, E.S., Liao, M., Chen, C.M., West, M., Nevins, J.R., and Huang, A.T. Gene expression predictors of breast cancer outcomes. *Lancet,* 361, 1590–1596, 2003.
5. Koopmann, J., Zhang, Z., White, N., Rosenzweig, J., Fedarko, N., Jagannath, S., Canto, M., Yeo, C., Chan, D., and Goggins, M. Serum diagnosis of pancreatic adenocarcinoma using surface-enhanced laser desorption and ionization mass spectrometry. *Clin. Canc. Res.,* 10, 860–868, 2004.
6. Li, J., Zhang, Z., Rosenzweig, J., Wang, Y.Y., and Chan, D.W. Proteomics and bioinformatics approaches for identification of serum biomarkers to detect breast cancer. *Clin. Chem.,* 48, 1296–1304, 2002.
7. Ma, X.J., Salunga, R., Tuggle, J.T., Gaudet, J., Enright, E., McQuary, P., Payette, T., Pistone, M., Stecker, K., Zhang, B.M., Zhou, Y.X., Varnholt, H., Smith, B., Gadd, M., Chatfield, E., Kessler, J., Baer, T.M., Erlander, M.G., and Sgroi, D.C. Gene expression profiles of human breast cancer progression. *Proc. Natl. Acad. Sci. USA,* 100, 5974–5979, 2003.
8. Petricoin, E.F., Ardekani, A.M., Hitt, B.A., Levine, P.J., Fusaro, V.A., Steinberg, S.M., Mills, G.B., Simone, C., Fishman, D.A., Kohn, E.C., and Liotta, L.A. Use of proteomic patterns in serum to identify ovarian cancer. *Lancet,* 359, 572–577, 2002.
9. Petricoin, E.F., III, Ornstein, D.K., Paweletz, C.P., Ardekani, A., Hackett, P.S., Hitt, B.A., Velassco, A., Trucco, C., Wiegand, L., Wood, K., Simone, C.B., Levine, P.J., Linehan, W.M., Emmert-Buck, M.R., Steinberg, S.M., Kohn, E.C., and Liotta, L.A. Serum proteomic patterns for detection of prostate cancer. *J. Natl. Canc. Inst.,* 94, 1576–1578, 2002.
10. Rai, A.J., Zhang, Z., Rosenzweig, J., Shih Ie, M., Pham, T., Fung, E.T., Sokoll, L.J., and Chan, D.W. Proteomic approaches to tumor marker discovery. *Arch. Pathol. Lab. Med.,* 126, 1518–1526, 2002.
11. Rosenwald, A., Wright, G., Chan, W.C., Connors, J.M., Campo, E., Fisher, R.I., Gascoyne, R.D., Muller-Hermelink, H.K., Smeland, E.B., Giltnane, J.M., Hurt, E.M., Zhao, H., Averett, L., Yang, L., Wilson, W.H., Jaffe, E.S., Simon, R., Klausner, R.D., Powell, J., Duffey, P.L., Longo, D.L., Greiner, T.C., Weisenburger, D.D., Sanger, W.G., Dave, B.J., Lynch, J.C., Vose, J., Armitage, J.O., Montserrat, E., Lopez-Guillermo, A., Grogan, T.M., Miller, T.P., LeBlanc, M., Ott, G., Kvaloy, S., Delabie, J., Holte, H., Krajci, P., Stokke, T., and Staudt, L.M. The use of molecular profiling to predict survival after chemotherapy for diffuse large B-cell lymphoma. *N. Engl. J. Med.,* 346, 1937–1947, 2002.

12. van de Vijver, M.J., He, Y.D., van't Veer, L.J., Dai, H., Hart, A.A., Voskuil, D.W., Schreiber, G.J., Peterse, J.L., Roberts, C., Marton, M.J., Parrish, M., Atsma, D., Witteveen, A., Glas, A., Delahaye, L., van der Velde, T., Bartelink, H., Rodenhuis, S., Rutgers, E.T., Friend, S.H., and Bernards, R. A gene-expression signature as a predictor of survival in breast cancer. *N. Engl. J. Med.,* 347, 1999–2009, 2002.

13. Tusher, V.G., Tibshirani, R., and Chu, G. Significance analysis of microarrays applied to the ionizing radiation response. *Proc. Natl. Acad. Sci. USA,* 98, 5116–5121, 2001.

14. Efron, B. and Tibshirani, R. Empirical Bayes methods and false discovery rates for microarrays. *Genet. Epidemiol.,* 23, 70–86, 2002.

15. Dudoit, S., Yang, Y., Speed, T., and Callow, M. Statistical methods for identifying differentially expressed genes in replicated cDNA microarray experiments. *Stat. Sinica,* 12, 111–140, 2002.

16. Nguyen, D.V. and Rocke, D.M. Tumor classification by partial least squares using microarray gene expression data. *Bioinformatics,* 18, 39–50, 2002.

17. Tibshirani, R., Hastie, T., Narasimhan, B., Eisen, M., Sherlock, G., Brown, P., and Botstein, D. Exploratory screening of genes and clusters from microarray experiments. *Stat. Sinica,* 12, 47–59, 2002.

18. Weston, J., Mukherjee, S., Chapelle, O., Pontil, M., Poggio, T., and Vapnik, V. Feature Selection for SVMs. In *Advances in Neural Information Processing Systems, 13,* Solla, S.A., Leen, T.K., and Muller, K.-R., Eds. MIT Press, Cambridge, MA, pp. 668–674, 2001.

19. Guyon, I., Weston, J., Barnhill, S., and Vapnik, V.N. Gene selection for cancer classification using support vector machines. *Mach. Learn.,* 46, 389–422, 2002.

20. Adam, B.L., Qu, Y., Davis, J.W., Ward, M.D., Clements, M.A., Cazares, L.H., Semmes, O.J., Schellhammer, P.F., Yasui, Y., Feng, Z., and Wright, G.L., Jr. Serum protein fingerprinting coupled with a pattern-matching algorithm distinguishes prostate cancer from benign prostate hyperplasia and healthy men. *Canc. Res.,* 62, 3609–3614, 2002.

21. Wald, A. *Statistical Decision Functions.* Wiley, New York, p. ix., 1950.

22. Fisher, A., Dickson, C., and Bonynge, W. The Mathematical Theory of Probabilities and Its Application to Frequency Curves and Statistical Methods, 2nd ed. The Macmillan Co., New York, p. v, 1922.

23. Vapnik, V.N. *Statistical Learning Theory.* Wiley-Interscience, New York, 736, 1998.

24. Zhang, Z., Page, G., and Zhang, H. Fishing expedition—A supervised approach to extract patterns from a compendium of expression profiles. In *Microarray Data Analysis II: Papers from CAMDA'01,* Lin, S.M. and Johnson, K.F., Eds. Kluwer Academic Publishers, Boston, 2002.

25. Zhang, Z., Page, G., and Zhang, H. Applying classification separability analysis to microarray data. In *Methods of Microarray Data Analysis: Papers from CAMDA'00,* Lin, S.M. and Johnson, K.F., Eds. Kluwer Academic Publishers, Boston, p. 125–136, 2001.

26. Burges, C.J.C. A tutorial on support vector machines for pattern recognition. *Data Min. Knowl. Discov.,* 2, 121–167, 1998.

27. http://www.netbeans.org/.

28. Li, J., White, C., Zhang, Z., Rosenzweig, J., Mangold, L., Partin, A., and Chan, D. Detection of prostate cancer using serum proteomic pattern. *J. Urol.,* 171, 1782–1787, 2004.

29. Hutchens, T.W. and Yip, T.T. New desorption strategies for the mass-spectrometric analysis of macromolecules. *Rapid Comm. Mass Spectrom.,* 7, 576–580, 1993.

30. Fung, E.T., Thulasiraman, V., Weinberger, S.R., and Dalmasso, E.A. Protein biochips for differential profiling. *Curr. Opin. Biotechnol.,* 12, 65–69, 2001.

15 Sample Characterization Using Large Data Sets

Brian T. Luke

CONTENTS

15.1 INTRODUCTION

Current proteomic investigations are able to generate large amounts of data for a relatively small number of samples representing different classes. These classes can represent diseased versus non-diseased patients or tumor cells from different organs. Computationally these data sets can be used to classify the samples. This chapter outlines some of the available characterization procedures. The emphasis is to show that, because of the over-determined nature of the data sets, it is very easy to *numerically* separate one class of samples from another, but creating a biologically realistic classification model is much harder. This requires choosing a small number of relevant features from the data set and using them to build the classification model. Described here are methods for scaling the data set, searching for outliers, choosing relevant features, building classification models, and then determining the characteristics of the models.

Modern experimental techniques are able to produce a large amount of data from a single sample. While the observations, or features, of each sample can number in the thousands, the number of samples is in the tens or hundreds. For example, microarray experiments are able to measure the fluorescence intensity in tens of thousands of wells to determine the concentrations of gene-specific mRNAs. These concentrations are assumed to be proportional to the expression levels of the particular proteins. Similarly, nuclear magnetic resonance or mass spectra of metabolites from intracellular samples (metabolomics) or blood or urine samples (metabonomics) also generate large numbers of features from a given sample.

Each feature set represents a *fingerprint* of the sample that, in certain cases, can be used to understand the biological processes that are taking place. In addition, each feature set may be able to identify the class of the sample. The class of a sample can represent its histological state or the particular organ that is affected.

Simply finding a computational procedure that distinguishes one class from another is a relatively easy numerical problem since the number of features is many times larger than the number of samples. A harder problem is to use samples that represent two or more classes and produce a concise model that can correctly classify the samples. One such procedure is to identify a small set of features that distinguish one class from another. The hope is that these relevant features can be used to learn about the underlying biochemical basis for a disease.

This chapter outlines some of the computational procedures that can be used to construct such a classification model. It starts with various methods for scaling the features from different samples so that all samples have an equal effect on the model. It is followed by a discussion of different procedures that can be used to search for outliers. Outliers are particular samples that are sufficiently different from the other samples of the same class that they must be treated as special cases; if they were included in the model generation, they would have an unusually large effect of perturbing the model to account for them.

Since only a small number of features will be used and the size of the feature set is so large that an exhaustive search is not possible, a feature selection method needs to be employed. Several heuristic and stochastic feature selection methods are described and their relative strengths and weaknesses are presented.

In order to classify two samples as representing the same state, some measure of their similarity or difference must be available. Therefore, after the discussion of various feature selection methods is presented, a brief description of several distance

TABLE 15.1
Notation Used in this Chapter

Symbol	Meaning
N	Total number of samples
L	Total number of features (wells, genes, peaks, bins, etc.)
J	Number of features used in a classification model ($J \ll L$)
K	Number of clusters or nearest neighbors used in a classification model
X	$N \times L$ matrix containing the data set
$x_{i,j}$	Value of a particular feature ($i = 1, N; j = 1, L$)
X_i	Sum of feature values for sample i [$\Sigma_{j=1,L}(x_{i,j})$]

metrics is given. This is followed by an overview of different classification procedures that can be employed. This section starts with a comparison of *crisp* vs. *fuzzy* classifications and then describes how either type of classification is possible using K-nearest neighbors, clustering, and neural networks.

Once a particular set of features, distance metric, and classification method are combined to produce a good classification model, this model needs to be numerically examined to determine its quality. To this end, a discussion of how a jackknife and/or a bootstrap analysis can be used to determine the robustness of the model is presented. For classification models that rely on clusters, including a self-organizing map, the average silhouette width and a Kelley analysis can quantitatively determine if one model is better than another and if the set of training samples is well described by this model. In addition, if a fuzzy classification is used, a receiver operating characteristic analysis can be used to determine an optimum threshold value for classification.

To assist in the presentation of various methods, the nomenclature listed in Table 15.1 will be used throughout.

Before the experimental results can be computationally examined, they may have to be processed first. For example, in a microarray experiment each gene product is represented by multiple wells. On-chip control wells can be used to correct differences in fluorescence from one region to another and the concentration/expression levels can then be summed or averaged for each gene.

In spectroscopic studies, the background levels need to be removed, as do peaks caused by a substrate, solvent, or other entities. Figure 15.1 represents an NMR spectrum of a urine sample.[1] The three major peaks represent the normalization standard, RF-damped water, and urine. These peaks need to be removed, and the fingerprint of this sample is really the collection of smaller peaks.

Two techniques can be used to reduce the raw data to a manageable size. The first fits each absorption peak with a Gaussian (Figure 15.2) and represents this with a single peak proportional to either the peak height or the total area. The second

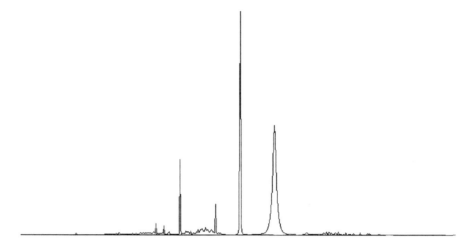

FIGURE 15.1 Example of an NMR spectrum of a urine sample. (From Lucas, D.A., et al., 2003. With permission.)

divides the spectra into bins and replaces the integrated intensity in the bin with a single peak (Figure 15.3).

At this point, each sample is represented by a set of L features, independent of the experimental source. For N samples, the full data set is represented by an $N \times L$ matrix **X**.

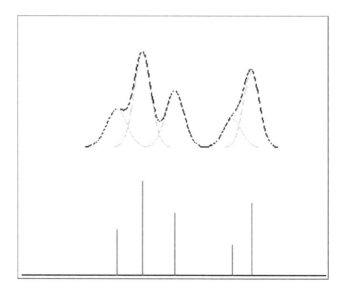

FIGURE 15.2 Example of replacing a spectral peak with a single line.

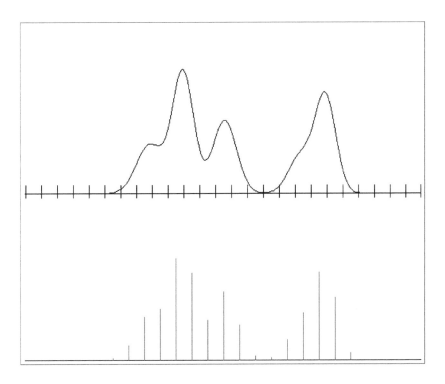

FIGURE 15.3 Example of replacing a continuous spectrum with discrete, binned intensities.

15.2 SCALING THE SAMPLES

It is highly recommended that each row of this data set be scaled (horizontal scaling or scaling by sample). This ensures that all samples, and any new samples, are treated equally. Common scaling methods include

X_i = constant (sum of feature values is a constant)

$\text{MAX}_{j=1,L}(x_{i,j})$ = constant (maximum feature value is a constant)

Care should be used to ensure that the scaling makes physical sense or is not improperly used. For example, the cosine similarity measure can be used in a distance metric (see below). To simplify the construction of a distance matrix, it is useful to normalize the selected feature values so that the sum of their squares equals 1.0. For this reason, the program CLUSTER[2] lets you scale the microarray expression levels so that

$$\Sigma_{j=1,L} \, x_{i,j}^2 = 1.0$$

This is done because the cosine similarity can be used with all genes to compare different samples, but it should not be used with any other distance metric since it destroys relative patterns and distances between samples (Figure 15.4).

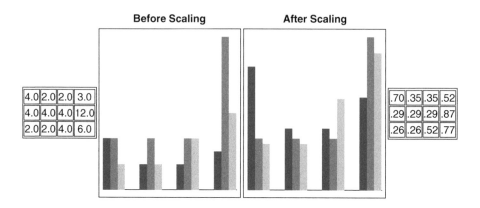

FIGURE 15.4 Effect of data normalization on relative sizes of feature values.

Similarly, the Pearson's correlation coefficient can also be used to determine the similarity of two samples.[3] To simplify the calculation of these pairwise coefficients the features for each sample can be standardized. Standardization sets the mean to zero and the standard deviation to one. This is also known as a standard normal distribution or converting the values to z scores, but again it should not be used unless this particular similarity metric is used since it also destroys relative patterns and distances between samples (Figure 15.5).

Microarray data can cluster the cell lines based on the gene expression profiles and/or can cluster the genes based upon expression levels in the different cell lines. The same scaling/distance metric does not have to be used in both. For example, Miki and coworkers[4] used standardization/Pearson's correlation to cluster the cell lines and normalization/cosine similarity to cluster the genes. This means that the expression level of a given gene in a particular cell line was given different scaled

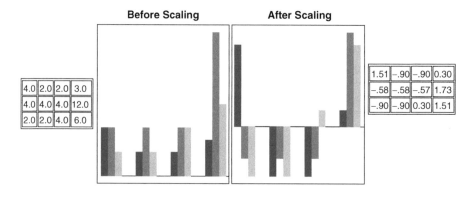

FIGURE 15.5 Effect of standardizing the data on relative sizes of feature values.

values in each clustering. There is nothing inherently wrong with this procedure; the experimentalists and anyone examining these results just need to be aware that this was done.

15.3 OUTLIER DETECTION

The next step is to examine the scaled data and remove any outliers. The standard technique for finding outliers is to use statistics to determine if an unexpected feature value occurs. For example, if a sample has a value that is more than two or three standard deviations from its mean, it can be considered an outlier. Underlying this are the basic assumptions:

1. The number of samples is large
2. Each sample has a small number of features associated with it
3. The values of a particular feature, when taken over all samples, form a normal, or Gaussian distribution

For the data sets examined here, none of these assumptions hold:

1. The number of samples is relatively small so that an accurate determination of the mean and standard deviation of a feature is not possible
2. There are a large number of features for each sample, and so being an outlier for a few of them should not make the sample an outlier.
3. The hope is that the set of values for one or more features is actually multi-modal so that it can be used to distinguish one state from another.

Therefore, methods that use the entire set of feature values (fingerprint or profile) for each sample will have to be used to find outliers.

A common practice is to perform a principal component analysis. Officially, the principal components are the eigenvectors of the variance/covariance, or dispersion, matrix that correspond to the largest eigenvalues. Principal component analysis is also known as factor analysis or Karhunen-Loeve transform or eigenanalysis. Unofficially, the principal components are linear combinations of features that do the best job of spreading out the data. The hope is that by examining the maximum spread of the data, outliers can be seen.

Since the number of features (L) greatly exceeds the number of samples (N), there are at most N non-zero eigenvalues. The magnitude of an eigenvalue divided by the sum of all N eigenvalues represents the fraction of the total variance in the data that is accounted for by its eigenvector (principal component). Figure 15.6 displays the first four eigenvalues of a particular data set and shows that the first principal component accounts for 49.6% of the total variance of the data set and the first four combine to account for 87.9% of the sample's variance.

Because the variance/covariance matrix is real and symmetric, the set of at most N eigenvectors (principal components) are orthogonal. Each principal component is a linear combination of all features that can be normalized to a unit vector, and by inserting the feature values into this expression, the projection of the sample onto this component is determined. Therefore, each sample that was

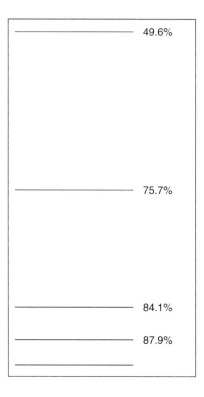

FIGURE 15.6 Percentage of the total variance contained within the first four principal components.

described by L features can now be described by (at most) N principal values. It also allows for Euclidean plots using the dominant principal components (Figure 15.7), although *the distances between the samples are only conserved if all components are used*. This figure definitely shows that the data set contains an outlier.

Another method of graphically searching for outliers is to use a Sammon map.[5] If there are L features for a given sample, the sample can be thought of as a point in L-dimensional space (assuming at least L samples). The Sammon map is a (nonunique) projection of the samples onto a lower M-dimensional space such that the distance between all pairs of points is preserved to the greatest possible extent (Figure 15.8). This map again shows the presence of a single outlier.

A third way to search for outliers is to return to its basic definition. By definition, an outlier is a sample whose feature values are significantly different from the others. A simple histogram plot of the distances to the closest three neighbors will show if there are one, two, or three outliers. Isolated outliers will show a large distance to all three of their closest neighbors. A close pair of outliers will show a small closest distance but large second and third closest distances. A close triplet of outliers will show small first and second closest distances but a large third closest distance.

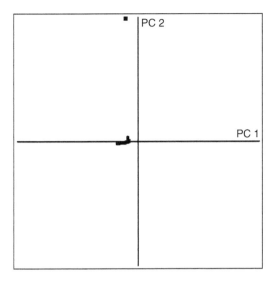

FIGURE 15.7 Plot of the first two principal components showing the presence of a single outlier.

Examples of this plot are shown in Figure 15.9 and Figure 15.10. In Figure 15.9, this procedure also shows that there is a single outlier, but unlike the previous two plots it identifies the 12th sample as the outlier. Figure 15.10 not only shows the presence of a single outlier, but also a pair of samples that are close to each other but far away from all other samples in the data set.

Two points must be emphasized about principal components: (1) the first few principal components are the linear combinations of features that result in the largest

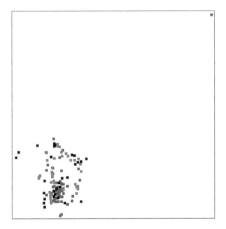

FIGURE 15.8 A two-dimensional Sammon map showing the presence of a single outlier.

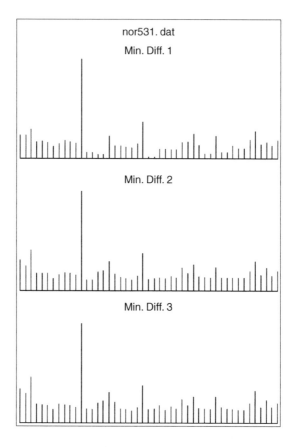

FIGURE 15.9 Histogram plot of the distances from each sample to its three nearest neighbors. This plot shows a single outlier.

variation (spread) in the data, and (2) distance is not conserved between samples unless all principal components are used.

If the data set contains the binned NMR/mass spectra of different samples, each sample will be composed of a few bins with large values (and therefore large absolute spreads) and many bins with small values (with small absolute spreads). The first few principal components will be mainly composed of the bins with large spreads. For example, Figure 15.11 shows 121 samples; 540 bins; 8 bins with large intensities; and all samples have equal summed intensities. This figure shows each bin's intensities centered about their medians. A black sample clearly has intensities that are above or below all others in all but the eight intense bins, while a dark gray sample is either high or low in seven of the eight intense bins but is never an outlier.

A plot of the first two principal components is shown in Figure 15.12. It suggests that the dark gray sample is an outlier while the black sample definitely is not.

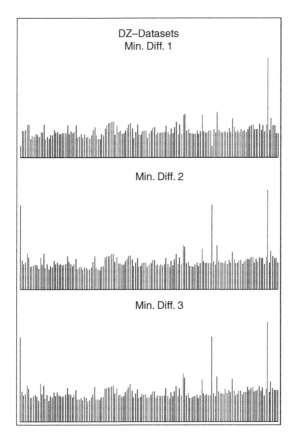

FIGURE 15.10 Histogram plot of the distances from each sample to its three nearest neighbors. This plot shows a single outlier and a closely spaced pair of outliers.

A Sammon map (Figure 15.13) shows that the black sample is an outlier while the dark gray sample is not. A display of the three nearest distances (Figure 15.14) also shows that the black sample is an outlier while the dark gray sample is not.

Since spectral data sets will have a small number of features with large intensities, and therefore potentially large variations, principal component analysis will only be able to locate an outlier if it is an outlier in one or more if these bins. If all of the outlier's intensity is located in another bin, it is possible that this bin will have a large enough variance that it will have a significant contribution to one of the first few principal components, but since the summed intensities are constant this also means that its intensity will be very low in one or more of the intense bins. If an outlier has a large distance from all other samples but this distance is spread across the large number of bins with small variances, it will not be identified by PCA.

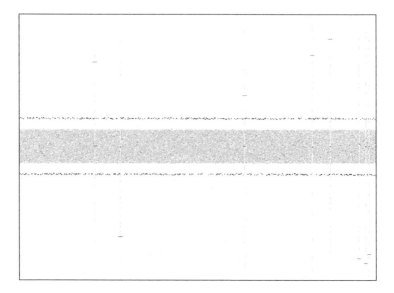

FIGURE 15.11 Plot of spectral intensities of 121 samples in 540 bins. Each bin has the intensities distributed about its mean value and all samples have a constant summed intensity. There are eight high-intensity (larger variation) bins. The black sample is an outlier in all but the eight high-intensity bins, while a black gray sample has intensities near the extremes in seven of the eight high-intensity bins but is never an outlier.

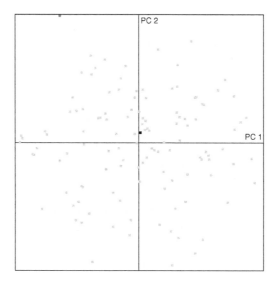

FIGURE 15.12 A principal component plot of the data shown in Figure 15.11 suggests that the dark gray sample is an outlier while the black sample is not.

FIGURE 15.13 A Sammon map of the data shown in Figure 15.11 suggests that the black sample is an outlier and the dark gray sample is not.

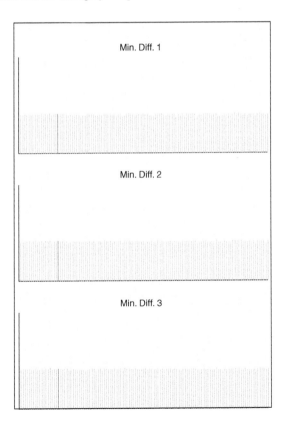

FIGURE 15.14 A nearest neighbor plot of the data shown in Figure 15.11 suggests that the black sample is an outlier and the dark gray sample is not.

15.4 CONSTRUCTING A CLASSIFICATION MODEL

It would be optimal if the overall fingerprint or profile of each sample can be used to classify it, and in some cases this is possible to a limited extent. In reality, most of the features have nothing to do with the biology that distinguishes one class from another. A good classification model would only use the features that amplify their differences.

When microarray or NMR data is used, a particular feature can be related to a particular protein or metabolite, and this can be used to establish a biological basis for the classification model. For this reason, principal components should not be used. Though they reduce the number of features from L to N, each component is a linear combination of up to L features and offers no biological information.

There are three basic steps in constructing a classification model:

1. Use a feature selection method to select sets of J features
2. Choose a distance metric that determines how similar/different two samples are with a given set of J features
3. Construct a classification model using these distances

The ability of the classification model to correctly determine the class of each sample is a measure of how good a set of features is. The goal is to determine the set of features that classifies the best for this distance metric and classification model. Changing either the distance metric and/or the classification model can change the optimum set of J features.

15.4.1 FEATURE SELECTION METHODS

Since there are L ways to choose one feature, $L(L - 1)/2$ ways to choose two unique features, and so on, choosing J features scales as $O(L^J)$. This means that trying all possible sets of J features is computationally impractical. A smarter way of searching for the best set of J features needs to be used.

There are three basic classes of feature selection methods:

1. Heuristic searches
2. Stochastic searches using a single solution
3. Stochastic searches using a population of solutions

15.4.1.1 Heuristic Searches

The simplest heuristic method is the Greedy search. The basic operation of a Greedy search is as follows:

1. Try each feature individually and find the one that classifies the samples the best
2. Keeping this "best" feature, sequentially try all of the others and find the best combination of two features

3. Continue this procedure of trying all remaining features with the set found best at the previous level until all J features are found

This process scales as $J \times L$, which is quite fast, and it generally leads to a good, but suboptimal, solution. The basic reason for this is that only the first selected feature acts directly on separating one class from another; the second feature's goal is to reduce any remaining errors (misclassifications). In general, better results are obtained if all pairs of features are considered since they can both be used to distinguish the classes. This argument can be extended to more than two features and in the limit of J features an exhaustive search is produced.

Branch and Bound is similar to a Greedy search, only instead of keeping the best set of features at each level, the best NBB feature sets are kept. To run this method, the following arrays are used:

FSOLD(J, NBB), QUOLD(NBB)
FSOLD(J,NBB) contains the previously selected NBB sets of features.
QUOLD(NBB) is the quality (ability to classify) of each feature set.
FSNEW(J,NBB), QUNEW(NBB)
FSNEW(J,NBB) is a temporary matrix holding the best NBB sets of new features found in a cycle.
QUNEW(NBB) are their quality values.

The search runs as follows:

1. Examine all $L(L-1)/2$ unique pairs of features and determine their quality. Store the best NBB sets, in order of decreasing quality, in FSOLD(J,NBB) and their quality values in QUOLD(NBB).
2. Set the counter j to 2.
3. Increment j by one and zero-out QUNEW(NBB).
4. For each set in FSOLD(J,NBB):
   a. Sequentially add a feature to this set to form a set of j features.
   b. Determine the quality of this feature set.
   c. Search down QUNEW(NBB) to determine if this feature set has a higher quality than any of the newly created sets. If so:
      i. Create a space by moving each set in FSNEW(J,NBB) and each quality in CNEW(NBB) down one, dropping the last entries off the list.
      ii. Place this feature set into FSNEW(J,NBB) and its quality in QUNEW(NBB).
5. Copy FSNEW to FSOLD and QUNEW to QUOLD.
6. If j is less than J, go to Step 3. Otherwise stop.

Care should be used in programming Step 4a to ensure that the same set of j features is not examined more than once. Each time a new set of features is created it should be compared to all previously treated sets in FSOLD(J,NBB). If all but one of the features is already present, this new feature set has already been examined.

When finished, the Branch and Bound method produces *NBB* unique sets of *J* features in decreasing quality. In addition, after completing Step 5, this process contains the best NBB containing 2 through *J* features. The best feature sets can be stored in each cycle and reported at the end.

The example above is used to maximize the quality of feature sets, but it could just as easily minimize the cost of feature sets. One disadvantage of this method is that finding the optimum set of *J* features is not guaranteed unless NBB is large enough that this becomes an exhaustive search. In addition, the computational time grows very rapidly with NBB. This is not only because many more combinations have to be tried but also the search through QUNEW and the updating of FSNEW and QUNEW becomes harder.

15.4.1.2 Stochastic Feature Selection Using a Single Solution

Three different stochastic search methods that modify a single solution are well suited to the feature selection problem. They are

1. Tabu search
2. Simulated annealing
3. Gibbs sampling

Tabu search[6,7] can be either a minimization (of the cost) or maximization (of the quality) procedure. This method starts with a randomly generated array of *J* features [FSOLD(J)]. The quality of this feature set is QUOLD.

Before the search starts, an array called the tabu list, TABU(J') [*J' < J*], is zeroed out and FSOLD(J) and QUOLD are copied to FSBEST(J) and QUBEST, respectively. In addition, the maximum number of search cycles, MAXCYC, is set. The Tabu search proceeds as follows:

1. Set NCYC = 0
2. Increment NCYC by one
3. Randomly choose an integer *I* between 1 and *J* to represent the member of the feature set to be changed
4. Look through TABU(J') to see if *I* is listed. If it is, return to Step 3
5. Sequentially try all other features in position *I* to create (*L − J*) new feature sets
6. Determine the classification quality of each set and save the best new feature set and quality in FSNEW(J) and QUNEW
7. If QUNEW is better than QUBEST, copy FSNEW(J) and QUNEW to FSBEST(J) and QUBEST, respectively
8. Copy FSNEW(J) and QUNEW to FSOLD(J) and QUOLD, whether or not QUNEW is higher than QUOLD
9. Place *I* into TABU(1) by pushing each index in this list down one position and dropping the last index off the Tabu list
10. If NCYC is less than MAXCYC, go to Step 2. Otherwise, stop.

When the search is finished, FSBEST(J) contains the best set of features. This method scales as MAXCYC $\times$ L, which is substantially less than the computational load of the Branch and Bound method. In addition, the replacement in Step 8 allows this method to accept a lower-quality solution in the hope of finding the best solution.

The Tabu list TABU(J') ensures that the same region of search space is not sampled too regularly. An alternative to Step 4 is known as the *aspiration criterion*. With this method, a position in the feature space vector is searched whether or not it is on the Tabu list. If it is on the list, Step 8 is only performed if Step 7 is true. In other words, it only becomes the current feature set if the search produced the "best to date" set. Whether or not this update occurs, index I is moved to the top of the Tabu list.

Simulated annealing[8] is a well-known minimization procedure, but it can also be used to maximize the quality of a feature set. Before this procedure is run, an initial and final effective temperature (Ti and Tf, respectively) needs to be selected, as well as MAXSTEP, which is the number of Monte Carlo steps to attempt at each temperature. Given these parameters, the following procedure is employed:

1. Randomly generate a set of J features, FSOLD(J), and store the quality of this feature set as QUOLD
2. Copy FSOLD(J) and QUOLD to FSBEST(J) and QUBEST
3. Set the effective temperature T to Ti
4. For MAXSTEP steps, do the following:
   a. Copy FSOLD(J) to FSNEW(J)
   b. Randomly change a feature (or a small number of features) in FSNEW(J) to create a new feature set
   c. Calculate the quality of this new feature set (QUNEW)
   d. If QUNEW is greater than or equal to QUOLD, then
      i. Replace FSOLD(J) and QUOLD with FSNEW(J) and QUNEW
      ii. If QUNEW is greater than QUBEST, replace FSBEST(J) and QUBEST
   e. If QUNEW is less than QUOLD, then
      i. Set ΔQ to (QUOLD-QUNEW) and calculate the Boltzmann probability $BP = e^{-\Delta Q/T}$
      ii. If BP is greater than a random number between 0.0 and 1.0, FSNEW(J) and QUNEW become FSOLD(J) and QUOLD; otherwise FSOLD(J) and QUOLD stay the same (the Monte Carlo step is rejected)
5. If T is greater than Tf it is decreased and the process returns to Step 4.

Because a change that decreases the quality can be probabilistically accepted, the hope is that this method can move away from a local maximum to find the global maximum. This also means that the search can find the global maximum and then "walk" away from it. Therefore, the best to date solution needs to be stored and this solution, not the last feature set, is the one that should be used.

Running a simulated annealing search is as much of an art as it is a science. The initial effective temperature (Ti) should be high enough to accept many steps that decrease the quality, while the final temperature (Tf) should be low enough to let the search settle on the (hopefully) global maximum. The path taken from Ti to Tf is known as the cooling schedule and, in general, different results will be obtained for linear vs. geometric decreases, or almost any other cooling schedule. In addition, MAXSTEP should be large enough so that the search can explore the entire available feature space at each temperature. In other words, the simulation should be ergodic at each temperature. This is not feasible in practice, so a compromised choice of Ti, Tf, the cooling schedule, and MAXSTEP needs to be made to get a good solution in a reasonable amount of computer time.

Gibbs sampling[9] can be thought of as a combination of a Tabu search and simulated annealing. Each time a position is selected, the old feature set and each of the new $(L - J)$ feature sets are given an unnormalized acceptance probability of

$$\text{UAP}_i = e^{Q(i)/T}$$

where $Q(i)$ is the quality associated with the ith feature set.

Once all feature sets are examined, dividing each by their sum normalizes the acceptance probabilities:

$$\text{NAP}_i = \text{UAP}_i/\Sigma_i\text{UAP}_i$$

A random number between 0.0 and 1.0 is chosen and each NAP_i is subtracted from this number until it becomes zero or negative; and that feature set becomes the base set [FSOLD(J)] for the next search.

As with simulated annealing, the sampling of the feature sets starts at a high temperature so that there is a good chance that any feature set will be chosen. As the temperature drops, only those sets that increase the classification quality or keep it constant will have a good chance of being chosen. Once again, the values of Ti, Tf, the cooling schedule, and MAXSTEP (the number of feature set positions scanned at each temperature) need to be set and the results may vary with different values.

Gibbs sampling differs from a Tabu search in that the starting feature set has a chance of being chosen for the next round (i.e., no change in FSOLD[J]). In addition, a Tabu list can be used, but it is not required here.

15.4.1.3 Stochastic Feature Selection Using a Population of Solutions

Instead of using a single set of features in the search, FSOLD(J), a population of features can be used, $\text{FSOLD}_i(J)$ ($i = 1$, NPOP, where NPOP is the size of the population). Four different stochastic search methods that use a population of solution can be applied to the feature selection problem. They are

1. Genetic algorithms
2. Evolutionary programming
3. Ant colony optimization
4. Particle swarm optimization

Genetic algorithms represent a large class of search heuristics about which several books[10,11] and reviews[12] have been written. This discussion is limited to the simple genetic algorithm (SGA).

This and the other search methods in this section start with a randomly generated population of feature sets $FSOLD_i(J)$. Each feature set has a corresponding quality $QUOLD_i$ and are used as parents to generate new solutions, or offspring.

In an SGA, combining the feature sets of two parents using a mating operator generates an offspring. These feature sets can be thought of as the parents' chromosomes, which are combined through crossover to generate the chromosome of their offspring. In general, this search runs for a fixed number of generations (MAXGEN), which is user-supplied. A "survival of the fittest" strategy is used in parent selection and to determine which feature sets become parents in the next generation.

One example of an SGA is outlined as follows:

1. Randomly generate NPOP feature sets, $FSOLD_i(J)$, and calculate their qualities, $QUOLD_i$. This becomes the parent population.
2. Set the generation counter IGEN to zero.
3. Increment IGEN by one.
4. Zero-out the arrays that will hold the feature sets, $FSNEW_i(J)$, and qualities, $QUNEW_i$, of their offspring.
5. Probabilistically choose two feature sets from the parent population such that parents with a higher quality are more likely to be selected.
6. Use these parents to generate an offspring using a 1-point crossover.
7. Calculate the quality of this offspring and place this feature set and its quality in $FSNEW_i(J)$ and $QUNEW_i$, respectively.
8. If $FSNEW_i(J)$ is not filled, return to Step 5.
9. Combine the parent and offspring populations and probabilistically choose NPOP feature sets to become parents in the next generation.
10. If IGEN is less than MAXGEN, return to Step 3.

The standard probabilistic selection procedure used in Steps 5 and 9 is a roulette wheel procedure. In Step 5, the probability that a particular parent is chosen is given by

$$P_i = QUOLD/\Sigma_{i=1,NPOP}\ QUOLD_i$$

A random number between 0.0 and 1.0 is chosen. The probability for each parent is subtracted from this number and the parent that causes this number to become zero or negative is the selected parent.

The 1-point crossover operator simply chooses a random integer Nr between 1 and $(J-1)$. Features 1 through Nr are taken from one parent and features $(Nr + 1)$ through J from the other. Since either parent can supply the first feature to the offspring a complimentary pair of offspring can be created:

```
A   B   C   D   E      Parent 1
            |          Crossover point
a   b   c   d   e      Parent 2

A   B   C   d   e      Offspring 1
a   b   c   D   E      Offspring 2
```

Both offspring can be examined and either the offspring with the highest quality or both feature sets can be placed in $FSNEW_i(J)$.

Because the next generation's parents are probabilistically chosen from the combined populations, an *elitist strategy* is used. This means that the feature set with the highest quality is automatically chosen and the rest are probabilistically chosen.

Since good parents generally create good offspring, some of the same features will become present in all members of the parent population. This is called *schema* formation and it reduces the dimensionality of the search space. The search will eventually converge onto a single set of J features for the entire population.

Again, this is only a simple example of a large number of algorithms available with this heuristic.[12] Other variations allow each parent to take part in a mating with a second parent that is chosen based on its quality, using a multipoint or uniform crossover procedure as the mating operator, allowing for mutations in the offspring, and using different probabilistic selection procedures in parent selection (Step 5) and forming a new parent population (Step 9). Each of these will affect the rate of schema formation and the quality of the final result.

Although evolutionary programming (EP) was independently developed,[13–15] it shares many features in common with a genetic algorithm. The major differences are that all parents are able to produce offspring and this offspring generation is asexual reproduction using a mutation operator.

An EP algorithm that is found to be an effective feature selection process can be outlined as follows:

1. Randomly generate NPOP feature sets, $FSOLD_i(J)$, and calculate their qualities, $QUOLD_i$. This becomes the parent population.
2. Set the generation counter IGEN to zero.
3. Increment IGEN by one.
4. Zero-out the arrays that will hold the feature sets, $FSNEW_i(J)$, and qualities, $QUNEW_i$, of their offspring.
5. Choose each parent to generate an offspring using the mutation operator.
6. Compare the offspring's feature set with all parents and offspring generated so far. If it is not unique, return to Step 5 and generate a new offspring.
7. Calculate the quality of this offspring and place this feature set and its quality in $FSNEW_i(J)$ and $QUNEW_i$, respectively.
8. If all parents have not created an offspring, return to Step 5.
9. Combine the parent and offspring populations and deterministically or probabilistically (with the elitist strategy) choose NPOP feature sets to become parents in the next generation.
10. If IGEN is less than MAXGEN, return to Step 3.

To create an offspring, the parent's feature set is copied to the offspring. One, or a small number, of the features is randomly chosen and replaced by a randomly selected feature.

The uniqueness criterion used in Step 6 is one of a broad class of operators called *maturation operators*.[12] Its basic effect is to stop the population from converging on a single solution. This can increase the computational time needed to complete the search, but is generally found to produce better results.

Ant colony optimization (ACO) is modeled after the movement of ants to and from a food source.[16] As each ant travels a route it lays down a pheromone trail, and the better routes will have a stronger trail and are more likely to be used again. Therefore, each new ant uses the information deposited by every member of the parent population and is not dependent upon the genetic information of a particular parent. This *social-only* model can be used in the feature selection problem.

At the start of the search, each of the L possible features is assigned a probability p_j of $1/L$ of being used. Each ant in the population is given a random set of features, $FS_i(J)$, and they are used with the distance metric and classification method to produce a classification model with a quality of Q_i. Once all members of the population are created and examined, the array that is used to update the feature selection probabilities, Δp_j, is set to zero. Each member of this array is updated using the formula

$$\Delta p_j = \Sigma_{i=1,NPOP} \ \delta_{ij} \ Q_i$$

In this expression, the sum is over all NPOP members of the population and δ_{ij} is 1.0 if feature j is used in set i, and 0.0 if it is not. β is a positive constant that is less than 1.0 and modulates the extent to which Δp_j increases as Q_i increases. The unnormalized probability of selecting the jth feature now becomes

$$p_j = (1 - \epsilon)_j + \epsilon \Delta p_j$$

In this expression, ϵ represents the evaporation rate and causes the importance of a suboptimal feature to decrease with time.

At the start of a new "generation" each ant is created from scratch. Each of its J features is selected using a random number r and a threshold value r_o. If r is less than or equal to r_o the unselected feature with the largest p_j is chosen, while if it is greater than this value a probabilistic selection procedure covering all unselected features (i.e., a roulette wheel selection) is used. Once all J features of an ant are determined its quality is measured. After NPOP feature sets are examined, p_j are recalculated and the process continues. Eventually, only J features will have significant values of p_j, and they represent the final set of features.

Particle swarm optimization (PSO)[17,18] is similar to ACO, but in this search each feature set is called a particle instead of an ant or parent/offspring. What differentiates PSO and ACO is that a good feature set used by a given particle also contributes to its search as well as the best feature set found by any particle. The J features that describe a particular particle, $FS_i(J)$, are floating point numbers instead of integers. To use a particular instance of $FS_i(J)$, they are first converted into an integer array using

$$IFSi(J) = NINT(FSi(J))$$

where NINT is the "nearest integer to" function. The integer feature set is then used to construct a classification model with a quality of Q_i. Also associated with each particle is a velocity vector $V_i(J)$ that determines its next step in feature space, and

$BFS_i(J)$ and BQ_i, which contains the (integer) best feature set found by this particle to date and its quality, respectively. $GFS(J)$ and GQ contain the best feature set and quality found by any particle during the search, and $RAN1_i(J)$ and $RAN2_i(J)$ are arrays of random numbers with each element between 0.0 and 1.0.

The PSO algorithm can be outlined as follows:

1. Randomly generate a set of numbers in $FS_i(J)$, $i = 1,NPOP$, such that each number is between 0.5 and FLOAT(L)+0.5 and the numbers are in ascending order for each i. Also randomly load $V_i(J)$ with random numbers.
2. For each particle load $IFS_i(J)$ use the expression above and use this integer set of features to calculate the set's quality Q_i. Save $IFS_i(J)$ and Q_i in $BFS_i(J)$ and BQ_i, respectively.
3. Find the largest value of BQ_i and store $FS_i(J)$ and BQ_i in $GFS(J)$ and GQ, respectively.
4. Initialize ICYC to zero.
5. Increment ICYC by 1.
6. Load $RAN1_i(J)$ and $RAN2_i(J)$ with random numbers.
7. Load $V_i(J)$ using the expression

$$V_i(J) = V_i(J) + C1 \times RAN1_i(J) \times [BFS_i(J) - FS_i(J)]$$
$$+ C2 \times RAN2_i(J) \times [GFS(J) - FS_i(J)]$$

8. Update $FS_i(J)$ using the expression

$$FS_i(J) = FS_i(J) + V_i(J)$$

9. For each particle load $IFS_i(J)$, use this integer set of features to calculate the set's quality Q_i. If Q_i is greater than BQ_i, update $BFS_i(J)$ and BQ_i with $IFS_i(J)$ and Q_i. If Q_i is greater than GQ, update $GFS(J)$ and GQ with $IFS_i(J)$ and Q_i.
10. If ICYC is less than MAXCYC, return to Step 5.

The net effect of the PSO algorithm is to allow each particle (feature set) to randomly move with a component of the motion directed toward the best solution it has found and toward the best solution found by all particles.

The constants C1 and C2 are user supplied and represent the maximum influence of these two controlling motions. If $C2 = 0.0$, this is a *cognition-only* model where each particle is only influenced by the best solution it has found. Conversely, if $C1 = 0.0$, this is a *social-only* model where each particle is only influenced by the best solution found by any particle. Literature results[17,18] suggest that $C1 = C2 = 2.0$ produces good results. With non-zero values for C1 and C2, all particles will eventually converge onto $GFS(J)$, and this feature set is the one that produces the best classification model.

Obviously an exhaustive search will give the best solution, but the size of the search space makes this method impractical. Branch and Bound is semiexhaustive and can give good results as long as NBB is large enough. Unfortunately, as NBB increases, the computational cost increases dramatically and a detailed enough search may not be possible.

Besides these two methods, evolutionary programming is the only other search heuristic discussed above that produces a collection (or population) of unique, good feature sets since the uniqueness operator can force population diversity at every step. The survival-of-the-fittest selection causes each feature set to be as good as possible, as long as it is unique.

Greedy search, Tabu search, simulated annealing, and Gibbs sampling only use a single solution and so only a single good solution is produced. Genetic algorithms, ant colony optimization, and particle swarm optimization are all designed to have the entire population converge on a single (hopefully) optimal set of features.

Obtaining a population of diverse, good solutions has a couple of advantages. The first is that this population can be searched to see if any features are present in a large number of the sets. If one or more features are used regularly, they are probably present because they do a reasonably good job characterizing the subjects and are the most likely candidates for biomarkers. The other features are added to take care of any minor errors.

The second advantage is particular to EP because of the nature of this search heuristic. As the experimental investigation proceeds, many new subjects could be added to the study and a search for a new classification model using the enlarged data set would be necessary. With all methods but EP the feature search would have to start from scratch, while here the new search can start with these old results. If adding new subjects causes a relatively small perturbation, the population of unique feature sets should have one or more members that are reasonably close to the new optimum set and can guide the search in that direction.

The same argument holds if it is found that one of the features is an artifact of different runs, or different laboratories, and this feature should be removed. In the EP search, any member of the population that has this feature can replace it with a randomly chosen one (again checking for uniqueness) and the search can continue. All other methods would have to start from scratch.

15.4.2 Distance Metrics

At this point, the feature selection method has chosen a particular set of J features that will be used to characterize the samples to the best possible extent. This classification procedure will need to know how similar or different a pair of samples is from each other. One way to do this is to calculate the distance between each pair with respect to this set of J features.

The most common distance metrics are known as L_N-norms. The L_N-norm between subjects i and k is defined by the expression

$$L_N(i,k) = [\Sigma_{j=1,J} |x_{i,j} - x_{k,j}|^N]^{1/N}$$

A plot of all points that have an L_N-norm that is less than or equal to a constant from a given point produces the results shown in Figure 15.15 for different values of N.

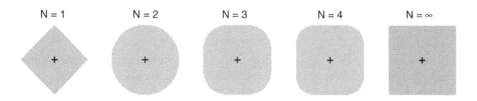

FIGURE 15.15 Plot of the points that have an L_N-norm distance that is less than or equal to a constant from a point for various values of N.

The L_1-norm is known as the city-block, or Manhattan, distance and is the sum of the absolute difference in each feature. The L_2-norm is the standard Euclidean distance and is the square root of the sum of the differences in each feature squared. The L_∞-norm is called the Chebyschev distance and is the maximum absolute difference in any single feature.

Each of these metrics represents a different way of "thinking" about distances. For example, if a building was four blocks down the street and then three blocks down a perpendicular street, the distance to the building depends upon the metric. If a Manhattan distance is used, the building would be seven blocks away (four down and three over). Conversely, the Euclidean distance would be five blocks "as the crow flies" and the Chebyschev distance would be no more than four blocks in either direction.

Other values of N in the expression above yield different results. In addition, several other distance metrics can be used. For example, the Canberra distance between two samples is given by the expression

$$D(i,k) = \Sigma_{j=1,J} (|x_{i,j} - x_{k,j}|/|x_{i,j} + x_{k,j}|)$$

This distance can be considered a *relative* Manhattan distance since the difference in each feature's value is divided by their sum. The squared chord distance between two samples is defined by

$$D(i,k) = \Sigma_{j=1,J} [(x_{i,j})^{1/2} - (x_{k,j})^{1/2}]^2$$

This metric obviously requires all feature values to be nonnegative and will have to be accounted for in the scaling of the data set for each sample. Finally, the squared chi-squared distance is given by

$$D(i,k) = \Sigma_{j=1,J} [(x_{i,j} - x_{k,j})^2/|x_{i,j} + x_{k,j}|]$$

This metric is basically a relative Euclidean-squared distance.

As mentioned in the section on data set scaling, distance metrics can also be obtained from similarity measures. For J features, each sample represents a point in J-dimensional sample space, or a vector that represents the line segment from the origin to each point.

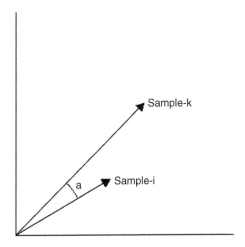

FIGURE 15.16 Plot showing how two samples can be viewed as vectors in J-dimensional space and the angle a between these vectors.

Both the cosine similarity and Pearson's correlation coefficient for a pair of samples can be defined in terms of the dot-product of their vectors. This, in turn, depends upon the angle between their (possibly scaled) vectors, a. A plot of this for $J = 2$ is shown in Figure 15.16. Using the definition of the dot product, the cosine similarity between two samples is given by

$$\cos(a) = \Sigma_{j\,=\,1,J}\ (x_{i,j}\ xk,j)/(|x_i|\ |x_k|)$$

The denominator is the magnitude of the two vectors. By normalizing the vectors for each sample relative to the J selected features, the denominator disappears. This is why normalization of expression levels is an option in the CLUSTER program. To obtain a distance metric from this similarity value one can use $[1 - \cos(a)]$ or just the absolute value of a. In the first case the distance is bounded by $[0.0, 2.0]$ while in the second it is bounded by $[0.0, \pi]$. The Pearson's correlation coefficient, r, can be defined by the expression

$$r = \Sigma_{j=1,J}\ [z_{i,j}\ z_{k,j}]/J$$

Here the J feature values for each sample are standardized to a mean of zero and a standard deviation of one (i.e., converted to their z scores). A distance metric can be represented by $(1 - r)$, and this distance is again bounded by $[0.0, 2.0]$.

Figure 15.17 represents a plot of the relative distances between two samples as the angle between them varies from 0.0 to π. The first sample has feature values of $(1.0, 0.0)$, and the second lies on a circle centered at the origin with a radius of 2.0. The Canberra, squared chord, and squared Chi-squared distances are obtained by adding 2.5 to the feature values in all cases so that all of the feature values are positive.

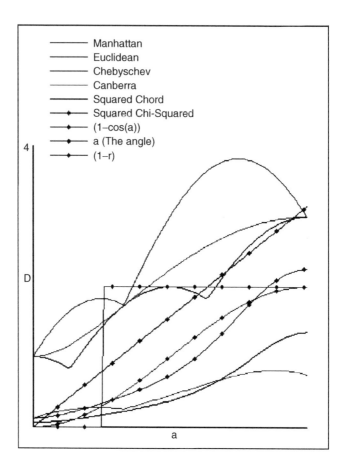

FIGURE 15.17 Plot of the distance between two samples as a function of the angle between their vectors for different metrics.

As this figure shows, the effective distance between the two samples varies significantly for different distance metrics. Of particular interest is the use of the Pearson's correlation coefficient r in a distance metric. The equation above shows that this is the dot-product of the vectors defined by the z score of each feature value, divided by the number of features. For $J = 2$, there are only two possible standard vectors defined by the endpoints; $(1,-1)$ and $(-1,1)$. The first vector represents all samples where $x_{i1} > x_{i2}$ and the second is for all samples where $x_{i1} < x_{i2}$. Therefore, there are only two possible r values, 1 and -1, depending upon whether the two samples have the same or different standard vectors. This means that all sample pairs must have a distance of either 0.0 or 2.0, and the inequalities show a sharp division between them. Samples with feature values of $(2.01,1.99)$ and $(-1.99,-2.01)$ will have a distance of 0.0, while samples with feature values of $(2.01,1.99)$ and $(1.99,2.01)$ will have a distance of 2.0.

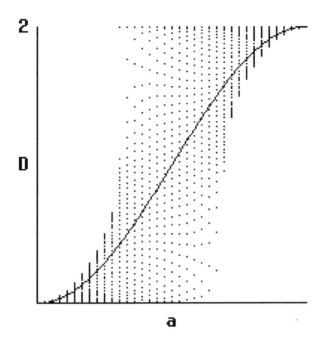

FIGURE 15.18 Plot if the [1 − cos(a)] distance (solid curve) and the (1 − r) distance (dots) between two samples as a function of *a* for different values of *t*.

To search for a more continuous change in distances, this examination is extended to *J* = 3. Here two sample points are placed on the unit sphere centered at the origin. The first sample is fixed with feature values of (0.0, 0.0, 1.0) and the second sample is moved around the unit sphere using spherical coordinates:

$$x_{i1} = \sin(a)\cos(p)$$

$$x_{i2} = \sin(a)\sin(p)$$

$$x_{i3} = \cos(a)$$

The distance between these samples is independent of the value of the angle *p* if the Euclidean distance is used as well as [1 − cos(a)], based on the cosine similarity, and a distance determined by the value of the angle *a*. Figure 15.18 is a plot of the distance between the samples as a function of the angle *a* for different values of the angle *p* for the [1 − cos(a)] distance, solid curve, as well as the (1 − r) distance, isolated dots. This plot shows that for many values of *a*, the (1 − r) distance can vary from 0.0 to 2.0 depending upon the value of *p*.

While the most logical metric for a classification problem depends upon the nature of the quantities and the same metric may not apply in all classification problems, these results suggest that the (1 − r) distance metric should be used with great care since points that have the same Euclidean distance may have very different values if the (1 − r) metric is used.

15.4.3 CLASSIFICATION MODELS

At this point, the feature selection method has chosen a particular set of J features that will be used to characterize the samples to the best possible extent and a metric has been chosen that determines how close two samples, or a sample and a point in J-space, are to each other. The next step is to construct a method that, given a set of samples whose class is known, can accurately determine the class of an unknown sample.

In all of the methods described here, the basic assumption is that if the unknown sample is close to a group of points, and all of the points are of the same class, the unknown sample must also be of this class. The extent to which this is true depends upon the set of features, the distance metric, and how a group of known points is constructed.

The classification methods described here are

K-nearest neighbors
Clustering
Neural networks

Before various classification models are described, it is important to distinguish the difference between a *crisp* and *fuzzy* classification. If the unknown sample is found to lie in a group of ten known samples and seven of the samples are of Class 1 and the other three of Class 2, these classification models yield different results. A crisp classification states that the unknown is definitely Class 1 since this is the majority class, while a fuzzy classification states that it is 70% Class 1 and 30% Class 2.

If a crisp classifier is used in a classification study, it is important to realize that even if the prediction is correct 100% of the time, the uncertainty in the results can approach 50% if there are two classes, 66.6% if there are three classes, and so on. A *maximum likelihood* classification is a "winner takes all" procedure where the unknown is assigned to the class that is present the most even if, as in the case of a multiclass problem, it is represented by less than 50% of the group of known samples.

15.4.3.1 K-Nearest Neighbors

For a given set of J features, each of the known samples represents positions in this J-dimensional space. The unknown sample also represents a point in J-space. As the name implies, this method finds the K known samples that are closest to the unknown sample (K is a user-supplied integer). By polling the classes of the known samples in this neighbor list, a crisp or fuzzy classification of the unknown sample is made.

Although this method appears to make physical sense, it is important to realize that in the standard K-nearest neighbors method the classification is independent of the distance to these nearest neighbors. This means that a sample that is close to some of its neighbors and relatively far from others will weigh them evenly in determining its classification. Similarly, this method does not distinguish between a sample that is close to all of its neighbors and one that is relatively far from them. For example, in the three cases shown in Figure 15.19, K = 3 and the unknown sample (shown as gray) needs to be fuzzy classified as fractionally black and white, based on the identity of its three neighbors.

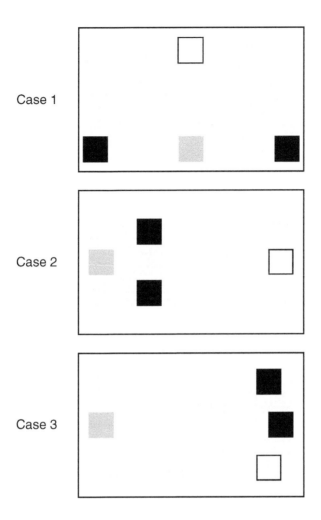

FIGURE 15.19. Different distributions of the three nearest neighbors of known samples (black and white) about an unknown sample (gray).

Case 1: In this case, the unknown is equidistant from two black and one white sample, and relatively close to all of them. Therefore, this unknown should be classified as 66.7% black and 33.3% white.

Case 2: In this case, the unknown is much closer to the black samples than it is to the white. Therefore, it should be considered more than 66.7% black and less than 33.3% white.

Case 3: Here the unknown sample is relatively far away from any of its neighbors. In this case, classifying the unknown as either black or white is suspect.

To handle the situation in Case 2, the unnormalized probability that the unknown sample is in the same class as its ith neighbor, $P(C_i)$, is taken to be a decreasing

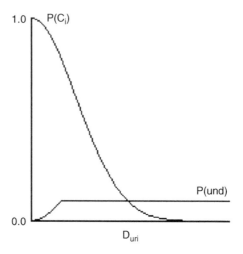

FIGURE 15.20 Probability distribution plots for belonging to the same class as neighbor i, $P(C_i)$, and belonging to the undetermined class as a function of the distance to the neighbor.

function of the distance between the unknown sample and this neighbor, $D_{u,i}$:

$$P(C_i) = f(D_{u,i})$$

To take care of Case 3 a new class called *undetermined* is added to the calculation. The unnormalized probability that the unknown sample belongs to the undetermined class, $P(und)$, is constant until $P(C_i)$ is sufficiently large, at which point $P(und)$ monotonically decreases to 0.0. An example of these probability functions is shown in Figure 15.20.

The probability of belonging to a particular class is then the sum of the unnormalized probabilities over all neighbors in that class divided by the sum of the unnormalized probabilities of being in any class, including the undetermined class. The undetermined class picks up a contribution from all of the neighbors, independent of their class, unless the distance to a neighbor is sufficiently small.

With these unnormalized probability functions, and setting $P(und) = 0.1$ if the distance to a neighbor is sufficiently large, the results now become (see Figure 15.19):

Case 1: Assuming that the neighbors are sufficiently close, the unnormalized probabilities become $P(black) \approx 2.0$, $P(white) \approx 1.0$, and $P(unk) \approx 0.0$. The unknown sample would then approximately be 66.7% black, 33.3% white, and 0.0% undetermined.

Case 2: In this case, the black neighbors are very close so $P(black) \approx 2.0$; the white neighbor is very far away so $P(white) \approx 0.0$, and $P(unk) \approx 0.1$. Therefore, this sample is 95.2% black, 0.0% white, and 4.8% undetermined.

Case 3: Since the unknown sample is relatively far away from any of its neighbors, $P(black) \approx P(white) \approx 0.0$ and $P(unk) \approx 0.3$. This means that the unknown sample is 100% in the undetermined class.

This distance-dependent K-nearest neighbors (DD-KNN) method yields classifications that are much more indicative of the local environment of the unknown sample and gives the researcher a better measure of the uncertainty of the classification.

15.4.3.2 Clustering

The classification of an unknown subject can also be determined by first clustering the known samples. Once the known subjects have been placed into K clusters, the unknown sample is added and placed in the appropriate cluster. From the known members of this cluster, the class of the unknown can be determined by either a crisp or fuzzy classification.

The goal is to find the feature set that, when used with the selected distance metric and clustering method, produces clusters that are as homogeneous as possible. This allows the prediction of the unknown to be as certain as possible.

Three different types of clustering algorithms are described here:

1. Agglomerative hierarchical clustering
2. Divisive hierarchical clustering
3. Nonhierarchical clustering

In agglomerative hierarchical clustering, each sample is initially placed in its own cluster. A rule is then used to determine which two clusters should be merged, reducing the number of clusters by one. This process continues until the desired number of clusters is obtained.

The well-known types of agglomerative hierarchical clustering are

1. Single linkage clustering: Each pair of samples from different clusters is examined and the pair with the smallest separation is located. The clusters containing these samples are merged. This allows a cluster to "snake" its way through the samples.
2. Average linkage clustering: Each pair of clusters is examined. For every sample in each cluster, the average intercluster distance is calculated. The cluster pair with the smallest average distance is merged. This keeps the average "density" of the clusters relatively constant, which means that regions with a higher density of points can have a larger cluster than those with a lower density.
3. Complete linkage clustering: Each pair of clusters is examined and the maximum intercluster distance is measured. The cluster pair with the smallest maximum distance is merged. This causes all clusters to have approximately the same size.
4. Ward's method: Each pair of clusters is temporarily merged and the centroid of this merged cluster is determined. The merged cluster with the smallest variance is the one that is kept. This also produces clusters that can be of various sizes depending upon the density of samples.
5. Jarvis-Patrick clustering[19]: A neighbor list of a given length is calculated for each sample. Each pair of samples from different clusters is examined

and the pair with the most number of samples in common in their neighbor list is selected and their clusters are merged. This is similar to single linkage in that a cluster can snake through the sample space, but distance to the neighbors or each other is not considered.

Divisive hierarchical clustering is the opposite of the agglomerative procedure in that all samples originally start in the same cluster. In each cycle, each cluster is examined and the sample pair that has the largest within-cluster distance is found. The cluster containing this pair is split in two using these samples as seed points. Each member of this pair is put into its own cluster and the other members of the original cluster are placed in the new cluster with the closest seed point.

Other options are available. In the first cycle the two samples that are furthest apart become the seed points for Cluster 1 and Cluster 2. After all samples are assigned to the cluster with the closest seed points the two clusters are individually examined to find the pair of samples with the largest within-cluster distance. If these points are in Cluster 2, for example, they become the seed points for Cluster 2 and Cluster 3. Instead of just dividing all samples that were in Cluster 2 among these new clusters, the three seed points can be used to separate all samples into clusters. This ensures that all samples are closer to their cluster's seed point than another cluster's. This can also be done in a nonhierarchical fashion by simply finding the set of K points that have the largest summed distance, or the largest minimum distance. They become the seed points and all samples are assigned to one of the K clusters.

The best known nonhierarchical clustering procedure is K-means clustering. It is also referred to as C-means clustering. For N samples with J selected features, the goal is to select K centroids in J-dimensional space such that when each sample is assigned to the cluster with the closest centroid (average coordinates), the sum of the distances squared to their centroid is a minimum. Finding the optimum locations of the K centroids is not an easy problem, and again the real goal is to find the set of J features that produce the most homogeneous clusters.

An outline of the K-means clustering algorithm[20,21] is as follows:

1. Randomly assign each of the N samples to one of the K clusters
2. Determine the centroids of each cluster and the sum of the distances squared from all samples to their centroid
3. For each sample,
   a. Sequentially place the sample in each of the other clusters
   b. Calculate the change in the centroid positions for the two affected clusters
   c. Calculate the new sum of distances squared. This only has to be done for the two affected clusters as long as the sum-of-squares distance is stored for each cluster.
4. Place this sample into the cluster that has the smallest sum of squares, adjust the positions of the centroids, and update the stored, intracluster sums
5. If any sample changes from one cluster to another, return to Step 3

To speed up the final assignment of samples in clusters, different algorithms have been proposed, which include H-means clustering,[22] J-means clustering,[23] and variable neighborhood search.[24–26]

Although K-means and its variants will generate a good set of clusters, they will not be the optimal set of K clusters for these J features. The reason is that in K-means clustering a final set of clusters depends on

1. The initial distribution of samples to clusters (i.e., the seed to the random number generator)
2. The ordering of the samples

Since there are K^N ways to initially distribute the points among the clusters and $N!$ orderings of the points, finding the optimal set of clusters is computationally impossible in most cases.

A modified H-means clustering algorithm has been found to yield good results in many cases. This procedure can be outlined as follows:

1. Select a set of K samples to act as initial centroids
2. Assign all samples to the cluster with the closest centroid
3. Use the samples in each cluster to recalculate the position of its centroid
4. If the cluster number of any sample changes, return to Step 2

In the first pass of this algorithm the samples do not have an assigned cluster so Step 4 automatically returns the process to Step 2. After a few iterations, no samples change clusters and the process stops.

This procedure is independent of the order of the samples since all samples are assigned to clusters at the same time. The problem is therefore reduced to finding the best set of K samples to act as initial centroids. The total number of unique sets of initial centroids is

$$N \times (N - 1) \times \cdots \times (N - K + 1)/K!$$

This can be a large number of initial sets, so an evolutionary programming algorithm, or any of the other search heuristics described above, can be used to search for the best set of initial centroids. Each set is then passed to this clustering algorithm and the cost of these clusters (total distance squared and/or the number of misclassified samples) is used to search for the best set of initial centroids.

If the number of samples is large, as would be the case of clustering the genes in a microarray study of normal and diseased cells, each of the final centroids should be close to a sample. If this is the case, the procedure outlined above can be stopped at Step 2 without iterating.

When a centroid is located on a sample it is called a *medoid*, and this is the basic idea behind **PAM** (partitioning around medoids).[27] This algorithm still searches through all unique sets of K samples and can be quite time consuming. Two non-EP search heuristics are used in the algorithms CLARA (Clustering LARge Applications)[27] and CLARANS (Clustering Large Applications based on

RANdomized Search)[28] where the search for the best medoids occurs by a series of reduced-dimensional searches.

15.4.3.2.1   Comparison of Clustering Methods

A given set of J features maps the samples onto points in J-dimensional sample space. The distance metric then defines how close or far they are from each other and from other points (centroids/medoids). The clustering methods described here separate the samples into K clusters but no two clustering methods will partition the sample space exactly the same way.

Single linkage and Jarvis-Patrick clustering are similar in that they can allow a single cluster to snake its way through sample space. They differ in that single linkage is based on distances, whereas Jarvis-Patrick is based on the similarity of neighbor lists, independent of distance.

Conversely, all of the other clustering methods divide the sample space into regions (which may seem more natural). Complete linkage and K-means clustering will generally form clusters of the same size, although high-density regions in sample space may actually have smaller K-means clusters since it is the sum of the distances squared and not the average squared distance. Average linkage clustering and Ward's method allow the clusters to be different sizes, but because average linkage is based on distances and Ward's on distances squared, the effects will be different.

Finally, only single linkage and Jarvis-Patrick guarantee that two points that are very close together will be placed in the same cluster. All of the other methods can place them in different clusters.

15.4.3.3   Neural Networks

Neural networks are learning algorithms that need to be trained using a set of samples, each with a given number (J) of known feature values. In general there are two types of training and networks of either type can be used to classify unknown samples. The types of training are called *supervised* and *unsupervised learning*.

In supervised learning each sample has a known response and the network is trained to produce this response to the best possible extent using the known feature values. The network of this type described here is called a feed-forward backpropagation multilayer neural network.

In unsupervised learning there is no known response and the network learns a pattern for each set of input values. The goal is to produce a comparable pattern for an unknown sample that has similar values. An example of this is handwriting analysis where a new person's t is understood as this letter because it is similar to other ts seen in the past. The network of this type described here is called a self-organizing map (SOM) or a Kohonen map.

Figure 15.21 is an example of a feed-forward backpropagation multilayer neural network. For each sample, three feature values ($J = 3$) are input, one to each of the input nodes. These nodes simply take the values and distribute them to each node in the hidden layer. Each connection, or wire, between nodes carries a weight $w_{j,h}$, where j is the index of the input node and h is the index of the node in the hidden layer. Each node in the hidden layer receives a total signal that is the sum of each

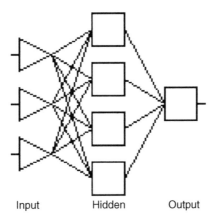

Input   Hidden   Output

FIGURE 15.21 Example of a feed-forward backpropagation multilayer neural network.

input value times the appropriate weight:

$$IN_h = \Sigma_{j=1,3}\ w_{j,h} \times x_{n,j}$$

The output from the hth node in the hidden layer is then a function (usually the sigmoid function) of the input signal:

$$OUT_h = f(IN_h)$$

The node in the output layer then receives the weighted sum of the outputs from hidden layer and produces a response for this nth sample using the same function:

$$IN_o = \Sigma_{h=1,4}\ w_{h,o} \times OUT_h$$

$$RESPONSE_n = f(IN_o)$$

This is obviously a multilayer network, but the counting of the layers has not been standardized. Many consider this a three-layer network (input, hidden, and output). Others only count processing layers, which means that the input (distribution) layer does not count and this is a two-layer network.

It is a feed-forward network because the signals travel from the input nodes to the hidden layer and then to the output layer.

The set of training samples are used to adjust the 16 weights in this network so that the error squared of the response is minimized. The error for a sample first adjusts the weights leading to the output layer, w_h, and then the weights between the input and hidden nodes, $w_{j,h}$. Therefore, the corrections to the weights are backpropagated through the network; hence its name.

To use this type of network in a classification problem, the class of each sample is simply given a number. In a two-class problem, the first class has a response of 0.0 and the second class a response of 1.0. Once the network is trained to give the best possible responses, an unknown sample is put into the network and its response can be used to assign the class of this sample using either a crisp or fuzzy classification.

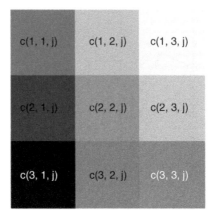

FIGURE 15.22 Example of a self-organizing map.

For example, a response of 0.75 can either be the second class (crisp or maximum likelihood) or 75% in the second class and 25% in the first (fuzzy).

A self-organizing map (SOM),[29] shown in Figure 15.22, is simply a collection of nodes arranged in a pattern (usually a rectangle). Each node is described by a triply indexed array $C(i,j,J)$. The first two elements represent the location of the node, and the third contains the J values representing the feature set coordinates of this node.

The J feature values of a given sample in the training set are compared with the coordinates of each node, and the sample is assigned to the node with the smallest distance squared.

At the start of the training, all nodes are adjusted so that their coordinates are stepped toward the sample by a relatively large amount. As the training proceeds the step-size decreases and not all nodes are adjusted. By the end of the training, the step-size is small and only the selected node and its adjacent neighbors have their coordinates moved.

The sum of the minimum distance squared should be recorded for each pass through the training samples so that the stabilization of the network can be verified. When the training is done, an unknown sample is checked and assigned to a node. The training samples associated with that node are used to make a crisp or fuzzy classification of the unknown.

An SOM is therefore very similar to a K-means clustering using this many clusters. Since adjacent nodes are more similar than distant nodes, it is akin to a K-means clustering followed by placing the centroids on a plot using a Sammon map. This classification method also suffers from the same problems as K-means clustering:

1. The final result depends upon the initial, random values assigned to the coordinates of each node
2. The final result depends upon the order of the training samples
3. Two closely spaced samples are not guaranteed to end up on the same node

15.5   EXAMINATION OF THE MODEL

Although the goal is to construct a classification model that correctly predicts the class of the known and unknown samples, it is also necessary to obtain some information about the model itself. This is particularly true when multiple models are obtained having nearly equivalent classification abilities.

To this point a classification model is constructed for a single set of training (and possibly testing) samples. It is important to determine the robustness of the model by trying other combinations of training and testing samples. Two procedures for doing this are a jackknife analysis and a bootstrap analysis. They will give you measures of the mean quality of the model and its standard deviation.

For models that generate clusters, including the SOM, a measure of the coverage and size of the clusters can be used to judge the relative merits of the models and can be used to determine the optimum number of clusters. Calculating the average silhouette width and running a Kelley analysis give useful information.

Finally, for models that give a fuzzy classification, finding a good threshold value for prediction is necessary. A receiver operating characteristic (ROC) analysis can be used to find the threshold value that maximizes the sensitivity and selectivity of the model.

15.5.1   CLASSIFICATION UNCERTAINTY

In studies with a small number of samples (N) all samples will have to be used to search for the best classification model. Although the ability of the model to correctly predict these training samples is important, this result is generated for the set it was optimized on. To obtain better estimates of the quality of this model (average quality and its standard deviation), a jackknife or bootstrap analysis can be done.

A jackknife analysis is simply a leave-one-out cross-validation. Each sample is sequentially removed from the set and the model is reoptimized on the ($N - 1$) remaining samples. The excluded sample becomes the test set and it is classified. The N classification uncertainties can then be used to generate a mean quality and an estimate of its standard deviation.

To increase the number of tests, an nth-order bootstrap analysis is used. Here, n samples are randomly selected and removed. The remaining ($N - n$) samples are used to construct the model and the n excluded samples are tested. This process is repeated a large number of times to give better estimates of the mean quality of the model and its standard deviation.

15.5.2   CLUSTER STATISTICS

Given a set of K clusters for the N samples, the average silhouette width (ASW)[30] gives an intercluster measure of how well the clusters cover the sample space, while a Kelley analysis[31] is an intracluster measure of their compactness.

For each sample n,

A(n) is the average distance between n and all other samples in its cluster.
C(n,k) is the average distance between n and all samples in each other cluster k.

$B(n) = MIN[C(n,k)]$ is the minimum average distance to samples in another cluster, and this cluster is called the *brotherhood* of n.

The Silhouette Width of n, SW(n), is then given by

$$SW(n) = [B(n) - A(n)]/MAX[A(n),B(n)]$$

The ASW is then defined as

$$ASW = \Sigma_{n=1,N} A(n)/N$$

The best clustering method and number of clusters K is then defined as the model that maximizes ASW. If SW(n) is near 1.0 the sample is well clustered; if it is near 0.0 the sample is between two clusters, while if it is negative it is probably in the wrong cluster.

A Kelley analysis[31] uses the average spread (AS) of clusters containing more than one sample. As the number of clusters decreases, their average spread increases, and this method minimizes a function of the average spread and number of clusters (including singletons).

The spread of non-singleton cluster k, SP_k is just the average of all intersample distances within this cluster. The average spread of K clusters with Kns non-singleton clusters is just the sum of the spreads divided by Kns:

$$AS(K) = \Sigma_{k=1,Kns} SP_k/Kns$$

The normalized average spread, ASnorm(K), is given by

$$ASnorm(K) = [(N-2)(AS(K)-ASmin)/(ASmax-ASmin)] + 1$$

where ASmin is the minimum distance between any two samples and ASmax is the average separation between all samples.

The Kelley Penalty Function is then

$$P(K) = ASnorm(K) + K$$

and the optimum clustering method and number of clusters minimize P(K). At this point the clusters are highly populated and compact.

15.5.3 Determining a Threshold in Fuzzy Classification

In a fuzzy classification, each sample has a Pi probability of being in Class i. From a diagnostic point of view, if a particular Pi is greater than a threshold value T the sample is diagnosed as being in this class. So if Pd is the probability of the subject being diseased, one can use

If Pd $\geq$ T, *the diagnosis is* "diseased"
If Pd $<$ T, *the diagnosis is* "nondiseased"

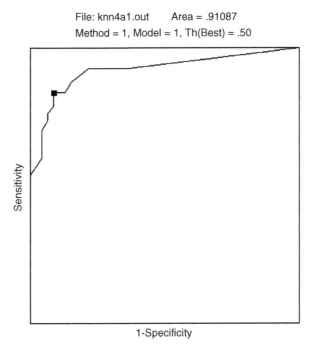

File: knn4a1.out   Area = .91087
Method = 1, Model = 1, Th(Best) = .50

Sensitivity

1-Specificity

FIGURE 15.23 Example of a receiver operating characteristic plot.

The sensitivity of the diagnosis is the fraction of diseased subjects that are correctly classified as such. This is also known as the true positive fraction (TPF). The specificity is the fraction of nondiseased subjects that are correctly classified, and this is also called the true negative fraction (TNF). The goal is to find the threshold such that both the sensitivity and specificity are maximized.

A receiver operating characteristic (ROC) analysis[32,33] can be used to graphically display the quality of the diagnosis as a function of T. The vertical axis of a ROC curve (Figure 15.23) is the sensitivity and the horizontal axis is (1-specificity). (1-Specificity) is the fraction of nondiseased subjects that are diagnosed as diseased and is therefore the false positive fraction (FPF).

For each value of T a point on the curve is determined and this can be used to find the value (range) of T that maximizes the sensitivity and specificity, or maximizes (sensitivity + specificity). The area under the curve is a measure of the quality of the model for all values of T, but due to the small sample size this curve is not smooth (monotonic) and the area might not be of much use.

15.5.4 VERIFICATION OF THE MODEL

Although one or more models can be generated that can be verified using a jackknife or bootstrap analysis and can have the quality of the clusters quantified, it is still uncertain whether the model(s) can be trusted in a real clinical setting. In other words, is each model valid, or is it simply a good numerical procedure that separates one class of subjects from another?

Since the number of features greatly exceeds the number of subjects, the latter is quite likely. The best verification of a model would be to have it yield a biological basis for the separation. Unfortunately, for a metabonomic study this is probably not possible and is definitely not possible from a mass spectral investigation.

Therefore, the only alternative is to produce a quantitatively good model and then test the model on a large number of blind subjects. As the number of tests increases, so does the confidence in the model.

15.6 CONCLUSIONS

Hopefully this discussion has stressed that there are a large number of classification models that can be built. In addition, because of the overdetermined nature of the data sets, most models will probably be nothing more than good numerical classifications of the existing data.

Great care must be used in the construction and verification of the model and it is up to both the experimentalist and the person constructing the model to ensure that any choices in the model (i.e., distance metric or classification method) make physical sense and that the verification/analysis of the model is sufficient.

ACKNOWLEDGMENTS

This work was funded in whole or in part with federal funds from the U.S. National Cancer Institute, National Institutes of Health, under contract no. NO1-CO-12400. The content of this publication does not necessarily reflect the views or policies of the Department of Health and Human Services, nor does any mention of trade names, commercial products, or organizations imply endorsement by the U.S. Government.

REFERENCES

1. Lucas, D.A., Luke, B.T., Collins, J.R., Van, Q., Klose, J., Chmurny, G.N., Conrads, T.P., Burt, S.K., Keay, S.K., and Veenstra, T.D. The use of urine metabonomic patterns for the diagnosis of interstitial and bacterial cystitis. 2003.
2. Eisen, M. Cluster and TreeView Manual (http://rana.lbl.gov/manuals/ClusterTreeView.pdf).
3. Eisen, M.B., Spellman, P.T., Brown, P.O., and Botstein, D. Cluster analysis and display of genome-wide expression patterns. *Proc. Natl. Acad. Sci. USA,* 95, 14863–14868, 1998.
4. Miki, R., Kadota, K., Bono, H., Mizuno, Y., Tomaru, Y., Carninci, P., Itoh, M., Shibata, K., Kawai, J., Konno, H., Watanabe, S., Sato, K., Tokusumi, Y., Kikuchi, N., Ishii, Y., Hamaguchi, Y., Nishizuka, I., Goto, H., Nitanda, H., Satomi, S., Yoshiki, A., Kusakabe, M., DeRisi, J.L., Eisen, M.B., Iyer, V.R., Brown, P.O., Muramatsu, M., Shimada, H., Okazaki, Y., and Hayashizaki, Y. Delineating developmental and metabolic pathways in vivo by expression profiling using the RIKEN set of 18,816 full-length enriched mouse cDNA arrays. *Proc. Natl. Acad. Sci. USA,* 98, 2199–2204, 2001.

5. Sammon, J.W. A nonlinear mapping for data structure analysis. *IEEE Trans. Comput.*, 18, 401–409, 1969.

6. Glover, F. Future paths for integer programming and links to artificial intelligence. *Comput. Oper. Res.*, 5, 533–549, 1986.

7. Glover, F. Tabu search: A tutorial. *Interfaces*, 20, 74–94, 1990.

8. Kirkpatrick, S., Gelatt, C.D., and Vecchi, M.P. Optimization by simulated annealing. *Science*, 220, 671–680, 1983.

9. Glover, F. Future paths for integer programming and links to artificial intelligence. *Comput. Oper. Res.*, 5, 533–549, 1986.

10. Holland, J. *Adaptation in Natural and Artificial Systems*. University of Michigan Press, Ann Arbor, MI, 1975.

11. Goldberg, D.E. *Genetic Algorithms in Search, Optimization, and Machine Learning*, Addison-Wesley, Berkeley, CA, 1989.

12. Luke, B.T. An overview of genetic methods, in *Genetic Algorithms in Molecular Modeling*, Devillers, J., Ed. Academic Press, London, pp. 35–66, 1996.

13. Fogel, L.J., Owens, A.J., and Walsh, M.J. *Artificial Intelligence through Simulated Evolutions*. John Wiley, New York, 1966.

14. Fogel, D.B., Fogel, L.J., and Porto, V.W. Evolutionary methods for training neural networks, in *IEEE Conference on Neural Networks for Ocean Engineering*, 91CH3064-3. pp. 317–327, 1991.

15. Fogel, D.B. Applying evolutionary programming to selected traveling salesman problems. *Cybernetic Systems (USA)* 24, 27–36, 1993.

16. Colorni, A., Dorigo, M., and Maniezzo, V. Distributed optimization by ant colonies, in *Proceedings of the First European Conference on Artificial Life*, Varela, F. and Bourgine, P., Eds. MIT Press, Cambridge, MA, pp. 134–142, 1991.

17. Kennedy, J. and Eberhart, R.C. Particle swarm optimization, in *Proceedings of the IEEE International Conference on Neural Networks, Perth, Australia, IV*. IEEE Service Center, Piscataway, NJ, pp. 1942–1948, 1995.

18. Kennedy, J. The particle swarm: Social adaptation of knowledge, in *Proceedings of the 1997 International Conference on Evolutionary Computation, Indianapolis, IN*. IEEE Service Center, Piscataway, NJ, pp. 303–308, 1997.

19. Jarvis, R.A. and Patrick, E.A. Clustering using a similarity measure based on shared near neighbors. *IEEE Trans. Comput.*, C22, 1025–1034, 1973.

20. Jancey, R.C. Multidimensional group analysis. *Aust. J. Bot.*, 14, 127–130, 1966.

21. MacQueen, J.B. Some methods for classification and analysis of multivariate observations. *Proceedings of 5th Berkeley Symposium on Mathematical Statistics and Probability*, 2, 281–297, 1967.

22. Howard, R. Classifying a population into homogeneous groups, in *Operational Research in the Social Sciences*, Lawrence, J.R., Ed. Tavistock Publishers, London, 1966.

23. Hansen, P. and Mladenovic, N. J-Means: A New Local Search Heuristic for Minimum Sum-of-Squares Clustering, (http://citeseer.nj.nec.com/107590.html).

24. Mladenovic, N. A Variable Neighborhood Algorithm—A New Metaheuristic for Combinatorial Optimization, Abstracts of papers presented at Optimization Days, Montreal, p. 12., 1995.

25. Mladenovic, N. and Hansen, P. Variable neighborhood search. *Comps. Opns. Res.*, 24, 1097–1110, 1997.

26. Hansen, P. and Mladenovic, N. "An Introduction to Variable Neighborhood Search," in Voss, S. et al. (eds.), *Proceedings of the 2nd International Conference on Meta-heuristics-MIC97*. Kluwer, Dordrecht, 1998.

27. Kaufman L. and Rousseeuw, P.J. *Finding Groups in Data: An Introduction to Cluster Analysis.* John Wiley & Sons, New York, 1990.

28. Ng, R.T. and Han, J. Efficient and Effective Clustering Methods for Spatial Data Mining. In *Proceedings of the 20th VLDS Conference.* Santiago, Chile, 1994.

29. Kohonen, T. *Self-Organization and Associative Memory.* Springer-Verlag, Berlin, 1989.

30. Rousseuw, P.J. Silhouettes: A graphical aid to the interpretation and validation of cluster analysis. *J. Comput. Appl. Math.,* 20, 53–65, 1987.

31. Kelley, L.A., Gardnew, S.P., and Sutcliffe, M.J. An automated approach for clustering an ensemble of NMR-derived protein structures into conformationally-related sub-families. *Protein Eng.,* 9, 1063–1065, 1996.

32. Lusted, L.B. Logical analysis of roentgen diagnosis. *Radiology,* 74, 178–193, 1960.

33. van Erkel, A.R. and Pattynama, P.M.Th. Receiver operating characteristic (ROC) analysis: Basic principles and applications in radiology. *European J. Radiol.,* 27, 88–94, 1998.

16 Computational Tools for Tandem Mass Spectrometry–Based High-Throughput Quantitative Proteomics

Jimmy K. Eng, Andrew Keller, Xiao-jun Li, Alexey I. Nesvizhskii, and Ruedi Aebersold

CONTENTS

16.1 INTRODUCTION

The objective of many proteomics experiments is to identify and quantify the proteins contained in complex samples such as body fluids, cell or tissue extracts, or fractions thereof. Increasingly, large-scale protein identification is achieved by a combination of (multidimensional) peptide chromatography, tandem mass spectrometry, and sequence database searchings.[1,2] To also quantify the proteins analyzed, they are imprinted with a stable isotopic signature prior to mass spectrometric analysis.[1,2] In a typical experiment, the proteins in a sample mixture are digested into peptides using a proteolytic enzyme. Such enzymes cleave the protein predictably at specific residue(s). The peptides are then separated by multidimensional, high-performance liquid chromatography (HPLC) and analyzed by a mass spectrometer. Sequence information on each peptide is generated by the mass spectrometer via its ability to select specific precursor ions out of a mixture of ions, to fragment the selected ions in a collision cell, and to record the precise masses of the thus generated fragment ions. These fragment ion spectra, also referred to as MS/MS spectra, contain the amino acid sequence of the selected precursor ion peptide. As it is challenging and time consuming to explicitly read the amino acid sequence from MS/MS spectra, computational analyses are employed to interpret spectra in order to derive a set of validated protein identifications. This involves searching the MS/MS spectra against sequence databases to identify the corresponding peptide sequences, validating the resulting assigned peptides, and using them to infer the proteins present in the sample mixture. In a typical liquid chromatography (LC) MS/MS–based proteomics experiment, tens of thousands of MS/MS spectra are collected, leading to the identification of thousands of peptides and hundreds to thousands of proteins.

The analysis and storage of these large data sets challenges the computer infrastructure of many research labs and institutions. In addition, if the data in or the conclusions of large proteomics data sets are to be published or entered into relational databases, it is essential that the quality of the data and conclusions, i.e., the sensitivity and error rate of the analyses carried out, are known and associated with the data. In this chapter, we describe and discuss currently available computer tools that support the data collection, analysis, and validation in a high-throughput LC-MS/MS–based proteome research environment.

16.2 COMPUTATIONAL METHODS TO IDENTIFY PEPTIDES AND PROTEINS USING MS/MS SPECTRA

Mass spectrometers have become the predominant tools for the identification of proteins and the volume and complexity of data collected has required the development of computer tools for the determination of protein sequences from mass spectrometric data. Initial mass spectrometry database search routines involved querying mass spectra of digests of purified proteins against sequence databases in a process termed "peptide mass fingerprinting" or "peptide mass mapping."[3–7] Each peak in the spectrum, which is typically acquired by a MALDI-TOF instrument, represents a peptide, and the whole of the spectrum represents the original protein. While each peptide mass by itself does not contain enough information to uniquely identify the target protein, the masses of the detected peptides collectively are usually sufficient for an unambiguous assignment of the protein digest. The utility of this approach is in the rapid identification of unknown, essentially pure proteins such as those excised from a 2D PAGE gel. However, the peptide mass mapping approach is generally unsuitable for analyzing proteins in mixtures and is therefore not compatible with LC/LC-MS/MS strategies for proteome analysis. When working with digests of mixtures of proteins, as simple as protein complexes or as complex as whole cell lysates, mass spectrometry methods that determine the sequence of individual peptides are necessary. Mass spectra that represent the peptide amino acid sequence via peptide-specific fragmentation information are typically acquired in a two-stage or tandem mass spectrometer. In tandem mass spectrometry data acquisition, the first mass spectrometry stage involves reading all peptide ions that are introduced into the instrument. The second mass spectrometry stage reads the masses of the fragment ion products of a peptide ion that has been isolated and fragmented in a collision cell in a process termed "collision induced dissociation" (CID). The acquired tandem mass spectrum, therefore, is a spectrum of the fragment ions from an isolated peptide ion. In principle, the identity of the peptide sequence can be deduced directly from the tandem mass spectrum either manually, with the potential aid of computational tools,[8] or via a sequence database search routine. Early computational methods for MS/MS spectra analysis were used to assist in the de novo sequencing process; they enhanced the ability of experts to read off peptide sequences from the complicated MS/MS spectra. Combining partial de novo sequencing and ion constraints, in the form of a peptide sequence tag,[9] with database searching also proved to be a viable method of peptide identification as long as the number of MS/MS spectra to be analyzed remained relatively low. However, the development of instruments and protocols that support the acquisition of thousands or tens of thousands of MS/MS spectra per experiment required the development of automated methods of searching uninterpreted MS/MS spectra against sequence databases and the development of such tools has been an important step in the development of a general proteomics technology.

Common MS/MS database search routines, such as SEQUEST,[10] Mascot,[11] and MS-Tag,[12] each take an uninterpreted MS/MS spectrum as input and identify a best-fit peptide for the output. Specifically, candidate peptides near the same nominal mass of the measured peptide mass are selected from the sequence database. Theoretical fragment ions are calculated for each of these candidate peptides using common dissociation rules. The calculated fragment ions, composed of both N-terminal ions and C-terminal ions, are specific for the mass spectrometer acquiring the data. Under the low-energy (a few keV) collision–induced dissociation conditions commonly encountered in mass spectrometers such as ion traps, triple quadrupole, and quadrupole TOF instruments, peptides primarily fragment at the amide bonds on the peptide backbone generating what are termed B-ions and Y-ions.[13] The high-energy CID conditions encountered in other types of instruments, including magnetic sector and TOF–TOF instruments, will also generate other ion types such as A, C, X, and Z ions resulting from breaking other bonds along the peptide backbone. Ions specific to residue side chain cleavages, such as D, V, and W ions, can also be generated at high-energy collision conditions. The database search routines compare the theoretically calculated fragment ions against the input spectrum using a score function. There are a wide variety of score functions used in uninterpreted tandem mass spectrum database search routines and they need to be adapted to the specific type of data (e.g., high- or low-energy CID) that are being searched. One class of algorithms uses spectral correlation functions[10] to measure closeness of fit between the input spectrum and theoretical spectra derived from candidate sequences in the sequence database. A second class of algorithms evaluates peptides using statistics based on fragment ion frequencies.[11] Yet a third class of routines uses a Bayesian statistics approach for comparing spectra to candidate peptides.[14,15] As this is a peptide identification method, unlike peptide mass fingerprinting, which identifies proteins, all types of sequence databases are suitable to be queried. These include genomic, expressed sequence tag, transcript, and protein sequence databases.

It should be noted that MS/MS search routines always return the best-fit peptide sequences in the database for each input spectrum whether or not the correct peptide sequence is present in the database. Therefore, irrespective of the class of score function used and the database searched, the significance of the search scores at the individual search result level needs to be further validated (see below). If the correct peptide sequence is not present in the sequence database that is searched, it will not be identified, and, potentially worse, a wrong assignment might be made. Thus, the choice of which sequence database to search is an important consideration. The notion of searching the largest sequence database available is attractive in terms of identifying unknown peptides since having comprehensive sequence data maximizes the chance of correctly matching peptide sequences. However, larger databases result in longer search times, which may be an analysis impediment depending on the available search options and computer hardware. Also, larger databases can result in a reduced number of correct identifications, as lower-quality spectra will match many more false-positive sequences. Whenever possible, smaller sequence databases, such as a species-specific database, offer the best chance at identifying peptides in an MS/MS database search as long as the peptides being identified are represented in the database.

Since the input sample of protein(s) is digested into peptides using a specific proteolytic enzyme such as trypsin, which cleaves after arginine and lysine (but usually not if either residue is followed by proline), it can be advantageous to only search against a subset of the peptides from a sequence database that is generated from the expected cleavage of the enzyme. By specifying the digestion enzyme as a parameter of the database search, this constraint reduces the number of peptides from the database that need to be analyzed, as opposed to considering any linear stretch of amino acids possible, which in turn reduces chance false-positive hits and greatly reduces search times compared to enzyme unconstrained searches. However, performing enzyme unconstrained searches does have its benefits. These include being able to identify peptides that exhibit unspecific cleavage, which can be significant (as high as 20%) depending on the quality of the enzyme and digestion protocol, that would otherwise be missed in an enzyme constrained search. Also, as in the case of trypsin, since lysines and arginines are only 2 of the 20 possible amino acid residues, tryptic peptides that are identified in an enzyme-unrestricted search have a 1 in 10 chance of exhibiting a tryptic cleavage at the N- or C-terminus and 1 in 100 chance of having tryptic cleavage sites at both the N- and C-terminus (assuming even distribution of the amino acid residues in the sequence database). Thus, the knowledge of the digestion enzyme might better be used after an enzyme unconstrained database search as an aide in the validation of the identifications.

16.3 VALIDATION OF IDENTIFICATIONS AT THE PEPTIDE LEVEL

For the majority of acquired MS/MS spectra, unfortunately even some of the best matching database peptides assigned to them by available database search tools are incorrect. For example, in a typical LC-MS/MS run on an ion trap mass spectrometer, less than 20% of all peptide assignments made by the database search tools are correct.

This means that, in order to derive meaningful information from the acquired data, the user has to evaluate each database search result and remove all or at least the majority of the incorrect peptide assignments. However, manual verification of peptide assignments can be achieved only in the case of small data sets. Manual validation is a very time-consuming approach and is simply not feasible in high-throughput analysis of large data sets containing tens of thousands of spectra. In addition, manual verification of peptide assignments to MS/MS spectra requires expertise in mass spectrometry and peptide fragmentation chemistry, which is often not available. Furthermore, even experienced researchers are not safeguarded from occasional spectrum interpretation errors and it is not clear if the manual interpretation process can be performed consistently and in an objective manner even by the same individual.

As an alternative to manual validation of the entire data set, researchers can attempt to separate the correct from the incorrect peptide assignments by applying filtering criteria based upon database search scores and properties of the assigned peptides; see Han et al. and Washburn et al.[16,17] However, the numbers of rejected correct identifications and accepted false identifications that result from applying such filters are not known.[18] The problem is further complicated due to diversity of experimental and computational methods (see Figure 16.1), which presents a challenge for interpretation of the data and for comparison of results of different research groups:

1. Peptides can be assigned to MS/MS spectra using a variety of available database search algorithms. Each database search tool scores candidate peptides using its own scoring scheme and the relationship between the different scoring schemes is not known.

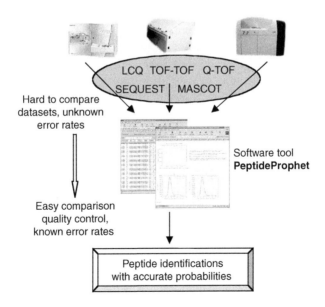

FIGURE 16.1 PeptideProphet facilitates comparison of data generated from different instruments and analyzed using different database search tools.

2. MS/MS data can be acquired using different types of mass spectrometers. Data sets of MS/MS spectra acquired on different mass spectrometers vary in terms of what peptides are selected for sequencing, mass accuracy, resolution, and signal-to-noise intrinsic to the acquired mass spectra, distributions of fragment ion types and intensities, and other factors. All these factors have influence on the search scores for correct results.

3. Overall data set quality, which can be defined as a fraction of the MS/MS spectra in the data set that got assigned the correct peptide, often varies significantly from experiment to experiment, even when the data is acquired on the same mass spectrometer and searched using the same database search tool.

4. The quality of the sample analyzed may be different from lab to lab and from experiment to experiment. Clearly, because of the first two reasons mentioned above, it would be impossible to define a set of filtering criteria based on the search scores provided by the database search tool, which, upon application to diverse data sets would result in uniform false-positive error rates (a fraction of all identifications passing the filter that are incorrect). Furthermore, even when the researcher uses the same type of mass spectrometer and the same database search tools, application of the same filtering criteria would likely result in different false-positive error rates in each new experiment, reflecting the differences in the overall data set quality. Thus, in order to be able to interpret the data consistently and reliably and to allow comparison of results of different experimental groups, peptide assignments to MS/MS spectra should be validated using robust statistical software tools.

One such tool, PeptideProphet™, has been recently described and made freely available to the scientific community (see Figure 16.1).[18] PeptideProphet is based on the use of the expectation maximization (EM) algorithm to derive a mixture model of correct and incorrect peptide identifications from the data, as illustrated in Figure 16.2. By employing the observed information about each assigned peptide in the data set, namely database search scores and peptide properties, the method learns to distinguish correct from incorrect peptide identifications and, in doing so, computes for each identification a probability of being correct. In a typical experiment, peptide properties such as the number of termini consistent with enzymatic cleavage, the number of missed enzymatic cleavages, and the difference between the measured and theoretical peptide mass is useful for discriminating correct from incorrect identifications and are employed by PeptideProphet. Any additional useful information when available, such as the presence or absence of a specific amino acid or sequence motif (e.g., cysteine in the case of quantitative isotope-coded affinity tag, or ICAT™ reagent experiments), can be easily incorporated in the model. If the database search tool outputs more than a single score useful for distinguishing correct from incorrect peptide assignments, as is the case with SEQUEST, all such scores are combined into one discriminant score that optimally discriminates between correct and incorrect peptide assignments. The discriminant function coefficients (weighting factors determining the relative contribution of each search score)

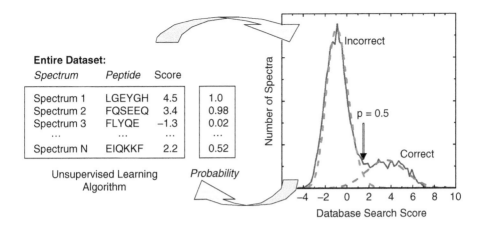

FIGURE 16.2 PeptideProphet learns the most likely distributions (dashed lines) among correct and incorrect peptide assignments given the observed data (solid line) and computes for each peptide assignment in the data set a probability of being correct.

are developed for each database search tool (SEQUEST, Mascot, etc.) and further optimized for different types of mass spectrometers (ion traps, quadruple time-of-flights, etc.) using training data with peptide assignments of known validity. Because PeptideProphet learns to distinguish the correct from incorrect peptide assignments from the data, it is robust toward variations in data quality, proteolytic digest efficiency, and other factors.

Probabilities computed by PeptideProphet are accurate measures of confidence that peptide identifications are correct. Probabilities can be considered accurate if, upon selection of all peptides in the data set having any given computed probability, the corresponding portion of them is correct. For example, if among all peptide assignments in a particular data set, 100 peptides are assigned by PeptideProphet a probability close to 0.9, then close to 90 of them should be correct. Extensive evaluation of the statistical model implemented in PeptideProphet demonstrated a very good agreement between the actual and computed probabilities in the entire 0.0 to 1.0 probability range.[18]

Computed probabilities are very efficient at separating the correct from incorrect peptide identifications. As a result, filtering data sets of peptide identifications using computed probabilities allows researchers to extract more correct identifications with no increase in the number of false identifications than otherwise would be possible using the database search scores. In addition, computed probabilities can be used to estimate the total number of correct identifications and the false-positive error rates resulting from filtering the data using a minimum computed probability as the filtering criteria. These model estimated parameters can serve as objective criteria for comparing data sets of peptide identifications obtained using different experimental protocols or different database search tools; see Figure 16.1. Most often in practice, however, no filtering is necessary at the peptide level. The entire list of peptide identifications and computed probabilities can be taken as input into a second

statistical analysis, ProteinProphet,[19] to compute probabilities that their corresponding proteins are present in the original sample, as described in the next section.

16.4 INFERRING PROTEINS IN THE SAMPLE

A tandem mass spectrometry database search generates a set of peptides assigned to
the MS/MS spectra in a data set. Most researchers, however, are interested in identifying proteins rather than peptides. This requires grouping the peptides assigned to
MS/MS spectra according to their corresponding protein. Various visualization tools
such as INTERACT,[16] DTASelect,[20] and CHOMPER,[21] facilitate such grouping.
Ultimately, the total peptide evidence observed for each protein must be determined
and evaluated to infer a set of protein identifications for the original sample.

Inferring protein identifications from MS/MS data is made more difficult by the
large numbers of incorrect peptide assignments to MS/MS spectra made by database
search algorithms, as discussed above. Several analyses have been described that
seek to assess the likelihood of the presence of a protein based upon database search
results. For example, Mascot computes an overall score for each protein that corresponds to identified peptides,[11] while Qscore estimates the likelihood due to chance
of an observed number of distinct peptides assigned to spectra in a data set, all
corresponding to a particular protein.[22] That score takes into account both the total
number of distinguishable peptides in the database used for a search and the number
possible for any particular protein. More recently it was described how probabilities
of peptides assigned to spectra that are correct can be combined together to compute
a probability that their corresponding proteins are present.[23]

The software ProteinProphet[19] computes protein probabilities using probabilities
that peptides assigned to MS/MS spectra are correct, such as those determined by
PeptideProphet. The method has been shown to compute accurate protein probabilities and hence enables the false-positive error rates (the fraction of results that are
incorrect) to be predicted for any data set. Importantly, it addresses two great
challenges to protein identification by MS/MS: nonrandom grouping of peptides
according to corresponding protein and the occurrence of peptides corresponding to
more than a single entry in the protein database.

Computing the probability that any particular protein is present requires grouping
together all peptides assigned to MS/MS spectra in the data set that corresponds with
that protein. Such grouping is not necessarily random, since correct peptides tend to
correspond to a small subset of correct proteins, whereas incorrect peptides tend to
correspond to "single hit" proteins, those to which no other peptide corresponds. This
effect becomes more pronounced as the number of spectra in a data set increases
relative to the number of proteins in the sample. This is illustrated in Figure 16.3,
where 10 peptides with computed probability of being correct equal to 0.5 are chosen
at random from the data set, 5 being correct and 5 being incorrect. When these peptides
are grouped according to their corresponding proteins, however, all 5 correct peptides
are found to correspond to only a single correct protein (Protein A), whereas each of
the 5 incorrect peptides corresponds to a different incorrect protein (Proteins B-F).
Therefore, a 50% false-positive rate at the peptide level (5/10 incorrect peptides)
translates into an 83% false-positive rate at the protein level (5/6 incorrect proteins).

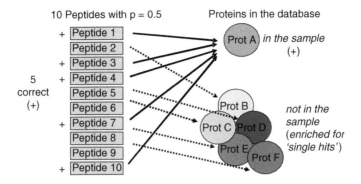

FIGURE 16.3 Nonrandom grouping of peptides according to corresponding protein.

Protein grouping information can be used to adjust peptide probabilities in advance so that accurate protein probabilities can be computed. In particular, peptides corresponding to single hit proteins should be penalized, and those not should be compensated. ProteinProphet learns from each data set by how much to penalize peptides corresponding to single hit proteins, and compensate the remainder. In the example in Figure 16.3, the probabilities that peptides 1, 3, 4, 7, and 10 are correct would be increased toward 1.0, and those of the remainder, toward 0.0, so that all 5 incorrect proteins would be assigned a probability close to 0.0.

Another challenge for inferring protein identifications is the occurrence of "degenerate" peptides, those corresponding to more than a single entry in the protein database. Such peptides are particularly prevalent in databases of large eukaryotes, such as human, which contain many homologues, splice variants, and redundant entries. Figure 16.4 illustrates an example of a degenerate peptide (Peptide 1) corresponding to two different proteins in the database, Proteins A and B, and a nondegenerate peptide (Peptide 2) corresponding only to Protein B. It is unclear whether Peptide 1, when assigned to an MS/MS spectrum in the data set, should be considered evidence for the presence in the sample of Protein A, of Protein B, or of both. ProteinProphet apportions each degenerate peptide among all its corresponding proteins in order to derive the simplest list of protein identifications that explain the observed data. In the

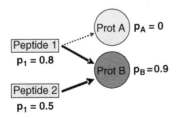

FIGURE 16.4 Peptide corresponding to two proteins in database is apportioned primarily to protein B.

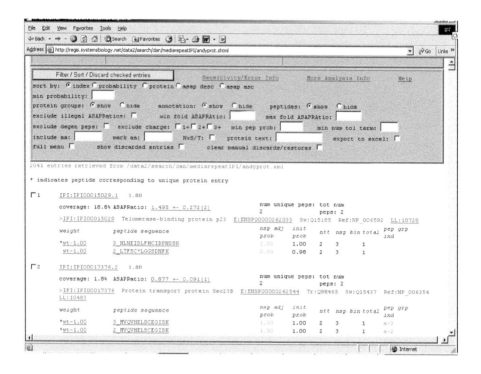

FIGURE 16.5 Screen capture of sample ProteinProphet™ output. Each protein entry is accompanied by its assigned probability of being correct, and when available, quantification information. In addition, annotation and peptide lists with links to the original MS/MS data help the user interpret the results of analysis.

case of the example in Figure 16.4, Peptide 1 would be apportioned primarily to Protein B since that protein on its own could explain both observed peptides.

The software ProteinProphet is easy to use and freely available. For each protein, it displays the computed probability of it being in the sample, along with annotation information and all corresponding peptides assigned to MS/MS spectra in the data set (Figure 16.5). Links allow users to access the original search results for each peptide, including the MS/MS spectra, map observed peptides onto the sequence of their corresponding proteins, and locate all other proteins that share any particular degenerate peptide. In order to make the results easier to interpret, ProteinProphet groups together all proteins that cannot be distinguished by the peptides assigned to MS/MS spectra in the data set and presents them as single identifications. Data can be exported to Excel format for further analysis.

Probabilities computed by ProteinProphet have been shown to be true reflections of the confidence of protein identifications. For example, using data sets for which the true protein contents of the sample are known, close to 50% of those proteins assigned probabilities of 0.5, and 90% of those proteins assigned probabilities of 0.9, were shown to be correct. These probabilities can serve as a standard for publishing protein identifications. If investigators report computed probabilities along with protein identifications, they could give the research community access to the maximal amount of data, including marginal identifications, while the accompanying probabilities serve as guides for interpretation. In addition, accurate probabilities allow the false-positive error rate to be predicted for any data set. That error rate can serve as an objective criterion for comparing large protein data sets of different research groups, generated using different mass spectrometer types, and even different database search algorithms.

16.5 QUANTITATIVE APPROACHES USING TANDEM MS

Mass spectrometry per se is not a quantitative technology. Detected ion intensities of individual peptides are determined not only by their abundance but also by their ionization efficiency, sample complexity, and other poorly reproducible parameters. To overcome these shortcomings, the technique of stable isotope dilution is applied in combination with mass spectrometry for protein identification and quantification.[1,2] In this method, proteins from different samples are labeled with different stable isotope tags of identical chemical structure but different mass. Labeled protein samples are then combined, enzymatically digested, separated, and concentrated by multidimensional HPLC systems and analyzed by tandem mass spectrometry. Peptides derived from proteins are identified by searching MS/MS spectra against a protein database, as described earlier. Since isotopically labeled peptides of the same sequence but from different samples differ only by their mass, their ion intensities are proportional to their relative abundance. Factors such as ionization efficiency and sample complexity, previously barriers to MS being a quantitative method, are addressed by this method of stable isotope dilution. The relative abundance of peptides can then be determined by the corresponding peptide ion intensities. Protein identification and quantification are deduced from the identification and quantification of the corresponding peptides.

There are several methods of implementing stable isotope tags into proteins.[24–30] Incorporation of stable isotope signatures by covalent chemical reactions as exemplified by the ICAT reagent method has some clear advantages over other methods.[26] Unlike metabolic labeling, which incorporates heavy or light isotopes such as 15N in place of 14N, 13C/12C, or 2H/1H, the ICAT reagent technology does not interfere with cellular processes and does not require live samples. It can be used on human samples from live cells as well as cell lysates and even samples such as serum in which cells are deprived. The ICAT reagent method specifically labels the cysteine residues of peptides with a reagent that incorporates an affinity tag with which labeled peptides can be purified by an affinity HPLC system. After release from the affinity column, only cysteine-containing peptides remain in the sample. This has the advantage of reducing sample complexity at the peptide level by tenfold and providing deep coverage of low-abundant proteins. One disadvantage of the ICAT reagent technology is that only cysteine-containing peptides are quantified, so proteins that do not contain a cysteine residue cannot be quantified by this method.

There are two complementary approaches to identify and quantify peptides in tandem mass spectrometry. In the first approach, peptides are identified first and quantified later. In the second approach, the two processes are reversed. Under the first approach, a reverse-phase HPLC system is coupled online with an electrospray ionization (ESI) mass spectrometer. Peptide samples are separated by reverse-phase HPLC and analyzed immediately by tandem mass spectrometry. Abundant peptide ions are then selected for fragmentation and identified by sequence database searching. Although most peptides are identified in only one of two isotopic forms, ions of the complementary isotopic form, the one not identified, can be easily determined from the expected mass shift between the two isotopic forms. Those ions can then be extracted from the MS spectra, the full MS scan spectra as opposed to the MS/MS spectra used for database searching, and used to reconstruct a single-ion chromatogram (SIC) for the two isotopic peptides. The relative abundance of the peptide in the two original samples is then determined by the ratio of the two corresponding SIC areas. A new software tool, ASAPRatio™,[31] has automated the analysis of obtaining robust and reliable quantification results. Since raw SICs are generally noisy, data smoothing and background subtraction are applied to improve the accuracy of quantification results and the dynamic range of detectable abundance changes. During ESI, most peptides are ionized into more than one charge state. Peptide ions of the different charge states can also be identified easily once the peptide sequence is known. A peptide abundance ratio can be determined from each detectable charge state and the ratios from all detectable charge states can then be used to calculate the mean and standard deviation of the corresponding peptide abundance ratio (see Figure 16.6).

Many peptides identified using this first approach belong to abundant proteins that show little change in abundance and normally play no causal roles in disease. To more selectively identify and quantify proteins of interest, a second approach was developed recently to quantify peptides before their identification, enabling the targeting of only peptides that exhibit a significant abundance change.[32] In this approach, peptide samples are separated by a reverse-phase HPLC system, collected in "wells" of a matrix-assisted laser desorption ionization (MALDI) plate, and mixed and crystallized with a matrix. Since samples are crystallized, they can be analyzed repeatedly by a mass spectrometer until all samples are exhausted. Unlike liquid-phase ionization, there is no real time constraint on the data acquisition of a MALDI sample, which allows for more flexible interrogation of the information. Peptides are ionized by a laser beam into singly charged ions and analyzed by a time-of-flight mass spectrometer. MS spectra are collected from each well of peptide samples.

A software tool then identifies the mass values of all potential peptide peaks within the spectra. Since peptides of the same sequence but in different isotopic forms have distinct mass shifts, ions corresponding to isotopically paired peptides can be identified. The relative abundance of the peptide pairs is determined by areas of their corresponding peaks in the MS spectra. The unpaired peptides likely belong to proteins existing in one sample but not in the other. The software tool then generates a list of ions potentially interesting for MS/MS data acquisition based on the calculated abundance ratios, including unpaired peaks, and user-defined thresholds. This mass list is sent back to the mass spectrometer, which is then directed to acquire MS/MS spectra on the ions in the list. These MS/MS spectra are searched against a sequence database to determine the identities of the corresponding peptides.

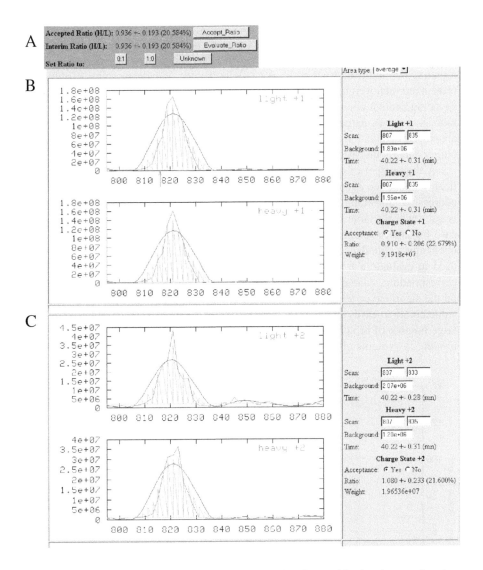

FIGURE 16.6 Screen capture images from CGI interface for peptide abundance ratio calculation. (A) Result on peptide abundance ratio. (B) Abundance ratio from ions in the [M+H]⁺ charge state. On the left are single-ion chromatograms reconstructed from LC-ESI-MS data. On the right are parameters used in abundance ratio calculations and ratio results from the charge state. (C) Same as B but from ions in the [M+2H]²⁺ charge state.

Similar to the situation of deriving protein identification from peptide identifications, it is a nontrivial step to derive protein abundance ratio from peptide abundance ratios. After the ProteinProphet program assigns all peptides to their corresponding proteins, the ASAPRatio program determines protein abundance ratios from the abundance ratios of their corresponding peptides. It is often the case that several peptides belong to the same protein, the same peptide is identified multiple times, and the abundance ratios from all peptide identifications of the same protein may not be consistent. The ASAPRatio program takes several steps to address these complexities. Abundance ratios of the same peptide sequence are first grouped together to calculate a unique peptide abundance ratio. All unique peptide abundance ratios of the same protein are then weighted by their standard deviations and used to calculate the corresponding protein abundance ratio. The standard deviation of the protein abundance ratio is also calculated; this provides an assessment on the reliability of the protein abundance calculation. Through each step, a statistical method is used to eliminate any outlier data (abundance ratios that disagrees with other data) from the calculation.[33]

A major goal of quantitative proteomics is to identify proteins showing significant abundance change. The ASAPRatio program adopts a statistical method to address the question of what changes should be considered significant. The method is valid when a large number of proteins do not change in abundance, which is the case for most protein samples. In this method, the ratio distribution of proteins of unchanging abundance is first estimated from the data set (see Figure 16.7) and then used to evaluate a p value for each protein based on its abundance ratio and the corresponding standard deviation. The p value provides a statistical measurement on the significance level of each abundance change value. One can use a suitable p value cutoff to easily distinguish proteins of significant abundance change from the large number of proteins that exhibit no abundance change. Users may use p values to identify proteins of interest very quickly.

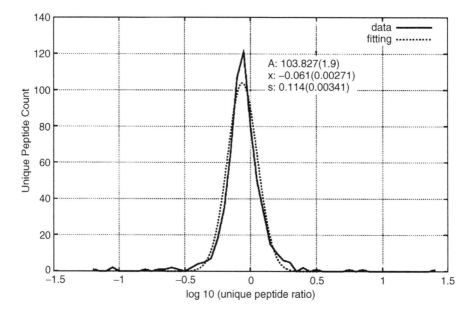

FIGURE 16.7 Distribution of unique peptide abundance ratios. The solid line is for the original data. The dotted line is for the corresponding normal distribution obtained by data fitting. The p value of a protein is determined by the distance between the protein abundance ratio and the center of the normal distribution.

16.6 CONCLUSION

Tandem mass spectrometry coupled with high-performance peptide separation tools enables high-throughput quantitative analysis of complex protein samples. Extracting useful information from large data sets generated by this method requires several steps of computational analysis. These include the assignment of peptides to MS/MS spectra, the validation of peptide assignments, the inference of protein identifications based upon the assigned peptides, and the quantification of peptides and proteins using stable isotope dilution methods. Software is currently available that automates this entire analysis pipeline and thus assists the researcher in identifying proteins that are of interest in the context of a specific experiment.

The statistical analyses of MS/MS database search results, such as PeptideProphet, ProteinProphet, and ASAPRatio, are particularly useful for lending some objectivity to protein identification and quantification. Accurate probabilities provided by the first two applications enable the false-positive error rate to be predicted for any data set. This can serve as an objective criterion by which any two data sets, generated by different research groups using different types of mass spectrometers and different database search algorithms, may be compared. Furthermore, if researchers publish computed probabilities and quanitification ratio errors along with their data, they can provide the research community with the greatest amount of data, while the probabilities and errors serve as guides for interpretation.

REFERENCES

1. Aebersold, R. and Goodlett, D.R. Mass spectrometry in proteomics. *Chem. Rev.,* 101, 269–295, 2001.
2. Aebersold, R. and Mann, M. Mass spectrometry-based proteomics. *Nature,* 422, 198–207, 2003.
3. Henzel, W.J., Billeci, T.M., Stults, J.T., Wong, S.C., Grimley, C., and Watanabe, C. Identifying proteins from two-dimensional gels by molecular mass searching of peptide fragments in protein sequence databases. *Proc. Natl. Acad. Sci. USA,* 90, 5011–5015, 1993.
4. Mann, M., Hojrup, P., and Roepstorff, P. Use of mass spectrometric molecular weight information to identify proteins in sequence databases. *Biol. Mass. Spectrom.,* 22, 338–345, 1993.
5. Pappin, D.J.C., Hojrup, P., and Bleasby, A.J. Rapid identification of proteins by peptide-mass fingerprinting. *Curr. Biol.,* 3, 327–332, 1993.
6. James, P., Quadroni, M., Carafoli, E., and Gonnet, G. Protein identification by mass profile fingerprinting. *Biochem. Biophys. Res. Commun.,* 195, 58–64, 1993.
7. Yates, J.R., III, Speicher, S., Griffin, P.R., and Hunkapiller, T. Peptide mass maps: A highly informative approach to protein identification. *Anal. Biochem.,* 214, 397–408, 1993.
8. Yates, J.R., III, Griffin, P., Hood, L., and Zhou, J. Computer aided interpretation of low energy MS/MS mass spectra of peptides. *Tech. Protein Chem.,* II, 477–485, 1991.
9. Mann, M. and Wilm, M. Error-tolerant identification of peptides in sequence databases by peptide sequence tags. *Anal. Chem.,* 66, 4390–4399, 1994.

10. Eng, J.K., McCormack, A.L., and Yates, J.R., III. An approach to correlate tandem mass spectral data of peptides with amino acid sequences in a protein database. *J. Am. Soc. Mass Spectrom.*, 5, 976–989, 1994.

11. Perkins, D.N., Pappin, D.J., Creasy, D.M., and Cottrell, J.S. Probability-based protein identification by searching sequence databases using mass spectrometry data. *Electrophoresis*, 20, 3551–3567, 1999.

12. Clauser, K.R., Baker, P., and Burlingame, A.L. Role of accurate mass measurement (+/– 10 ppm) in protein identification strategies employing MS or MS/MS and database searching. *Anal. Chem.*, 71, 2871–2882, 1999.

13. Johnson, R.S., Martin, S.A., and Biemann, K. Collision-induced fragmentation of (M+H)+ ions of peptides, side chain specific sequence ions. *Intl. J. Mass Spectrom. Ion Proc.*, 86, 137–154, 1988.

14. Bafna, V. and Edwards, N. SCOPE: A probabilistic model for scoring tandem mass spectra against a peptide database. *Bioinformatics*, 17, S13–S21, 2001.

15. Zhang, N., Aebersold, R., and Schwikowski, B. ProbID: a probabilistic algorithm to identify peptides through sequence database searching using tandem mass spectral data. *Proteomics*, 2, 1406–1412, 2002.

16. Han, D.K., Eng, J., Zhou, H., and Aebersold, R. Quantitative profiling of differentiation-induced microsomal proteins using isotope-coded affinity tags and mass spectrometry. *Nat. Biotechnol.*, 19, 946–951, 2001.

17. Washburn, M.P., Wolters, D., and Yates, J.R., III. Large-scale analysis of the yeast proteome by multidimensional protein identification technology. *Nat. Biotechnol.*, 19, 242–247, 2001.

18. Keller, A., Nesvizhskii, A.I., Kolker, E., and Aebersold, R. Empirical statistical model to estimate the accuracy of peptide identifications made by MS/MS and database search. *Anal. Chem.*, 74, 5383–5392, 2002.

19. Nesvizhskii, A.I., Keller, A., Kolker, E., and Aebersold, R. A statistical model for identifying proteins by tandem mass spectrometry. *Anal. Chem.*, 75, 4646–4658, 2003.

20. Tabb, D.L., McDonald, W.H., and Yates, J.R., III. DTASelect and contrast: tools for assembling and comparing protein identifications from shotgun proteomics. *J. Proteome Res.*, 1, 21–26, 2002.

21. Eddes, J.S., Kapp, E.A., Frecklington, D.F., Connolly, L.M., Layton, M.J., Moritz, R.L., and Simpson, R.J. CHOMPER: A bioinformatic tool for rapid validation of tandem mass spectrometry search results associated with high-throughput proteomic strategies. *Proteomics*, 2, 1097–1103, 2002.

22. Moore, R.E., Young, M.K., and Lee, T.D. Qscore: An algorithm for evaluating SEQUEST database search results. *J. Am. Soc. Mass. Spectrom.*, 13, 378–386, 2002.

23. MacCoss, M.J., Wu, C.C., and Yates, J.R., III. Probability-based validation of protein identifications using a modified SEQUEST algorithm. *Anal. Chem.*, 74, 5593–5599, 2002.

24. Oda, Y., Huang, K., Cross, F.R., Cowburn, D., and Chait, B.T. Accurate quantitation of protein expression and site-specific phosphorylation. *Proc. Natl. Acad. Sci. USA*, 96, 6591–6596, 1999.

25. Pasa-Tolic, L., Jensen, P.K., Anderson, G.A., Lipton, M.S., Peden, K.K., Martinovic, S., Tolic, N., Bruce, J.E., and Smith, R.D. High throughput proteome-wide precision measurements of protein expression using mass spectrometry. *J. Am. Chem. Soc.*, 121, 7949–7950, 1999.

26. Gygi, S.P., Rist, B., Gerber, S.A., Turecek, F., Gelb, M.H., and Aebersold, R. Quantitative analysis of complex protein mixtures using isotope-coded affinity tags. *Nat. Biotechnol.,* 17, 994–999, 1999.

27. Mirgorodskaya, O.A., Kozmin, Y.P., Titov, M.I., Korner, R., Sonksen, C.P., and Roepstorff, P. Quantitation of peptides and proteins by matrix-assisted laser desorption/ionization mass spectrometry using (18)O-labeled internal standards. *Rapid Commun. Mass Spectrom.,* 14, 1226–1232, 2000.

28. Yao, X., Freas, A., Ramirez, J., Demirev, P.A., and Fenselau, C. Proteolytic 18O labeling for comparative proteomics: Model studies with two serotypes of adenovirus. *Anal. Chem.,* 73, 2836–2842, 2001.

29. Goodlett, D.R., Keller, A., Watts, J.D., Newitt, R., Yi, E.C., Purvine, S., Eng, J.K., von Haller, P.D., and Aebersold, R. Differential stable isotope labeling of peptides for quantitation and de novo sequence derivation. *Rapid Commun. Mass Spectrom.,* 15, 1214–1221, 2001.

30. Ong, S.E., Blagoev, B., Kratchmarova, I., Kristensen, D.B., Steen, H., Pandey, A., and Mann, M. Stable isotope labeling by amino acids in cell culture, SILAC, as a simple and accurate approach to expression proteomics. *Mol. Cell Proteomics,* 1, 376–386, 2002.

31. Li, X.-J., Zhang, H., Ranish, J.A., and Aebersold, R. Automated statistical analysis of protein abundance ratios from data generated by stable-isotope dilution and tandem mass spectrometry. *Anal. Chem.,* 75, 6647–6648, 2003.

32. Griffin, T.J., Lock, C.M., Li, X.J., Patel, A., Chervetsova, I., Lee, H., Wright, M.E., and Ranish, J.A., Chen S.S., and Aebersold R. Abundance ratio-dependent proteomic analysis by mass spectrometry. *Anal. Chem.,* 75, 867–874, 2003.

33. Dixon, W.J. Processing data for outliers. *Biometrics,* 9, 74–89, 1953.

17 Pattern Recognition Algorithms and Disease Biomarkers

Ben A. Hitt, Emanuel Petricoin, and Lance Liotta

CONTENTS

17.1 INTRODUCTION

The publication of our successes[1–4] in using mass spectral patterns to identify patients with ovarian and prostate cancers and animals with premalignant pancreatic cancer has led to increased interest in using these methods to characterize a variety of disease and biological states. Adoption of a pattern discovery and recognition

---

The views expressed here are solely by the authors and should not be construed as representative of those of the Department of Health and Human Services, the U.S. Food and Drug Administration. Moreover, aspects of the topics discussed have been filed as U.S. government and company owned patent applications. Drs. Hitt, Petricoin, and Liotta are coinventors on these applications and may receive royalties provided under U.S. law.

approach to biomarker identification requires the acceptance of two basic premises—
that there is a need and that a pattern of features in a signal constitute a biomarker—
as well as choosing an adequate set of tools. This chapter will present a case for
acceptance of both premises, describe some of the methods available, and present
the pitfalls associated with the process.

17.2   WHAT IS THE NEED?

17.2.1   THE PROBLEM WITH SINGLE BIOMARKERS

The search for single tumor markers appears to be reaching a point of diminishing
returns. The number of new diagnostic tumor markers released for use has been in
steady decline for the last five years.[4] While the underlying cause for this decline is
unclear, one can speculate on what factors may be involved.

The first point to consider is that useful specific tumor markers may only exist in
vanishingly small amounts. For real clinical utility, a tumor marker must be measurable
at the earliest stage of the disease. Markers that appear during late-stage cancer offer
little chance in triggering any effective treatment plan that alters the ultimate course
of the disease. Early stage cancers will have a small tumor mass with only a relatively
few cells. When only a small number of tumor cells exist, the amount of any specific
marker secreted into the extracellular media, or produced within the microenvironment
at the tumor–host interface is unlikely to be large enough to be found.

The second is that a single tumor biomarker depends on quantitative measurement
for success. For a quantitative relationship to exist, the tumor type would likely have
to be homogeneous. An example is PSA. Prostate cancer is a single-tissue type cancer
and thus as a prostate cancer progresses a predictable increase in PSA should ensue.
However, even under this circumstance, PSA is not as reliable as one would like it to
be. Indeed its usefulness has been called into question.[5] An opposing example is CA125
and its use for ovarian cancer. Ovarian cancer has many different tissue types and a
strict quantitative relationship does not exist. CA125 is not acceptable as an indicator
of the onset of ovarian cancer, and is only FDA approved for monitoring indications.

Finally, for a single specific marker to be uniformly effective, its expression
would have to be independent of other genotypic and phenotypic factors in the host.
Of course there are markers that are specific to infectious agents that demonstrate
this quality. However, in these cases the marker expression is dependent on the
immune system for expression, the protein produced is a unique entity not found at
any time within the human population, and there is significant variation in expression.

17.2.2   THE PATTERN ADVANTAGE

Mathematically it should be obvious that a pattern of multiple biomarkers will contain
a higher level of discriminatory information compared to a single biomarker alone,
particularly for large heterogeneous patient populations. Patterns are not dependent on
any single biomolecule. Unlike single-marker approaches, pattern recognition spreads
its dependency on all the features in the pattern. This allows for increased robustness
compared to dependency on a single feature. Consider the effect of the degradation
of a single feature in a ten-feature pattern. In true pattern recognition, each feature

contributes equally to the pattern. In our ten-feature pattern, each feature contributes 10% to the pattern. If a single feature were to degrade, say, 50%, 95% of the information needed to recognize the pattern would remain.

A true pattern recognition model is able to recognize multiple patterns based on a single-feature set. It is suspected that diagnostic serum proteomic patterns are the result of tumor interaction with the host. Thus it is possible that given a specific tumor type, two genetically different hosts may generate distinct patterns. Each of those patterns signals the presence of the tumor.

This ability illustrates the underlying capacity of pattern recognition modeling to capture nonlinear and chaotic effects. It is perhaps this ability that is the real strength of pattern-based modeling. Biology is intrinsically nonlinear. While on a macro level an organism's behavior appears continuous, at the cellular level most biological systems are nonlinear. For example, nerve transmission is an all or none phenomenon as is muscle contraction. Furthermore, biological systems are chaotic.[6] For any chaotic process, unless the beginning point is precisely known, the end result is unpredictable. The consequence of chaos is that tumor markers cannot be expected to be uniform from patient to patient. However, host response to a tumor is dependent on a number of regulated mechanisms. Pattern recognition in sera can be expected to detect changes resulting from perturbation of any or all of these regulated systems, in effect capturing the chaos of tumor host system.

17.3 PATTERN RECOGNITION METHODS

17.3.1 NONLINEAR FEED-FORWARD NEURAL NETWORKS: THE BASICS

Nonlinear feed-forward neural networks, most commonly known as back-propagation[7] neural networks (ANN), were the premier pattern recognition algorithms used in a variety of industrial settings. Generally, an ANN consists of three layers: an input layer, a hidden layer, and an output layer (Figure 17.1). Each node in a layer is connected to every node in each successive layer. There is also a single node — the bias node — that is connected to every node in the hidden layer and the output layer. The input layer is inactive in that all it does is receive input data. The hidden layer nodes each perform two functions. The first is a summation of the products of each input value and its corresponding connection weight. The sum is called the internal activation:

$$A = \sum (w_i I_i) \qquad (17.1)$$

where A is the internal activation, w is a weight and I is an input associated with the weight. The second is a nonlinear transformation of the internal activation, e.g., the Sigmoid function:

$$O = \frac{1}{1 - e^A} \qquad (17.2)$$

where O is the node output and A is the internal activation. A plot of the Sigmoid function (Figure 17.2) demonstrates an S-shaped curve.

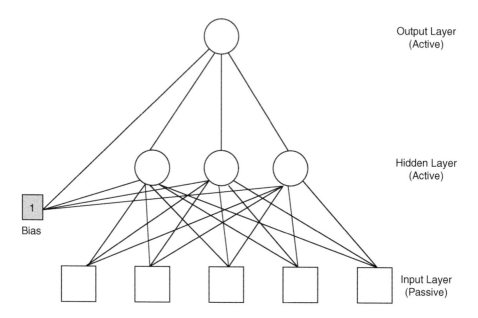

FIGURE 17.1 A nonlinear feed-forward neural network comprised of a passive input layer, an active hidden layer, and an active output layer. The network is said to be fully connected in that each node in a layer is connected to every node in the layer immediately above it. There is a bias node that is always equal to 1 and is connected to every active node in the neural network.

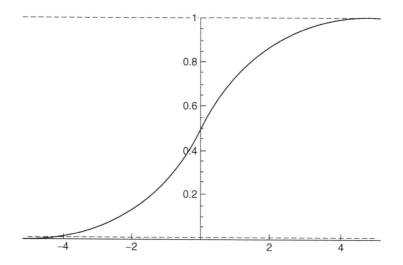

FIGURE 17.2 A plot of the Sigmoid function. The plot is an S-shaped curve that approaches 0 and 1. The dynamic range of the curve falls in the range $-2 \le x \le 2$.

The output layer nodes perform the same two functions. The output of an ANN is a score between 0 and 1 when the transformation is through the Sigmoid function as above or between −1 and +1 when a function such as the hyperbolic tangent or sine is used. A training cycle consists of

1. Presenting a data record or exemplar to the input layer
2. Computing the internal activation sum for each node in the hidden layer
3. Computing the nonlinear transform of the internal activations in the hidden layer nodes
4. Computing the internal activation sum for the output layer node(s)
5. Computing the nonlinear transform of the internal activations in the output layer node(s) for output
6. Determining the error of the computed output as different from the known outcome associated with the input data vector
7. Backpropagating the error through the first derivative of the transformation function
8. Adjusting the connection weights proportionately to the backpropagated error

This process is iterated over and over again until the training error over all the exemplars ceases to decrease. Because the training process requires knowledge of expected outcome and is essentially an error minimization process, ANNs are an example of supervised learning.

17.3.2 INTERPRETATION OF RESULTS

ANN results are generally interpreted by ranking the input record's score and desired outcome. A successful ANN model is one where the scores of those records where the desired outcome is 1 approximately 1. A cutoff or threshold score is chosen so that the majority of the records with outcomes of 1 fall above it and, conversely, the majority of records with outcomes of 0 fall below it. In general ANN use, two kinds of errors are reported:

1. Type I, or false negatives
2. Type II, or false positives

The goal is to minimize both error types while maintaining model robustness. For biological data the results are reported as sensitivity (1-Type I rate) and specificity (1-Type II rate). The general goal remains the same.

Reliable robust results must be ascertained through the use of at least two sets of data, a training set and a testing set. Ideally, a third set of data, a validation set, is employed. The training set is only used to adjust the connection weights to minimize the overall error. The testing set is used to determine if the model has been overtrained or overspecified and to set the threshold score. The validation set is used to test for robustness and serves as the source of the final metrics of model performance. For an ANN model to be robust, the overall error rate in the training, testing, and validation data sets should be approximately the same. Overtraining and overspecification are indicated when the training error is lower than either the testing or validation errors.

The makeup of the three data sets is crucial. The training set needs to be balanced; that is to say, there needs to be an equal representation of records with each of the desired outcomes. In the example of a diagnostic model, there should be an approximately equal population of exemplars of diseased cases and case controls. The reason for this is that an ANN will adjust the connection weights in direct response to errors it makes on the data. If there are many more of one type of exemplar, the adjustments will be predominantly in favor of that one exemplar type. However, since the testing and validation sets are not used to adjust connection weights, they do not need to be balanced. There should be a sufficient number of records of each type to ensure reliable metrics.

It is also imperative that input data to ANN be scaled appropriate to the nonlinear transformation function employed. If the function used is the Sigmoid function, then the input data should be scaled to fall in the range $-2 \leq I \leq 2$. If either the sine or hyperbolic tangent is used, the range should be $-0.8 \leq I \leq 0.8$. The rationale behind the scaling is that the dynamic range of the transformation function lies in this range. The goal is to scale both the input data and the connection weights such that the internal activations fall within the transformation functions range.

17.3.3 ANN PROS AND CONS

The major benefit of an ANN is that there is a proof[8] that states that a nonlinear feed-forward neural network with a single hidden layer can approximately solve any measurable relationship. While one can argue at some length as to what constitutes a measurable relationship, the practicality is that ANNs can be thought of as universal approximators.

On the downside is the observation that ANN requires large amounts of data for training, testing, and validation. The wholly interconnected nature of ANN provides for increased risk of overfitting. There are only two ways to avoid this: one is to limit network size, i.e., number of nodes and connections, or to have sufficient data to overcome the problem. A rule of thumb is that seven training cases are needed for each connection in the network. Thus it should be obvious that there needs to be a balance between network size and data availability. This may limit the usability of ANN in biological work. The strength of an ANN to unravel complex relationships resides in the hidden layer. The more nodes there are in the hidden layer, the more complex a function can be approximated, but more data are required. The bottom line is laboratory and clinical studies where ANN use is anticipated require larger numbers of subjects than usually anticipated.

17.4 SELF-ORGANIZING MAPS: THE BASICS

Self-organizing maps (SOM) were developed by Stephen Grossberg[9] and Tuevo Kohonen[10,11] in the 1970s and have constantly improved since then. They have found significant use in a variety of settings. A basic SOM consists of two layers as shown in Figure 17.3. Similar to ANN, there is an input layer. The second layer is a two dimensional grid, or map, of nodes. Each node in the input layer is connected to

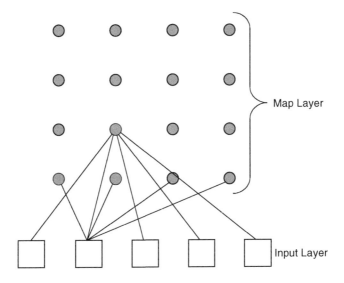

FIGURE 17.3 A self-organizing map. The SOM is comprised of an input layer and a map layer. The map layer is a two-dimensional grid of nodes. Each node in the input layer is connected to each node in the map layer (most connections are not shown for clarity). No map node is connected to another map node.

every node in the map layer. While the ANN connections are coefficients like one finds in multivariant regression equations, the connections in SOMs are directly related to the inputs. A map layer node's connections represent a coordinate in N-dimensional space where N is the number of connections coming into each node, or the number of inputs to the map.

SOM training is competitive in nature. The competition is based on Euclidean distance according to an expansion of the Pythagorean Theorem into N dimensions:

$$D = \sqrt{\sum (w_i - I_i)^2} \qquad (17.3)$$

where D is the Euclidean distance, w is a weight, and I is the input associated with the specific weight.

Each node's vector of weights constitutes a coordinate in the space, as does the vector of inputs. The distance between the input vector and every node in the map is computed. The node whose weight vector is closest to the input vector wins and the weight vector is adjusted to be closer to the input vector. The process does not stop here. The winning node's neighbor's weight vectors are also adjusted to be closer to the input vector. The shape and size of the neighborhood is one of the critical user-determined parameters. Typically, three types of neighborhoods are employed:

1. The diamond shaped neighborhood is smallest containing, at a minimum four adjacent nodes, one each to the north, south, east, and west of the winning node

2. The square neighborhood contains, at a minimum, eight adjacent nodes: three across the top and bottom, and one on either side
3. The hexagonal neighborhood contains at least six adjacent nodes, one at each apex

There are no rules of thumb to determine which of these neighborhoods to use under which circumstance. However, the hexagonal neighborhood seems to have come into vogue of late.[11]

SOM training also includes a conscience mechanism so that training tends to spread out evenly over the entire map. The object is to bias the Euclidian distance in such a way that the more a node wins, the more difficult it becomes for it to win. This is accomplished by adding a fraction of the distance back to the distance where the size of the fraction is proportional to the number of times the node has won:

$$D_c = \sqrt{\sum (w_i - I_i)^2} + aD_o \qquad (17.4)$$

where D_c is the current distance, w and I are defined as in Eq. (3), a is the conscience factor and is given in Eq. (5) below, and D_o is the original or last distance,

$$a = c_n/T \qquad (17.5)$$

where c_n is the count of the number of times node n has won and T is the total number of iterations through the map.

The net effect of this is that data dense areas tend to spread out, whereas sparse areas tend to concentrate. The effect can be visualized as a wadded-up piece of gauze unfolding and ultimately spreading out evenly.

17.4.1 INTERPRETATION OF RESULTS

Interpretation of SOM results depends on their use. One of the primary attractions of SOM is the ability to view the vector of weights for each node in the map as a prototype of a pattern or putative pattern in the input data. This makes SOM useful as a front end to a supervised technique such as ANN. The goal is to filter out some of the noise in raw data by assigning a prototype vector and using the prototype as the input to the ANN or other supervised techniques. If this technique is employed, the SOM must be organized first and then the supervised technique follows after the SOM is fixed. The SOM organization step is, of course, unsupervised.

The prototypical nature of node weights can also be used to determine the features or components of the pattern that discriminate one pattern from another. One of the major shortcomings of ANN is the inability to explain which features in the pattern contribute to the outcome. With SOM, the weight vectors of two nodes can be compared directly by subtracting one from the other and the key differences will emerge.[12] The utility of this property of SOM is illustrated in the following example. Consider the transition of a normal cell to an abnormal cell.

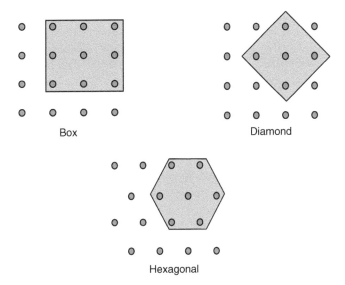

FIGURE 17.4 Three SOM neighborhoods. The neighborhood defines which nodes adjacent to the winning nodes will be adjusted along with the winning node. The box neighborhood has eight adjacent nodes, the diamond neighborhood has four adjacent nodes, and the hexagonal neighborhood has six adjacent nodes.

The investigator has the ability to measure a number of specific proteins known to be involved in the transition but does not know how the transition actually involves those proteins. He or she measures the proteins in a number of cases and controls sufficient to support SOM development and organizes an SOM (Figure 17.4). Three different kinds of nodes emerge.

1. Nodes that associate with cases
2. Nodes that associate with controls
3. Nodes that either have a weak or no association with either

Now he or she can compare the patterns in the case nodes against the patterns in the control nodes and see which proteins actually discriminate between the two states. However, what is interesting is three types of nodes that lie between the cases and controls. Their weight vectors represent putative patterns in the transition from normal to abnormal. An indication of these transitional states may be of great value in the treatment and management of disease.

17.4.2 SOM PROS AND CONS

The primary advantage of SOM is its ability to organize data based on the data itself and not its relationship with some known outcome; i.e., SOMs are unsupervised. Furthermore, since an SOM node's weight vectors reflect real-world input values, they can be used as prototypical patterns of the clusters found in the data. An organized node is said to be a centroid whose coordinates in N-dimensional space are the weights. As suggested in the example above, these features make SOM very useful in knowledge discovery, where there is a wealth of data but the information is masked by that very wealth.

On the downside, SOM requires a moderate amount of trial and error to achieve the desired results. There are a number of interacting critical parameters that must be balanced for the desired result. These are number of nodes in the map, geometry of the map, dimensions of the map, neighborhood configuration, learning parameter for the winning node, learning parameter for the neighboring nodes, and sensitivity of the conscience mechanism. For example, too few nodes will not distribute the data correctly, while too many nodes may separate clusters into inappropriate subclusters.

17.5 ADAPTIVE PATTERN RECOGNITION: THE BASICS

As noted above, SOMs require significant trial and error to find the correct number and geometry of nodes in the map layer. Furthermore, ANN and SOM share a common shortcoming: once they have been trained, they cannot recognize or adapt to data that lie outside the envelope of that used during training. They will score and/or classify novel data as though they had been represented in the training corpus. Adaptive pattern recognition algorithms address this issue.

There are a variety of algorithms that are adaptive pattern recognition devices: Fuzzy ARTMAP,[13] the Adaptive Fuzzy Feature Map (AFFM),[14] and the Lead Cluster Map (LCM)[15] are three examples. While they differ in the way that they learn and adapt, they all basically do the same thing. That is, they try to match an incoming pattern to a prototypical pattern they have already learned. If they find a match, they update the prototype; if not, they create a new prototype from the incoming pattern. Of the three listed, LCM is the simplest and it flows directly from SOM, so it will be used to explain the process.

LCM begins with an empty N-dimensional space, where N is number of features used in the desired pattern mapping. Values in vectors presented to an LCM must be scaled to the range $0 \leq x \leq 1$. The result of this is the bounding of the space by zero and one in all dimensions. Under this circumstance, the maximum distance in a single dimension is 1. Therefore, the maximum distance allowed in the space is the square root of N.

The first vector presented to the space simply fixes a centroid at the coordinates specified by the vector. A decision boundary is drawn around the centroid, the radius of which is a fraction of the maximum distance allowed in the space. For example, if the user wishes to achieve a pattern match of 90%, they would set the boundary radius to be $0.1N^{1/2}$. The second vector is presented and its distance to the centroid is computed.

If the distance between the points specified by the centroid and the incoming vector is less than the distance from the centroid to the decision boundary, then the incoming vector is said to match the pattern specified by the centroid. That being the case, the centroid is moved fractionally closer to the vector. If the distance is greater than the decision boundary radius, the vector is used to fix a new centroid at the coordinates specified by the elements in the vector. When the third vector is presented, the process is repeated with two distance calculations and comparisons. The process is repeated until every vector in the data set is processed. Each vector in the data set is processed only once.

17.5.1 Interpretation of Results

At the end of training an LCM will contain only the nodes/clusters specified by the data. The population count of each cluster is a direct indication of the abundance of records in a data set that conform to the pattern described by the cluster centroid. For any given naturally generated training set, one may expect LCM nodes with a broad range of population count. An LCM that generally describes a data set should have relatively few nodes. Of those, some, perhaps two or three, should be highly populated. Nodes containing population counts of only one or two should also be expected. An LCM containing a large number of nodes with a small population count has learned to differentiate between individual records and is not good for general use.

The centroids developed during LCM training represent prototypical patterns for the data belonging to that cluster. This permits direct comparison of the clusters to determine which features in the patterns are driving the difference. For example, one can simply subtract two centroid vectors and display the difference vector as a histogram (Figure 17.5). Those distinctive features will show up as significantly different from zero either in a positive or a negative direction. In proteomics this kind of information can be used to guide protein/peptide sequencing efforts, for example.

A cluster's population is relatively homogenous with respect to the features defining the pattern. If the clusters were organized in a truly unsupervised fashion, examination of a cluster's population could reveal whether the pattern features have any relevance to a biologic state, such as a disease. Consider a data set containing data collected from subjects with a particular cancer and subjects without the disease. If a cluster is formed that contains only subjects with the cancer, then it would be interesting to compare the centroid pattern of the cluster to centroids of clusters without or thinly populated with diseased subjects. The resulting differences may have bearing on the oncogenic process or identify potential therapeutic targets.

17.5.2 Pros and Cons

The major advantage of the LCM, along with other adaptive pattern recognition algorithms, is its ability to recognize and track novel and rare events in a data stream. An LCM, after its initial training or organization, can be fielded in static or adaptive mode. In static mode the centroids are fixed and the reported results would include cluster membership and distance to the centroid. If a vector of data does not match any known pattern in the space, the nearest cluster is reported but the distance is reported as being outside the decision boundary of any known node/cluster. An ANN or SOM, once trained, will report a classification but have no means of recognizing novel data vectors. In a biomedical setting, this is an important advantage. An LCM

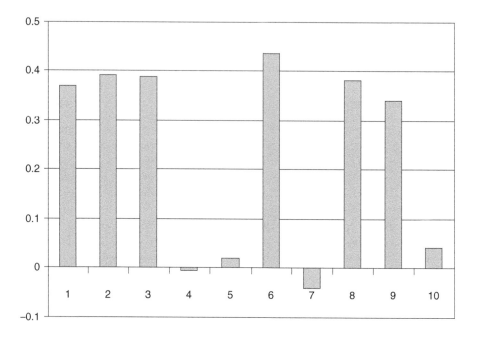

FIGURE 17.5 A comparison of two lead cluster map nodes. The differences in the features shows as a change from zero. Those features that are similar in the two patterns do not differ greatly from zero, whereas the features that separate the two patterns do.

classifying patients can identify patients whose patterns are novel. This may be valuable to the public health in general. For example, if a novel pattern appeared and then started to show up in a specific locale, then an epidemiologic investigation may be warranted.

On the downside, LCMs require a great deal of trial and error to find the optimum set of parameters that direct their learning. One has to balance the radius of the decision boundary with the adjustment to new member vector along with specifying the correct number and selection of pattern features. Lead cluster mapping is prone to overfitting on two fronts, too many features in the pattern, and too small a decision radius. The number of features is limited by the amount of data in the training set. The number of features one can use in the pattern increases as the amount of training data increases. However, this does not mean that one should search for the largest number of features allowed by the data. Any problem probably has a natural limit to the number of features required to create a signature pattern. Therefore, it is possible that if the size of the feature set grows beyond a certain point, unnecessary features will be included. More importantly, data sets common to biologic investigation are usually limited in size. In this case the inclusion of too many features in the pattern will result in overspecification to the point of being able to specify each record in the data set. Overspecification can also result when the decision radius is set too small; thus the need for rather extensive trial and error using true training and testing data sets.

17.6 FEATURE SELECTION

Any modeling method, be it ANN, SOM, LCM, or other, requires that the correct set of features be used. The search for an optimal feature set represents the real effort in finding useful robust models. As the number of possible features grows, the combinatorial numbers become astronomical very quickly. With mass spectrometric data, for example, one can expect something on the order of 10,000 to 1 million data points. Assuming a 10,000 data point string, if a five-feature pattern is expected, there are $10,000^5$, or 1×10^{20} possible combinations. An explicit search of a combination set of this magnitude is virtually impossible, even using massively parallel computers. It is clear that a systematic means of streamlining the search is required.

There are a number of ways available for feature selection. Classically, the statistical methods of stepwise regression and correlation have been with reasonable success. However, it is not clear that a statistical method is truly compatible with the nonlinear methods described here. Therefore, it seems desirable to use a method that uses these nonlinear techniques in an integral fashion.

One such method uses the genetic algorithm[1-3,16] in conjunction with an LCM to find an optimal set of features. The genetic algorithm is a computer simulation of natural evolution. The natural processes of mating, reproduction, crossover mutation, and population limitation are used to evolve a set of features that an LCM can use to cluster data such that it can distinguish biological states. All the processes are probabilistic, with the more fit members of the population having a greater chance of mating and surviving than those less fit. The process guarantees that each successive generation of candidate solutions will be, on average more fit, i.e., better able to solve the problem, than the previous until a near optimal solution is found.

REFERENCES

1. Petricoin, E.F., Ardekani, A.M., Hitt, B.A., Levine, P.J., Fusaro, V.A., Steinberg, S.M., Mills, G.B., Simone, C., Fishman, D.A., Kohn, E.C., and Liotta, L.A. Use of proteomic patterns in serum to identify ovarian cancer. *Lancet,* 359, 572–577, 2002.
2. Petricoin, E.F., III, Ornstein, D.K., Paweletz, C.P., Ardekani, A., Hackett, P.S., Hitt, B.A., Velassco, A., Trucco, C., Wiegand, L., Wood, K., Simone, C.B., Levine, P.J., Linehan, W.M., Emmert-Buck, M.R., Steilhberg, S.M., Kohn, E.C., Liotta, L.A. Serum proteomic patterns for detection of prostate cancer. *J. Natl. Canc. Inst.,* 94, 1576–1578, 2002.
3. Hingorani, S.R., Petricoin, E.F., Maitra, A. Rajapakse, V., King, C., Jecobetz, M.A., Ross, S., Conrads, T.P., Veenstra, T.D., Hitt, B.A., Kawaguchi, Y., Johann, D., Liotta, L.A., Crawford, H.C., Putt, M.E., Jacks, T., Wright, C.V., Hruban, R.H., Lowy, A.M., and Tuveson, D.A. Preinvasive and invasive ductal pancreatic cancer and its early detection in the mouse. *Canc. Cell,* 6–21, 2003.
4. Anderson, N.L. and Anderson, N.G. The human plasma proteome: History, character, and diagnostic prospects. *Mol. Cell. Proteomics,* 1(11), 845–867, 2002.
5. Punglia, R.S., D'Amico, A.V., Catalona, W.J., Roehl, K.A., and Kuntz, K.M. Effect of verification bias on screening for prostate cancer by measurement of prostate-specific antigen. *New England Journal of Medicine,* 349, 335–342, 2003.

6. Kauffman, S.A. *The Origins of Order: Self-Organization and Selection in Evolution.* Oxford University Press, New York, 1993.

7. Rummelhart, D.E. and McClelland, J.L. *Parallel Distributed Processing.* MIT Press, Cambridge, MA, 1986.

8. Hornik, K., Stinchcombe, M., and White, H. Multilayer feedforward networks are universal approximators. *Neural Network.*, 2, 359–366, 1986.

9. Grossberg, S. Adaptive pattern classification and universal recoding, I: Parallel development and coding of neural feature detectors. *Biol. Cyber.*, 23, 121–134, 1976.

10. Kohonen, T. Analysis of a simple self-organizing process. *Biol. Cybern.*, 44, 135–140, 1982.

11. Kohonen, T. Self-organized formation of topologically correct feature maps. *Biol. Cybern.*, 43, 59–69, 1982.

12. Kohonen, T. *Self-Organizing Maps.* Springer-Verlag, New York, 2001.

13. Carpenter, G.A., Grossberg, S., Markuzon, N., Reynolds, J., and Rosen, D. Fuzzy ARTMAP: A neural network architecture for incremental supervised learning of analog multidimensional maps. *IEEE Transactions on Neural Network.*, 3, 698–713, 1992.

14. Hitt, B.A. Adaptive Fuzzy Feature Mapping. U.S. Patent # 6,249,779, 2001.

15. Kohonen, T., Kaski, S., and Lappalainen, H. Self-organized formation of various invariant-feature filters in the adaptive-subspace SOM. In *Self-Organizing Map Formation, Foundations of Neural Computing.* MIT Press, Cambridge MA, pp. 354–368, 2001.

16. Holland, J.H. *Adaptation in Natural and Artificial Systems.* MIT Press, Cambridge, MA, 2001.

18 Statistical Design and Analytical Strategies for Discovery of Disease-Specific Protein Patterns

Ziding Feng, Yutaka Yasui, Dale McLerran, Bao-Ling Adam, and John Semmes

CONTENTS

18.1 INTRODUCTION

The rapid advances in proteome technology offer great opportunities for cancer researchers to find protein biomarkers or protein patterns for early detection of cancer.[1] Since many types of cancer are curable if treated early and are incurable at later stages, early detection is an effective way for fighting against cancer.

Bioinformatic and statistical methods for genomic data, in particular the data from Affimatrix microarray or c-DNA spot array experiments, have matured over the past five years and have made contributions to biology and medicine. To make similar headway in proteomics, we need to understand two challenges in proteomic data analyses. First, the number of proteins, due to posttranslational modifications, is even larger than the number of genes. This high dimensionality leads to difficulties in identifying protein biomarkers or protein patterns truly diagnostic for cancer. Not only computational demand, but also the chance of false findings, is high because the dimension of the data is usually much larger than the number of samples under investigation. This is parallel to genomic data but at a larger scale. Second, in the Affimatrix microarray or c-DNA spot array data, the complementary double-helix structure of DNA greatly facilitates the fidelity and reproducibility of the expression data. Such an advantage does not apply to proteomics due to proteins' three-dimensional structures. Measurement variation occurs not only on protein expression intensity but also on the protein mass quantification. Specifically, the same protein or peptide can appear at different mass values on different proteomics platforms and, worse, in different runs on the same instrument, due to the limitations of the instrument. This creates great difficulties in protein identifications and measurements for using protein patterns as a disease diagnostic tool.

Statistics has contributed greatly in agriculture, industry, technology, and biomedicine. Regardless of the settings, good statistical practice follows three ordered principles. First, we need to understand the data-generating process, the sources of variations, and systematic biases in the process. Second, based on these understandings, we need to design experiments that eliminate or reduce the biases and variations (noises), or at least enable their control and measurement so that rigorous analyses and inference could be made. Third, analysis methods must take into account the data-generating process and must be congruent to the experimental design. This means a good analysis is impossible without correct execution of the first two principles. It also emphasizes the importance of close interactions between laboratory/bench scientists and statisticians from the early experimental planning to the final reporting of findings.

The objective of this chapter is to provide operational guidelines to meet these three principles for biologists and biostatisticians in planning proteomics experiments and analyzing and interpreting their data.

The remaining part of this chapter is organized into three sections corresponding to the above three principles, plus a summary section. We use surface enhanced laser desorption/ionization–time of flight (SELDI-TOF) data as an example, but the principles apply to data from other proteomic platforms, such as matrix-assisted laser desorption/ionization–time of flight (MALDI-TOF). Our discussions focus on using protein/peptide peaks to form disease classifiers without knowing protein identities. SELDI technology does not directly provide protein identification. Therefore, quantitative methods for peptide/protein identification are not discussed here. We also exclude the statistical design and analysis for *formally validating* a given disease classifier (e.g., a protein marker or a panel of protein markers/peaks with an explicit rule to combine them for diagnosis). The statistical experimental design and analysis for formal validation of a given classifier are very different from that for classifier construction and would require a whole chapter to discuss them. This chapter answers two questions: (1) how to plan an experiment for studying protein/peptide patterns for disease classification, and (2) after obtaining such data, how to construct a classifier.

18.2 STATISTICAL PROPERTIES OF PROTEIN EXPRESSION MEASUREMENTS

SELDI-TOF utilizes the ProteinChip Technology developed by Ciphergen Biosystems, Inc. Biological samples are first placed on the surface of wells on a protein chip. Proteins bind to the surface with certain chemical (e.g., hydrophobic, hydrophilic, ionic) or biochemical (e.g., antibody, enzyme, receptor) properties. Nonbinding proteins, salts, etc., are then washed away. Matrix, or energy-absorbing molecule (EAM), is then applied to the surface that will absorb the energy from a laser beam and transfer the energy to ionize proteins. The protein mass and its intensity are determined by the time-of-flight technology. Heavier proteins/peptides take a longer time to reach the detector, while lighter ones require less time. These different times to reach the detector are translated into different protein masses. The number of protein particles hitting the detector at a given time point is translated into a protein intensity measure.

There are 8 wells per protein chip, 96 chips per manufacture batch. Assays may be run on different days and/or different machines. Therefore, the sources of variations are well-to-well variation, chip-to-chip variation, batch-to-batch variation, day-to-day variation, machine-to-machine variation, and sample-to-sample variation. Only the last variation associated with a certain disease phenotype is the signal of scientific interest. All the others are noise that we want to reduce so that the signal-to-noise ratio is increased.

Figure 18.1 shows two spectra from the same serum sample. A visual inspection suggests that the protein mass/charge points are very consistent across spectra. However, the magnified segments reveal that the seemingly aligned peaks are not actually aligned. Typically, the magnitude of SELDI mass accuracy error is about 1,000–2,000 ppm (0.1–0.2% of mass per charge values). This misalignment makes assessment of protein intensity variation at a specified mass/charge point not very meaningful. Therefore, for the materials discussed in this section, we have performed peak identification and alignment of spectra. Baseline subtraction and normalization have been applied on these data at the laboratory and are not discussed here. A detailed discussion of our peak identification and alignment method is given in Section 18.4.

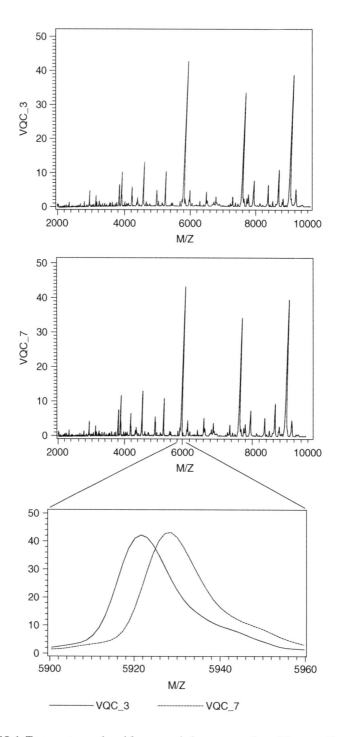

FIGURE 18.1 Two spectra produced from a pooled serum sample and the magnified segment from 2950 to 2980 m/z.

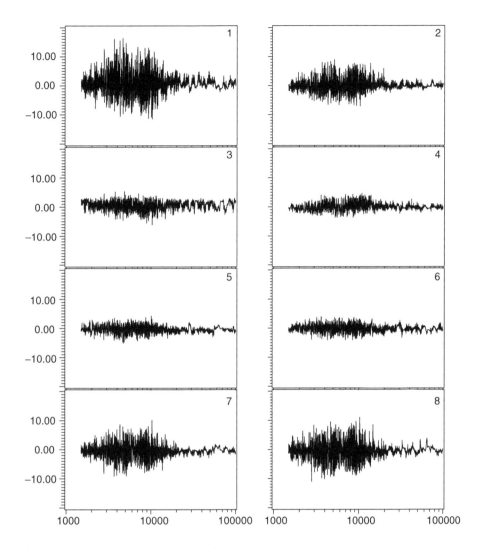

FIGURE 18.2 t-Statistic for departure from grand mean.

In an experiment investigating the sources of biases and variations, a pooled serum sample was measured repeatedly using one SELDI machine. Six chips were taken from each of two batches and assays were run on five different dates, 2 to 4 chips per day. This yields 96 spectra (12 chips × 8 wells per chip).

Figure 18.2 depicts, for each well's expression intensity, the t-statistic for departure from the grand mean over all chips and wells. It indicates that there are some systematic differences in intensity variability among wells, with wells 1 and 8 (the top and bottom wells on a chip) having more deviations from the grand mean intensity values. The systematic variation becomes smaller toward the middle wells. This could occur for many potential reasons; for example, if chips are not in a perfect

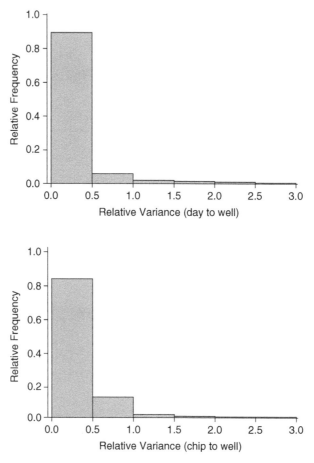

FIGURE 18.3 Histogram of day-to-day to well-to-well variance ratio and chip-to-chip to well-to-well variance ratio.

perpendicular position in the machine. In planning a new experiment, either this systematic problem must be corrected technologically, or wells 1 and 8 should not be used for analyzing samples. It also suggests that samples should be assigned to wells at random to avoid systematic bias. Section 18.3.2 will discuss experimental design in depths.

Figure 18.3 shows histograms of the ratio of day-to-day variance to well-to-well variance, and the ratio of chip-to-chip variance to well-to-well variance, for all mass per charge points. It indicates that the chip-to-chip variance and the day-to-day variance are usually much smaller than the well-to-well variance. Note that "variance" refers to the degree of random measurement errors and should be distinguished from "bias," the systematic measurement errors shown in Figure 18.2. This is a desirable property because the well is the smallest measurement unit and it is relatively easy to reduce the well-to-well variance by using multiple wells per sample and taking their mean.

18.3 STUDY DESIGN

18.3.1 THE ISSUE OF OVERFITTING

One central challenge in study design is the fact that the number of candidate protein peaks under examination for their diagnostic potential is usually substantially larger than the number of biological samples, even after peak identification and alignment that reduce the number of candidate protein peaks dramatically. In a classic regression problem, a model would fit the data extremely well if the degrees of freedom of the model are large enough, or fit the data exactly if the model has the degrees of freedom equal to the sample size; yet such a model does not approximate well the underlying mechanism that generated the data (e.g., proteins and their interactions that specify the observed disease status). This is because the large degrees of freedom tailor the model too finely to the features of observed data; the fitted model describes not only the systematic features of the underlying data-generating mechanism, but also the random features that are unique to the observed set of data. When such an overfitted model is used to predict a new observation, it fails. This phenomenon is called "overfitting." This issue is so critical for high-dimensional genomic and proteomic data analyses and deserves more discussions below.

The degrees of freedom for a model is a statistical concept. It refers to the number of independent pieces of information used by the model. The more complex a model is, the more degrees of freedom it has. A set of n independent observations has n pieces of independent information. Let's assume that we have a total of n independent samples, some are from disease cases and the rest are from normal controls, and the biomarker measure is a continuous variable with distinct values for each sample. A simple model using the mean for each of the case and control groups has two degrees of freedom. A model that classifies a sample as "disease" if its value equals any of the values of the case samples and "normal" if its value equals any of the values of the control samples has n degrees of freedom. It classifies the n samples perfectly but it is useless for future prediction. This is an extreme example of overfitting because it does not uncover any underlying mechanism of biology and did not filter out any noise from the data. A good model would approximate the underlying mechanism of biology to a degree consistent to the amount of information in the data; that is, it is not too simple or too complex relative to the amount of information at hand. An overfitting occurs when a model is inappropriately too complex and flexible, relative to the amount of information available. The result of fitting such a complex flexible model to the finite data at hand is a model excellent for describing the data at hand but poor for approximating the underlying mechanism of biology and, consequently, poor for predicting disease classes of new observations.

The large number of protein peaks allows analysts to construct a complex model with degrees of freedom possibly near or equal to the number of samples, leading to overfitting. This overfitting and the resulting overly optimistic claim/interpretation could occur implicitly and the analysts or readers may not realize them.

van't Veer et al.[2] examined microarray data with 23,881 genes measured over 78 breast cancer cases, clinically, 44 cases were in the good prognosis group and 34 were in the poor prognosis group. The goal of their analysis was to use gene expressions to predict prognosis. A microarray predictor was generated by the following process: approximately 5,000 genes were selected initially from 23,881 genes using a fold-change and p value criterion, of which 231 genes were selected if they had large absolute correlation with the disease class labels. Then, they used a "leave-one-out" cross-validation method to select an optimal subset of 70 genes, constructed a 70-dimensional centroid vector for the 44 good-prognosis cases, and used the correlation of each case with this centroid to choose a cutoff value to form a binary microarray predictor. This classifier was produced by the multistage modeling process described above from a large number of genes. Therefore, overfitting is a potential concern. Did it really overfit?

Tibshirani and Efron[3] did an interesting analysis to address whether there is an overfitting. If one naively uses the 70-gene predictor as a covariate in a logistic regression model together with six other clinical covariates, the odds ratio (OR) of the disease associated with the microarray predictor is 60, while the largest OR from the six clinical predictors is 4.4. The OR of 60 indicates a remarkably strong association between the disease and the predictor. In contrast, Tibshirani and Efron[3] used a "prevalidation" procedure; that is, they set aside 6 cases at a time when performing the above gene/cutoff point selection using the remaining 72 cases. A binary microarray predictor from the 72 cases was used to predict the 6 cases, and the predicted values for the 6 cases were saved as their binary microarray predictor. They repeated this process 12 additional times and obtained a microarray predictor for each of the 78 cases. When they used this predictor in the same logistic regression model, the OR for the microarray predictor dropped from 60 to 4.7, less than the largest clinical OR, 4.9. A more rigorous full cross-validation showed that adding the microarray predictor to the six clinical predictors in logistic regression models only decreased the prediction error from 29.5 to 28.2%. van't Veer et al. also performed some sort of cross-validation (but details were not provided in their paper) and the OR for the microarray predictor was 18. The large difference in the ORs for the microarray predictor suggests that an overfitting might have occurred.

The key message from this example is the importance of having an unbiased assessment of true future prediction error of a predictor. Future prediction error is defined as the prediction error of a predictor constructed from training samples on (a large set of) independent new samples from the same population. This can be assessed unbiasedly by separating samples into a training set and a test set at the design stage (Section 18.3.2), or by rigorously estimating it from all available data (Section 14.4.5). The practice of not estimating the future prediction error by an independent test set or a careful statistical approach should be avoided.

18.3.2 USE OF TRAINING AND TEST DATA SETS AND SAMPLE SIZE CONSIDERATIONS

Section 18.3.1 illustrates the importance of having an independent test data set, not involved in the construction of a predictor, to evaluate the true future prediction error. We recommend this as the first choice for the estimation of future prediction error when there are a large enough number of samples to split into training and test data sets. This should be an essential part in the planning of experiments. When an independent test data set is not feasible, a careful and thorough cross-validation is necessary to guide the construction of a predictor and assess its performance. The cross-validation is the second choice because there are many ways of performing cross-validation, as seen from the above example, and analysts could use the approach that leads to seemingly "better" results, without realizing that an overfitting has occurred. This section discusses the setting when we have enough samples to split into training and test sets.

How large should each data set be? There is no formal simple rule to determine the size of a training set. This is because the training sample size depends on signal-to-noise ratios in the data and the complexity of a final predictor (the number of protein peaks to be combined and how they are combined). These are unknown before the analysis of data. Because of the high dimensionality and the potential complexity of predictors that could arise, our suggestion is to get as many samples as feasible, usually at least 100 samples for each disease class.

The prediction rule usually has its potential clinical application in mind. Therefore, the test sample size can be determined using a joint confidence region for sensitivity and specificity. Sensitivity is the true positive fraction, the proportion of the disease cases that are classified correctly by the prediction rule. Specificity is the true negative fraction, the proportion of the normal controls that are classified correctly by the prediction rule. Sensitivity and specificity are binomial proportions so the sample size calculation follows the standard one for binomial proportions with additional considerations of the two jointly. The null hypothesis is that either sensitivity or specificity of the rule is lower than a respective predefined unacceptable value. The sample size should be large enough such that if the classification performance is truly better than the specified unacceptable cutoffs, the joint confidence region of sensitivity and specificity will not include the specified unacceptable cutoffs. For example, if we hope a new test has sensitivity and specificity both at 92% and we want above 90% power to rule out the unacceptable sensitivity and specificity of 84% using a 95% joint confidence region, we need 250 diseased and 250 nondiseased samples. Details of this procedure are described in Section 8.2 of Pepe.[4]

The unacceptable sensitivity and specificity depend on the intended clinical application. For example, for a general population screening for ovarian cancer, due to the very low prevalence of the disease the specificity has to be very high, say above 99%. Otherwise, a lot of false positives will be generated for unnecessary work-up and worry. However, if it will be used for a high-risk population surveillance, for example, a prostate cancer screening for men with moderately elevated PSA levels, a test with a lower specificity can be applied to decide whether the subjects should be evaluated further by biopsy. In the latter situation, high sensitivity is more important than high specificity because the current practice sends all men with moderately elevated PSA levels to biopsy.

The sample size consideration for a test set discussed above is for the situation where we have a well-defined predictor for an intended clinical use and want to assess the predictor's performance for the intended use. Often we just want to use the test set to check potential overfitting and overinterpretation. In this setting, the majority of the sample, say 70 to 85%, should be used for the training set, especially when there is not a large number of samples to adequately split to the two sets. This is because at the training stage the construction of a predictor is the main objective and should use as many available samples as possible.

The test set should be kept by a person who is not involved in constructing the classifier. Only the final predictor should be tested on the test set. If an analyst modifies the predictor after seeing the test set performance, or chooses a predictor from multiple predictors based on their test set performance, the final test set performance is overestimated, sometimes extremely, because the test set is becoming the training set when it is used for selecting predictors.

There is another situation where an estimate of future prediction error from even a rigorous assessment differs from the prediction error in an independent test set. This happens when the test samples and the training samples are not from the same population. Differences in subject selection criteria, sample processing, and storage conditions could lead to this situation. The definition of "population" must, therefore, include subject selection, sample processing, and storage. Samples are from the same population if the subject selection criteria, the sample processing, and storage conditions are consistent.

When samples are collected from multiple laboratories, these conditions must be consistent across all laboratories. If samples are processed under different conditions in one of the labs, bias could occur and a predictor constructed from this lab's data will not be consistent with data from the other labs because the samples are not from the same population. If samples are from multiple labs, one way to perform a cross-validation is to leave one lab's data out, using the remaining data to construct a predictor that is then applied to the data from the lab left out to estimate the prediction error. Perform this in turn for all labs and then combine the prediction errors ("leave-one-lab-out cross-validation"). If all predictors are similar and performed similarly well, that means the data were collected in a consistent way or at least any inconsistency did not affect protein profiles. We could then combine all data to construct one predictor. Otherwise, it is necessary to identify which lab's samples appear abnormal and potentially down-weight or discard their data. In a prospective study requiring sample collection from multiple sites, it is absolutely necessary to develop a formal protocol that details subject selection, sample processing, storage, and assay procedures, and all sites must follow the protocol closely. This is a general good practice for all multicenter studies, but it is particularly important when the measurements of interest are high-dimensional because small systematic bias could accumulate in multidimension and therefore has a big impact on future prediction error.

18.3.3 CONTROLLING FOR POTENTIAL CONFOUNDERS BY FREQUENCY MATCHING

In comparing control and cancer subjects with respect to protein profiles, it is advantageous to select controls such that they have the same distributions as cancer cases for all factors that might distort the associations of protein profiles with the disease, such as smoking status, age, race, family history of cancer, etc. Such factors are called potential confounders and they may be correlated with both disease status and diagnostic protein markers. Imbalance between cases and controls with respect to confounders will introduce bias. Often an individual match is not necessary and a frequency matching will be adequate; e.g., two groups have the same percent of smokers, age distribution, racial mix, etc. If the individual matching is used, it would be advantageous to account for it in the analysis procedure (e.g., by using conditional logistic regression).

Potential confounders could also arise from sample collection, processing, storage, and assay procedures. Even when samples are collected, processed, and stored in the same way, if case and control samples are assayed at different times but are not consistent, potential confounding could occur. As implied by Figure 18.2, even the allocation of samples to the wells on ProteinChips could introduce bias. Because we can never know all confounding factors, the safest way is to randomly assign cancer and control samples to wells, chips, processors, and assay dates, balancing the number of control and cancer samples in each chip. Assay technicians should be blinded on disease status when performing the assays.

18.3.4 REPLICATED ASSAYS

Each sample should be assayed at least twice, preferably three or more times repeatedly. As we will see in Section 18.4.1, poor spectra with extremely low signal-to-noise ratios do occur. With three measurements, we have information to judge whether an observed abnormal spectrum is a poor-quality spectrum due to instrument/assay, or a correct spectrum from a poor-quality sample. In addition, an average of three replicated intensity measurements would have a standard deviation that is $1/\sqrt{3}$ times (less than 58% of) the standard deviation of a single measurement. This is a simple statistical scheme to reduce the noise level in the intensity measurements. The exceptional setting where multiple measurements cannot be taken is when the assay cost is far more than the cost of obtaining specimen. This is not the case for SELDI and MALDI experiments.

18.3.5 BALANCE BETWEEN COMPARISON GROUPS AND REPLICATE OVER THE MAJOR SOURCES OF VARIANCE COMPONENTS

The variance components analysis in Figure 18.3 could help us in deciding how to allocate control and cancer samples in the experimental design. The principle is to balance the control and cancer sample allocation on the major source of variance components. If the well-to-well variation is very small, not achieving balance between cancer and control samples on well positions is not critical. Figure 18.3 indicates, however, that this variance component (well-to-well) is indeed large. Therefore, balancing the well positions between two disease groups is important. In particular, allocating a specific type of samples (e.g., QC samples, disease samples, etc.) to a specific well position in each chip would be a bad idea.

Another use of variance component analysis is to guide where to place replicates. We often observe laboratory scientists taking duplicates in the same assay run for convenience, e.g., on the same chip. By doing it this way, however, they are likely to get consistent measurements on the same sample; this is not a good practice in taking replicates. In fact, this approach leads to the largest variance of an average intensity of the replicate measurements because the replicates are taken at the well level and they could only reduce the well-to-well variance component. If the nesting structure is in the order of well, chip, assay day, and chip batch, for example, the average of replicate measurements would reach the smallest variance when replicates are placed at the chip batch level. The reason is that when we replicate at the chip batch level, we automatically replicate all factors nested within it, as long as we have a random assignment. If we replicate wells within chip, only wells are replicated but nothing above it is replicated. If, for practical reasons, we want to replicate at a level lower than the top, say at the chip level, it is then important to make sure that the levels with large variance components get replication. For example, from Figure 18.3, the well-to-well variance is the largest one. Replication at the well level, though not optimal, is not unreasonable if the logistic for a higher-level replication is very difficult.

18.4 DATA ANALYSIS

18.4.1 IDENTIFYING POOR-QUALITY SPECTRA

Sometimes it is easy to determine a poor-quality spectrum by visual inspection but often it is hard to decide (Figure 18.4). This is particularly true when they are at the borderline. We developed a logistic regression model to quantify the probability that a spectrum is of poor quality using three predictors:

1. Square root of mean square errors of intensity (poor quality spectra tend to have higher noise)
2. Autocorrelation (poor quality spectra tend to have low correlation between adjacent points)
3. Maximum intensity at a prechosen range of mass-per-charge values where almost all spectra showed a strong peak (poor-quality spectra tend to lack this strong peak)

The first two calculations are restricted to low molecular weight (2,000 to 3,170 Daltons) where high-intensity values are typically observed. Figure 18.5 shows the classification of "good" vs. "bad" spectra by this model. This approach is more objective and systematic than the subjective visual inspection and one can easily change the cutoff point to be more or less conservative.

18.4.2 PEAK IDENTIFICATION AND ALIGNMENT

As Figure 18.1 indicated, the error in mass accuracy poses a challenge in data analysis because the same protein or peptide could appear at slightly different mass-per-charge values across samples. Although many analysts use SELDI-TOF data without defining and aligning peaks for biomarker-profile identification, we believe these are essential procedures before using the data for the marker identification.

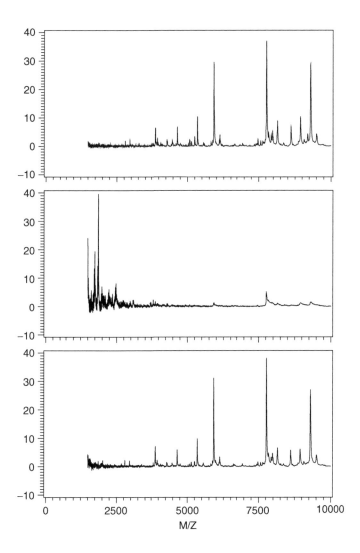

FIGURE 18.4 Triplicate spectra from same serum sample. The middle one indicates a bad spectrum.

To align peaks, we first need to define them. Peaks can be identified by using the manufacturer's internal algorithm. Our alternative simple definition calls a point a "peak" if it is the maximum intensity in its nearest $\pm N$ mass per charge points. By trial and error with visual inspection, we found $N = 10$–20 as reasonable.

After peaks are identified by the simple rule, we look at every mass per charge point and a window around it: the window consists of all points with mass per charge values within $\pm 0.2\%$ (or $\pm 0.1\%$; this is equal to the mass accuracy of the instrument) of the point under inspection. We count the number of peaks in the window across all spectra, pick up the mass per charge point that has the largest number of peaks in its window, and use this point as the first aligned mass per charge point with an

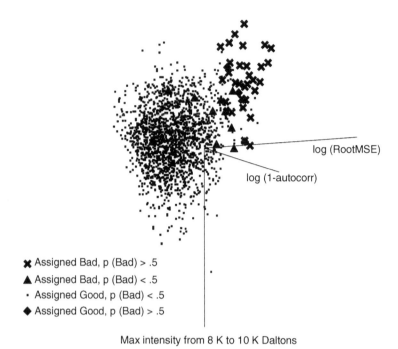

log (RootMSE)

log (1-autocorr)

✖ Assigned Bad, p (Bad) > .5
▲ Assigned Bad, p (Bad) < .5
· Assigned Good, p (Bad) < .5
◆ Assigned Good, p (Bad) > .5

Max intensity from 8 K to 10 K Daltons

FIGURE 18.5 A three-dimensional plot indicated that the bad spectra were separated from the good spectra by three predictors via a logistic regression model.

aligned peak value of each spectrum set at its maximum intensity value within the window. Note that a spectrum without any peak in the window still gets an aligned peak value, but it would be relatively low. Now this first window is taken away from the mass-per-charge axis and we repeat the process (i.e., pick a mass per charge point that has the largest number of peaks in its window, use this point as the aligned mass per charge point, and take out the window from the remaining mass per charge axis) until no peak is left to be aligned in any spectrum. Figure 18.6 indicates four spectra before and after the peak alignment. See Yasui et al.[5] for details of this procedure.

18.4.3 REDUCTION OF DIMENSIONALITY

Now we are ready to construct a classifier. Depending on the peak identification and alignment methods, the number of potential predictors (aligned peaks) could still be on the order of a few dozen to a few thousand. If the number is big, it is hard to construct a classifier from the pool of potential predictors directly. One way to reduce the dimensionality is to filter out most of the peaks and only use the peaks with some promising features to be combined for a powerful classifier. In microarray analyses, fold changes, *t*-statistics or correlations between two disease groups are often used for filtering. However, as Pepe, et al.[6] pointed out, it makes more sense

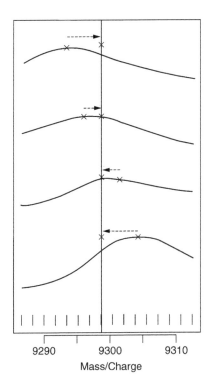

9290   9300   9310

Mass/Charge

FIGURE 18.6 Spectra before and after peak alignment.

to use sensitivity and specificity to rank the candidate genes for further investigation for disease diagnosis and early detection purposes. We have used a similar idea. Sensitivity and specificity are two measures and they change with the cutoff point. It would be easier to use a one-dimensional measure as a filtering criterion. Two sensible one-dimensional measures are partial area under the receiver operating characteristic (ROC) curve (PAUC) and Yuden distance (Figure 18.7). Yuden distance is defined by the sum of sensitivity and specificity minus 1. The range of Yuden distance is from 0 to 1, with 1 indicating zero classification error and 0 indicating no diagnostic capacity at all. Thus, filtering peaks by Yuden distance would give an equal weight to sensitivity and specificity.

An ROC curve is the functional curve of sensitivity against (1-specificity). It summarizes the operating characteristics of a predictor with continuous values. Each point on the curve corresponds to a cutoff point for a predictor (e.g., test positive for cancer if the predictor value is above the cutoff point, otherwise test negative). PAUC can be defined as the area under the ROC curve from specificity 1 to a predefined specificity, say 0.9. Choosing protein peaks that have larger PAUC values will preselect protein peaks that have high specificity. Similarly, we can define PAUC as the area under the ROC curve from the specificity corresponding to a predefined high sensitivity, say 0.9, to specificity 0 (i.e., sensitivity 1), and select protein peaks with large

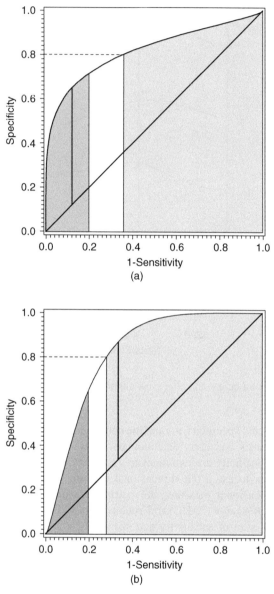

FIGURE 18.7 The shaded area containing the bolded vertical bar represents the criteria for filtering predictors. (a) PAUC based on high specificity. (b) PAUC based on high sensitivity. (c) Yuden distance.

PAUC values. This preselects protein peaks that have high sensitivity. We could predetermine the top percentiles for the two types of PAUC to be selected to control the number of protein peaks filtered out. Note that the multiple criteria (i.e., Yuden distance and the two types of PAUC) can be combined for filtering by taking the union of the selected peak set by each criterion. This is advantageous since we wish to keep

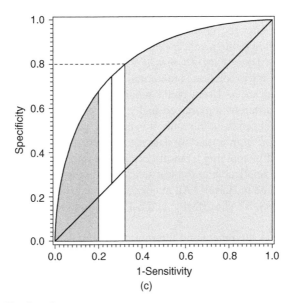

FIGURE 18.7 (Continued).

the peaks that are not necessarily discriminatory globally, but either highly specific or sensitive, to capture some smaller subgroups of cancer cases or controls.

18.4.4 CLASSIFIER CONSTRUCTION

There are many classifier construction algorithms: most of them are from either statistical science or computer science/machine learning. A good discussion of this topic can be found in a recent book by Hastie, et al.[7] We describe here two algorithms that we found to have good performance and interpretations. They are logistic regression and boosting.

18.4.4.1 Logistic Regression

Logistic regression is a standard statistical regression algorithm to model a binary outcome Y (e.g., $Y = 1$ for cancer vs. $Y = 0$ for noncancer) using an array of candidate predictors (continuous, binary, categorical, or mixture of them). It is one of the most popular analysis methods in the field of epidemiology. Logistic regression assumes the following model:

$$\text{Logit}(E[Y]) = b_0 + b_1 X_1 + b_2 X_2 + \cdots + b_M X_M, \quad Y \sim \text{Independent Bernoulli}(E[Y])$$

where $\text{Logit}(a) = \ln(a/1 - a)$, X_i is ith predictor (i.e., ith peak in the spectra), and b_i is the regression coefficient associated with the ith predictor. The value of b_i has an interpretation as the increase in log odds of cancer associated with one unit increase of X_i. For a binary predictor (i.e., having peak or not having peak at a specific mass per charge point), it is the log odds ratio of cancer for a patient with a peak at this mass-per-charge point compared to a patient without a peak there, holding all other factors constant. After the model is finalized, we can predict the probability of a spectrum in question being a cancer rather than noncancer using the formula: probability of cancer = $1/[1 + \exp\{-(b_0 + b_1 X_1 + b_2 X_2 + \cdots + b_M X_M)\}]$. If the probability is larger than 0.5, the prediction is more indicative for cancer. Otherwise, it is more indicative for noncancer. Users could modify this cutoff point if sensitivity and specificity have different importance under their particular setting. Note that the intercept term b_0 depends on the ratio of cancer vs. noncancer cases in the data set: if the ratio changes from a training data set to a test data set, for example, caution must be exercised in applying a cutoff because the intercept term from the training set is not directly applicable to the test set. More discussions on logistic regression models can be found in Harrell.[8] All major commercial statistical software packages, such as SAS, SPSS, STATA, BMPP, and S-Plus include logistic regression analysis capability.

Even with peak alignment and filtering, there may still be a few hundred peaks as potential predictors for logistic regression. We can use a forward variable-selection procedure: we first identify, among all X_js, the peak that has the best prediction power. That means that, among all logistic models with just one peak in the model, the model with this peak has the minimum misclassification error. After selecting the first peak for the model, we identify and add the best peak of all remaining peaks, given the first peak in the model. We continue the selection by adding the best peak, given the previously selected peaks in the model.

When shall we stop the selection? As we noted before, we could add more and more predictors and eventually overfit the data. We could use a cross-validation procedure to determine when to stop and minimize overfit. The idea of a tenfold cross-validation, for example, is as follows. We divide the data into ten equal parts randomly and only use nine parts to construct a classifier with the best p predictors. We then use the classifier to predict the remaining 10% of the data not used in classifier construction. This 10% data act like a validation data set. We continue this process with a fixed p for all ten parts. By the end, there is precisely one classification result for each sample that can gauge the performance of the classifier-building procedure with the fixed p on new independent data. After trying a set of different values for the number of predictors in the model, the optimal p could be determined by the one that gives the smallest cross-validation error. Alternatively, we can fix the significance level p of a newly added protein peak, instead of the number of peaks in the model, to determine the optimal stopping rule.

There is an important implication: if we use cross-validation only once for a prespecified number of predictors in the model, the cross-validation error is a good estimate of future prediction error. However, if we use cross-validation for model selection, such as described above, we still could overfit slightly. This is because, although such cross-validation discourages analysts from choosing too complicated models (too many predictors), the "optimal" number of predictors that minimizes the cross-validation error depends on the particular data at hand. If we draw another set of data from the same population, the "optimal" number may differ. Thus, selecting an optimal number of predictors itself is also overfitting but at a smaller order compared to that from a classifier with too many predictors.

18.4.4.2 Boosting

Boosting is an ingenious method developed in computer science during the 1990s[9–11] for combining multiple "weak" classifiers into a powerful committee. There are variations of boosting algorithms. For a comprehensive discussion, see Hastie, et al.[7] We will describe two popular boosting algorithms here: Discrete AdaBoost and Real AdaBoost. The applications of boosting for SELDI data were described in Qu[12] for Discrete AdaBoost and Yasui[13] for Real AdaBoost.

18.4.4.2.1 Discrete AdaBoost

Discrete AdaBoost uses a classifier $f_m(x)$ that has value 1 (to indicate disease) or -1 (to indicate nondisease) for a given protein peak indexed by m. Suppose we use a simple "stump" classifier that classifies a sample as disease if the peak intensity is at the mth protein peak above (or below) a certain level. This can be done by simply dividing the peak intensity into a number, say 50, of equally spaced segments and finding the cutoff point that best differentiates disease and nondisease samples. If we are not comfortable about the quantitative measure of intensity, we can simply use $X = 1$ (peak) or $X = 0$ (no peak). We examine all peaks this way and choose the peak with a particular cutoff that gives the smallest misclassification error. This forms our first classifier $f_{m=1}$. In this first classifier, we use an equal weight (weight $=1/N$) for all N observations.

This classifier is usually a weak classifier. Now we update the weights by assigning larger weights for the observations that were misclassified by this classifier. For observation Y_i ($Y_i = 1$ for disease, $Y_i = -1$ for nondisease), the new weight is $w_i =$ previous weight $\times \exp\{c_m \times I(y_i f_m(x_i) = -1)\}$, where $c_m = \log[(1 - \text{err}_m)/\text{err}_m]$, where err_m is the proportion of all N samples that are misclassified by the current classifier and $I(.)$ is an indicator function and equals 1 if the statement in parenthesis is correct, 0 otherwise. Note that if the prediction is correct, then $y_i f_m(x_i) = 1$ and $\exp\{c_m \times I[y_i f_m(x_i) = -1]\} = 1$. The new weight is then equal to the old weight. If the prediction is incorrect, then $\exp\{c_m \times I[y_i f_m(x_i) = -1]\} = \exp(c_m)$ is larger than 1 as long as c_m is positive: this always holds with a high probability if the weak classifier is better than flipping a coin. This is the first important feature of boosting: assign larger weights to those difficult to classify. Logistic regression uses the same weight for each observation.

We apply our favorite classifier algorithm again, the same way as described above, except now to the weighted data. We repeat this process M times and the final classifier is

$$f(x) = sign\left[\sum_{m=1}^{M} c_m f_m(x)\right]$$

a weighted sum of all M classifiers: the classifier is $f(x) = 1$ if the sign of the sum is positive and -1 otherwise.

We can think of this final classifier $f(x)$ as a committee formed by M members (classifiers). Each member gives a vote 1 or -1. A member that has better knowledge (a smaller err_m and therefore a larger c_m) carries larger weight in voting. This is the second important feature of boosting: a committee voting. This is analogous to b's (regression coefficients) in logistic regression.

The third feature of boosting is that the same protein peak can be repeatedly selected as a committee member, likely with different values of c_m, while in logistic regression each protein peak can appear in the final predictor at most once.

How shall we choose the number of iteration M? We can use the cross-validation described above to determine M that gives the smallest cross-validation error. If we let M be too big, we will eventually overfit the data.

18.4.4.2.2 Real AdaBoost

Real AdaBoost has the following differences from the Discrete AdaBoost:

1. Instead of using a discrete classifier $f_m(x)$, Real AdaBoost uses a continuous classifier that produces a class-probability estimate p_m, the predicted probability of disease based on mth protein peak. A natural choice for constructing class-probability estimates is via logistic regression with a single covariate.
2. Calculates a quantitative classifier $f_m(x) = 0.5 \log[p_m(x)/(1 - p_m(x)]$
3. Updates the weight by new weight = previous weight $\times \exp[-(y_i f_m(x_i)]$, $i = 1, 2, \ldots N$, and renormalizes so that the sum of weights over N samples equals to 1. The initial weight is $1/N$.
4. Repeat this process M times and the final classifier is

$$f(x) = sign\left[\sum_{m=1}^{M} f_m(x)\right]$$

Note that in Real AdaBoost, $f_m(x)$ will be positive if $p_m(x) > 1/2$ and will increase as $p_m(x)$ increases. Therefore, in each iteration it assigns weights to each observation not only according to whether it is correctly classified but also to the confidence of correct specification or the extent of misclassification. In each iteration, Discrete AdaBoost assigns one weight for all correctly classified and one weight for all incorrectly classified. Therefore, Real AdaBoost uses "confidence-rated" prediction in weights and we expect it will "learn faster" and have better predicting power. See Friedman et al.[14] for discussion and a simulation experiment on this point.

18.4.4.2.3 Why Boosting?

Among many data mining algorithms, why do we prefer boosting? The key is the observation that it appears to be resistant to overfitting, yet it still has good predicting power. This consideration is important because any high-dimensional data provides ample opportunities to overfit and therefore it is the main trap we need to avoid. It is still not completely clear why boosting is resistant to overfitting. One possibility is due to its "slowness" in learning. It adds one predictor at a time and when a new predictor is added, boosting does not try to optimize all predictors in the model simultaneously. This reduces the variance of the model. Another explanation is that the impact of each weak classifier $f_m(x)$ reduces when iteration proceeds as the correctly classified observations with high confidence have the sum of $f_m(x)$ far away from zero and are less likely to change their signs by the new added weak classifier. Therefore, the bulk of the data are not affected much. That increases the stability of the model.

Another consideration is the ease of interpretability. Because each weak classifier is a very simple model, their weighted voting is easy to interpret. This is the advantage of boosting over other methods like neural network.

The logistic regression with cross-validation may be preferred in some instances over the boosting if their performance difference is not appreciable and the former uses considerably fewer number of peaks in the classifier. This is because further scientific investigations following the SELDI/MALDI-based biomarker exploration would study specific protein peaks in the classifier as potential biomarkers.

18.4.5 ASSESSMENT OF PREDICTION ERROR

Assessment of prediction error serves two main purposes: selecting a classifier with the minimum future prediction error and estimating future prediction error of the classifier.

The number of iterations in boosting and the number of predictors to be included in logistic regression are called "tuning parameters." They specify the complexity of the classifier. We can select the value of these tuning parameters such that the classifier has the smallest prediction error in cross-validation. The cross-validation error mimics the future prediction error. Therefore, we can use the estimated future prediction error to select our classifiers.

As described in Section 18.4.4.1, when the cross-validation procedure is used more than once, as we have to in classifier selection, the minimum cross-validation error among several classifiers will generally underestimate the future prediction error of the classifier in an independent new data set. Therefore, we want to have an assessment of the future prediction error after we finalized our classifier. This can be achieved by using bootstrap.

Bootstrap is a powerful yet general statistical strategy for making inference on quantities that is difficult otherwise. For good explanations of the bootstrap method, see Efron and Tibshirani.[15] In our situation, we want to estimate a future prediction error for the final model. It goes like this:

We randomly draw, with replacement, N observations from original N observations to form a bootstrap sample. We repeat this B times. We suggest B to be at least 100.

For each bootstrap sample, we repeat our whole model selection process. That is where we start using disease status information, i.e., the reduction of dimensionality in Section 18.4.3. That means that, for each bootstrap sample, we will select candidate peaks, use cross-validation to determine the tuning parameters, and construct a classifier using a favorite algorithm. By the end, we will have B bootstrap classifiers, one for each of B bootstrap samples.

Use each of B bootstrap classifiers to classify the observations in the original sample that are not selected in this particular bootstrap sample and compute the prediction error (misclassification error, sensitivity, specificity, etc.). The average of B prediction errors is our estimate of future prediction error. We call this "validation prediction error."

It is important to note that we have three types of prediction errors: training-set prediction error (the observed prediction error for the final model on the original training data), validation prediction error, and test-set prediction error, with increasing generalizability. The validation prediction error is the most you can extract from a training data set on future prediction error. It cannot completely replace the need for a test data set, but it could be a good compromise when it is not feasible to have a large enough test data set. At a minimum, analysts need to use cross-validation to select a classifier. Letting a model become too complex, either explicitly or implicitly, to achieve an apparent lowest training-set prediction error without cross-validation easily leads to overfitting.

18.5 CONCLUSION

Proteomics offers great hope as well as great challenges in biomedical research. There are, and will be, many statistical and bioinformatics algorithms to analyze such data. However, the following general principles will apply:

1. Need to understand sources of bias and variation of the data
2. Design experiments that eliminate or reduce bias and/or variance, and meet study objective
3. Make the analysis strategy consistent with the experimental design and the study objectives and also resistant to overfitting; rigorously assess future prediction errors

REFERENCES

1. Hanash, S. Disease proteomics. *Nature*, 422, 226–232, 2003.
2. van't Veer, L.J., Dai, H., van de Vijuer, M.J., He, Y.D., Hart, A.A.M., Mao, M., Peterse, H.L., van der Kooy, K., Marton, M.J., Witteveen, A.T., Schreiber, G.J., Kerkhoven, R.E.M., Roberts, C., Linsley, P.S., Bernards, R., and Friend, S.H. Gene expression profiling predicts clinical outcome of breast cancer. *Nature,* 415, 530–536, 2002.
3. Tibshirani, R. and Efron, B. Pre-validation and inference in microarrays. *Statistical Applications in Genetics and Molecular Biology,* 2002, http://www.bepress.com/sagmb/vol1/iss1/art1.
4. Pepe, M.S. *The Statistical Evaluation of Medical Tests for Classification and Prediction.* Oxford University Press, New York, 2003.
5. Yasui, Y., McLerran, D., Adam, B.L., Winget, M., Thornquist, M., and Feng, Z. An automated peak-identification/calibration procedure for high-dimensional protein measures from mass spectrometers. *J. Biomed. Biotechnol.,* Special Issue, 1–7, 2003.
6. Pepe, M.S., Longton, G., Anderson, G.L., and Schummer, M. Selecting differentially expressed genes from microarray experiments. *Biometrics,* 59, 133–142, 2003.
7. Hastie, T., Tibshirani, R., and Friedman, J. *The Elements of Statistical Learning,* Springer, New York, 2001.

8. Harrell, F.E. *Regression Modeling Strategies*. Springer, New York, 2001.

9. Schapire, R. The strength of weak learnability. *Mach. Learn.*, 5, 197–227, 1990.

10. Freund, Y. Boosting a weak learning algorithm by majority. *Inform. Comput.*, 121, 256–285, 1995.

11. Freund, Y. and Shapire, R. A decision-theoretic generalization of online learning and an application to boosting. *J. Comput. Syst. Sci.*, 55, 119–139, 1997.

12. Qu, Y., Adam, B., Yasui, Y., Ward, M., Cazares, L., Schellhammer, P., Feng, Z., Semmes, J., and Wright, G. Boosted Decision Tree Analysis of SELDI Mass Spectral Serum Profiles Discriminates Prostate Cancer from Non-Cancer Patients. *Clin. Chem.*, 48, 1835–1843, 2002.

13. Yasui, Y., Pepe, M., Thompson, M., Adam, B., Wright, G., Qu, Y., Potter, J., Winget, M., Thornquist, M., and Feng, Z. A data-analytic strategy for protein-biomarker discovery: Profiling of high-dimensional proteomic data for cancer detection. *Biostatistics*, 4, 449–463, 2003.

14. Friedman, J., Hastie, T., and Tibshirani, R. Additive logistic regression: A statistical view of boosting (with discussion). *Ann. Stat.*, 28, 337–407, 2000.

15. Efron, B. and Tibshirani, R. *An Introduction to the Bootstrap*. Chapman & Hall, New York, 1993.

19 Image Analysis in Proteomics

Stephen Lockett

CONTENTS

19.1 INTRODUCTION*

Proteomics includes not only the identification and quantification of proteins but
also the determination of their localization, modifications, interactions, activities,
and, ultimately, their function. Initially encompassing just two-dimensional (2D) gel
electrophoresis for protein separation and identification, proteomics now refers to
any procedure that characterizes large sets of proteins.[1]

The mainstream analytical techniques used in proteomics involve extraction of
proteins from samples followed by determination of their abundance, interactions,
and modifications. These techniques have the advantage that large numbers of
proteins can be rapidly analyzed, but such measurements do not provide complete
information about the proteome of the sample for the following three key reasons:

1. Protein abundance normally varies from cell to cell in a sample; thus it
   is necessary to analyze cells in the sample individually.

---

* Excetpted with permission from *Science* 291: 1221–1224, 2001. Copyright 2001 AAAS.

2. In the molecularly and structurally complex environment of the cell, protein concentrations, interactions, and modifications vary within the cells.

3. The cells of multicellular organisms are constantly interacting with their neighbors and are heterogeneous in their structural, molecular, and behavioral properties. Thus, to determine the proteome of a cell, tissue, or organism, it is necessary to analyze multiple proteins in individual cells, as well as in cells in the natural spatial and temporal context of their neighbors. Integration of knowledge about cell proteomes will enable us to understand the complex signaling networks that drive cell behavior. [2]

This chapter describes how quantitative analysis of fluorescence microscope images augments mainstream proteomics by providing information about the abundance, localization, movement, and interactions of proteins inside cells. Emphasis is given to methods that facilitate the analysis of several different proteins in the same cell simultaneously.

The chapter is organized as follows: Sections 19.2, 19.3, and 19.4 present methods for fluorescence labeling of proteins in cells, the acquisition of optical microscope images of fluorescence-labeled cells, and quantitative analysis of images, respectively. In Section 19.5 we present methods to analyze the dynamics and interactions of proteins in live cells, and in Section 19.6 we outline emerging techniques in the field of fluorescence microscopy.

19.2 FLUORESCENCE LABELING PROTEINS IN CELLS

19.2.1 USE OF FLUORESCENCE DYES

In optical microscopy, two types of labels may be used: colorimetric dyes that absorb light, creating dark contrast in the image, and fluorescent dyes that absorb light and then re-emit light at a longer wavelength, thus creating bright contrast in the image. Fluorescent dyes have wider utility for the following five reasons:

1. Higher sensitivity enabling detection of lower concentrations
2. An emitted signal that is proportional to the concentration of dye, thus facilitating quantification of protein concentrations
3. The ability to simultaneously detect multiple proteins, by labeling each protein with dyes that have different excitation and emission spectra
4. The ability to collect true three-dimensional (3D) images of samples (using a confocal microscope), because the sample remains transparent
5. Applicability in live cells

Consequently, in this chapter we will only describe techniques that use fluorescent dyes. There are two principal techniques for fluorescence labeling proteins: antibodies or green fluorescent protein (GFP). Generally, the labeling of specific proteins by these methods is combined with additional fluorescent markers of cell structures.

FIGURE 19.1 (Color insert follows page 204) Example of antibody labeling in fixed tissue. The picture is a triple label confocal image of the cerebellum of the mouse mutant ataxia (axJ). Granule cells and other nuclei are labeled with DAPI (blue), Purkinje cells are labeled with antibody to calbindin (red), and synaptic contacts are identified by an antibody to syntaxin (green). (Courtesy of Dr. Rivka Rachel, NCI-Frederick).

19.2.2 FLUORESCENCE LABELING OF PROTEINS WITH ANTIBODIES

The most common method for localizing proteins in cells is by indirect immunofluorescence antibody labeling. Such labeling has the advantages that the protocol is usually simple; antibodies specific for many different proteins are available, and the endogeneous protein is being detected. However, the major disadvantage of antibody labeling is that the sample must be fixed. Antibody labeling involves the following steps:

1. Fixation of cells
2. Exposure to the primary antibody directed against the protein of interest
3. Removal of unbound antibody
4. Incubation with a fluorescently tagged secondary antibody directed against the primary antibody.

Simultaneous double and triple labelings are possible, but care must be taken to use primary antibodies generated from different species (e.g., mouse and rabbit) in order to avoid cross-reaction. A wide variety of protocols for antibody labeling are described in "Current Protocols"[3] and Figure 19.1 shows an example of a fluorescence microscope image of an antibody-labeled sample.

19.2.3 FLUORESCENCE LABELING OF PROTEINS WITH GREEN FLUORESCENCE PROTEIN (GFP) AND DsRED

GFP and DsRed are proteins isolated from jellyfish *Aequoria victoria* and *A. discosoma*, respectively, which, upon excitation by the appropriate wavelength of light, fluoresce brightly from inside cells in a species-independent fashion. Moreover, cloning

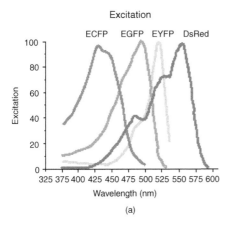
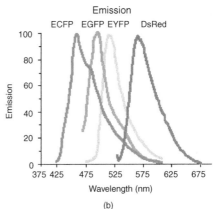

FIGURE 19.2 Excitation (a) and emission (b) spectra of enhanced GFP (EGFP) and its spectral variants. (©Copyright 2004 Becton, Dickinson and Company)

of GFP into an expression construct with a gene of interest directly ligated to the GFP gene followed by transfection into cells results in stable expression of the GFP (or DsRed) fused to either the N- or C-terminal of the protein of interest. This critical aspect of GFP-based reporter systems provides the ability to monitor the locations, dynamics, and interactions of proteins of choice in real time in live cells.

A variety of GFP and DsRed variants are commercially available that not only yield enhanced fluorescence compared to their natural counterparts but, more importantly, have different excitation and emission spectral properties (Figure 19.2). This latter property enables multiple proteins to be followed simultaneously and independently in live cells. In addition, destabilized GFPs are available that have a short half-life in cells, enabling kinetic studies of regulated proteins, as well as photoactivatable GFP that has applications in studies of protein dynamics.[4]

When using GFP chimeras, precautions should be taken to ensure that the chimeras behave similarly to the natural protein, which can be checked by comparing the localization pattern of the chimera with that of the antibody-labeled protein in the same cells. The main reasons for differences in behavior between the GFP-tagged protein and the endogeneous protein are often due to the high level of expression of the chimera relative to endogeneous protein, or from an increase in mass due to attachment of the GFP.

Protocols for expression of GFP chimeras are available in "Current Protocols"[3] and the Clontech Web site.[5]

19.2.4 Other Fluorescence Labels: Cell Structures and Organelles, and Indicators of Cell Activity

A wide variety of fluorescent labels are available for labeling cell structures and organelles and specific nucleic acid sequences. It is very common practice to "counterstain" cell nuclei in fixed samples with a DNA-specific dye, such as DAPI or Hoechst, that fluoresce at wavelengths significantly shorter than most dyes used to label proteins. In addition, fluorescent indicators are available for detecting a wide variety of activities in cells (e.g., calcium release, mitochondrial membrane potential, caspase activation). Further information is available from Fluorescence Probes.[6]

19.3 ACQUISITION OF OPTICAL IMAGES

19.3.1 CONVENTIONAL FLUORESCENCE MICROSCOPY

Images of fluorescence-labeled cell samples are acquired using an optical microscope in the reflected (epi) configuration (Figure 19.3). The source of excitation light in the microscope is a high-intensity arc discharge lamp emitting white light as well as ultra-violet and infrared. The light is collimated by the condenser lens in order to fill the back aperture of the objective lens, and en route it passes through an excitation filter to select the wavelength range appropriate for exciting a particular fluorescent dye followed by reflection into the objective by a dichroic (two-color) mirror. The objective lens focuses the light through the immersion media and coverslip onto the sample. Emitted light is collected by the objective lens, and being a longer wavelength than the excitation light, passes through the dichroic mirror. Emitted light in the wavelength range of the fluorescent dye is selected by an emission filter and is then refocused onto a camera by the tube lens. The excitation filter, emission filter, and dichroic mirror are a readily interchangeable unit (the "filter cube"), enabling rapid selection of the appropriate cube for different fluorescence labels in the same specimen.[7] When imaging fluorescence dyes, it is best to use a high numerical. aperture (NA) objective lens in conjunction with a high refractive index immersion media (oil or water) between the

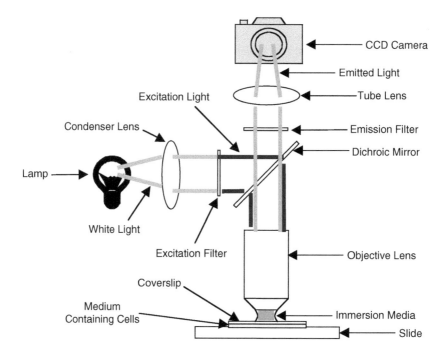

FIGURE 19.3 Basic configuration of an upright epifluorescence microscope.

Combined Use of DIC and Fluorescence Illumination

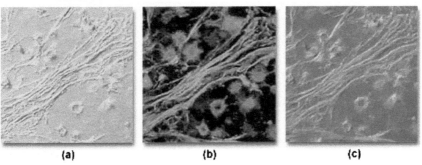

(a)   (b)   (c)

FIGURE 19.4 (Color insert follows page 204) (a) A thin section of cat brain tissue infected with cryptococcus and imaged using DIC optics and a full-wave retardation plate. Note the pseudo three-dimensional appearance of the photomicrograph. (b) The same field of view, but imaged with fluorescence illumination and an Olympus WIB filter cube. The cells were stained with a combination of fluorescein-5-isothiocyanate (FITC) and Congo red (emission wavelength maxima of 520 and 614 nanometers, respectively). (c) The two techniques are used in combination, illustrating the infected cat brain tissue in both fluorescence and DIC illumination.

lens and coverslip over the sample in order to collect as large a solid angle of the emitted light as possible, as well as to achieve high spatial resolution.

The camera is usually a charge-coupled device (CCD), which provides low background noise, very high sensitivity, a linear response over three decades of intensity (i.e., very high dynamic range), and direct coupling to a computer for digital image storage.

Fluorescence microscopy, which conveys molecularly specific information, is often combined with differential interference contrast (DIC or Nomarski) or phase contrast microscopy, which provides structural information about the cells (Figure 19.4).

The configuration shown in Figure 19.3 is an upright microscope; however, generally for live cell applications, an inverted microscope is preferred so that cells can be in an open, coverslip-bottomed dish and imaged from below. Increasingly, microscopes are available with an environmental chamber, motorized filter cubes, and focusing and stage positioning. These enhancements facilitate imaging of live cells for extended time periods (hours) and automation of the image acquisition process in high-throughput screening applications. References 8, 9, and 10 provide further information about fluorescence microscopy.

19.3.2 THREE-DIMENSIONAL FLUORESCENCE MICROSCOPY

For quantitative imaging of the spatial and temporal distribution of proteins in cells, it is necessary to acquire three-dimensional (3D) images, since the cell/tissue samples are themselves inherently 3D. This is achieved by acquiring a series of two-dimensional (2D) slice images through the cell at increasing or decreasing focal depths of fixed increments. However, this method suffers from a severe limitation when performed with a conventional epifluorescence microscope, because the microscope is in focus only for a narrow layer of the specimen (<0.5 μm for a high NA objective

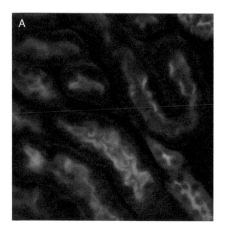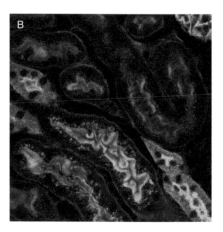

FIGURE 19.5 (Color insert follows page 204) Tissue section of mouse kidney 15 μm thick, labeled with Alexa Fluor 488 wheat germ agglutinin (green) and Alexa Fluor 568 phalloidin (red). Images were acquired with an LSM 410 confocal microscope (Carl Zeiss Inc., Thornwood, NY). (A) Image acquired with the pinholes open to mimic conventional microscopy. (B) Image acquired with a small pinhole for confocal microscopy.

lens), whereas a cell is often at least 3 μm in thickness. The consequence is that a conventional image of a cell always consists of the sum of an in-focus image of the layer of the cell in focus plus an out-of-focus image of the majority of the cell above and below the focal layer. This is the case regardless of where the microscope is focused in the cell, and hence the image is always blurry (Figure 19.5A).

There are two methods for removing the out-of-focus contribution from a 3D image while retaining the in-focus contribution. Deconvolution (see Section 19.4.3.3), an offline computational method, may be applied,[8,11,12] but confocal microscopy is much more commonly used.

19.3.3 CONFOCAL MICROSCOPY

Confocal microscopy uses a combination of an excitation light source focused to one point in the sample, and a pinhole in the emission path to collect emitted light only from the same point (Figure 19.6). In order to obtain a 2D slice image over the focal plane, the excitation light and the projection of the pinhole at the sample are simultaneously scanned across the sample (Figure 19.7). A 3D image of the specimen is obtained by mechanically moving the focal plane between the acquisition of each 2D slice. Figure 19.5B is an example of an image from a confocal microscope. Emission light is recorded with a photomultiplier tube and stored digitally in a computer for subsequent visualization and analysis. For further information, refer to references 8 and 13.

19.3.4 NEAR-FIELD MICROSCOPY

There are two near-field optical microscopy techniques for imaging fluorescence labels specifically on the cell surface: total internal reflection microscopy (TIFM) and scanning near-field optical microscopy (SNOM).

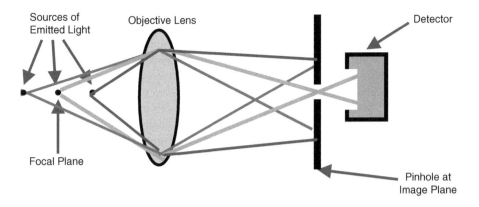

FIGURE 19.6 Emission path of a confocal microscope. Only light emitted from the light source at the focal plane is efficiently passed through the pinhole and collected by the detector. Emitted light originating from points in front of and behind does not focus at the pinhole and is therefore not efficiently collected.

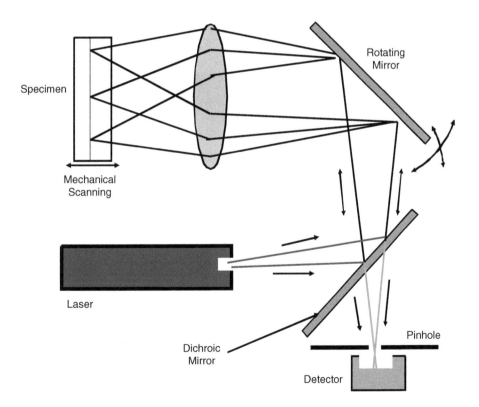

FIGURE 19.7 Schematic of a laser scanning confocal microscope. Generally a bright excitation source, i.e., a laser, is employed. In practice, the rotating mirror is a pair of mirrors that together scan the entire area of the field of view. Excitation and emission filters are not shown.

TIFM[8,14] images fluoresce only on the cell surface in contact with the glass coverslip. This is achieved by directing the excitation light toward the cells and through the coverslip at a high angle of incidence so that total internal reflection takes place at the glass–cell interface. In this situation, only an evanescent wave, which exponentially decays as a function of distance from the glass surface, enters the cell and results in only excitation of fluorescence molecules less than 0.15 μm from the glass.

SNOM[15,16] images fluoresce only on the exposed cell surface, by mechanically scanning the surface with a fine-tipped, single-mode optical fiber. The tip is coated with aluminum and has a subwavelength aperture of approximately 50 nm diameter. Laser excitation light is directed through the fiber and, when it exits the tip, diffraction causes the light to spread out in all directions. This results in the excitation intensity rapidly decreasing as a function of distance from the tip, and consequently only fluorescence molecules that are very close (50 nm) to the tip are excited.

19.4 ANALYSIS OF FLUORESCENCE IMAGES

19.4.1 THE DIGITAL IMAGE

A 2D gray-level digital image is a rectangular array of elements ("pixels"), where each pixel is assigned a value between 0 and 255 (or between 0 and 4095) (Figure 19.8A). The value assigned to a pixel by digitization of the signal from the detector is

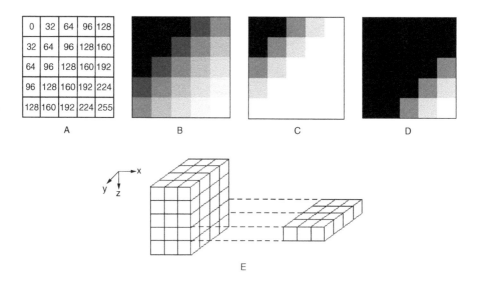

FIGURE 19.8 (A) Digital images of 5 × 5 pixels showing pixel intensity values. (B) The image in A displayed as a grey-scale image such that 0 is black and 255 is white and intermediate values are different shades of grey. (C) B after application of contrast enhancement to the lower intensity pixels. The image has been rescaled so that 0 is still black, but 128 and above is white. (D) B after contrast enhancement to the higher intensities. 128 and below is black and 255 is white. (E) 3D image of 3 × 5 × 4 voxels. A 3D image can be considered as consisting of a stack of 2D images.

approximately proportional to the intensity of light emitted from the corresponding position in the sample. The range of intensities over which proportionality holds is called the "dynamic range." For visual interpretation, 0 is assigned to black and 255 to white (Figure 19.8B). Images typically consist of very large numbers of pixels (e.g., $512 \times 512 = 262{,}144$), and thus the individual pixels are not discernable when viewing the image. 3D digital images are stacks of 2D images (Figure 19.8E), and a typical size for a 3D confocal image would be $512 \times 512 \times 30$. For 3D images, pixels are also called "voxels." Each pixel in a 3D image represents a small volume of the sample; for example, $0.2 \times 0.2 \times 0.5$ µm, which would be typical for a high NA objective lens. It should be noted that in this example the physical distance represented by the voxel in the depth (z) dimension (0.5 µm) is greater than the physical dimension in the lateral (x and y) dimensions (0.2 µm). This condition is known as anisotropy and must be taken into account during image processing, analysis, and visualization.

Binary images are a type of image where each pixel can only have values 0 or 1. They most commonly arise following the segmentation of gray-level images (see below), where an intensity of 1 is used to represent pixels in objects and intensity 0 represents background.

For specimens labeled with multiple fluorescent dyes, each dye is imaged separately using different filters and stored as separate gray-level images.

19.4.2 COMMON DISTORTIONS IN MICROSCOPE IMAGES

Images acquired by digital optical microscopy can suffer from a number of common distortions.

19.4.2.1 Saturation

Saturation occurs when the light emitted from the specimen is either more intense than the dynamic range of the detector, or so dim that it is below the detection threshold. Saturation thus results in the image no longer being a quantitative representation of the sample. Significant saturation in an image can be easily detected by observing whether a significant but small fraction (>0.1 %) of pixels in the image has an intensity of 0 or 255 (or 4095). It can usually be removed by adjusting the detector's gain and black level settings, and in the case of CCD cameras the integration time used to collect the image.

19.4.2.2 Linearity

A detector is said to be linear when the pixel values are proportional to the input light intensity. CCD cameras are usually highly linear providing there is saturation, but linearity is less accurate for other types of detectors, such as photomultiplier tubes (PMTs).

19.4.2.3 Spatial Resolution and Point Spread Function

Microscopes are limited in their precision for detecting the originating locations of photons emitted by the specimen. This imprecision is largely unavoidable (however, see Section 19.4.3.3) and is primarily determined by the numerical aperture of the

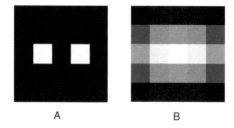

A   B

FIGURE 19.9 (A) Two closely spaced spots. (B) Image of the two spots after blurring because of the finite spatial resolution of the microscope.

objective lens and the wavelength of the light. The practical effect is a limited ability to detect small details in the specimen. Spatial resolution is the term given to this imprecision and it has several definitions. One of these definitions is the closest distance separating two point sources (of fluorescence) while the sources still appear as separate spots in the image (Figure 19.9A and Figure 19.9B). For optical microscopy, spatial resolution is worse in the depth (z) dimension versus the lateral (x and y) dimensions.

The point spread function (PSF) is the image obtained from a point source of light, and an alternative definition of spatial resolution is the full width at half maximum intensity (FWHM) of the point spread function. Mathematically, optical images are a convolution (see Figure 19.10C) of the actual distribution of fluorescence molecules in the sample and the PSF, and often images are said to be "blurred" by the PSF.

19.4.2.4 Background

Background is artifactious signal added to the true signal in the image, which leads to a loss of quantitative accuracy of the image and loss of contrast. Its main causes are nonspecific labeling by the fluorescence dye, autofluorescence from the sample, extraneous light from the room, and extraneous signal generated by the detector. In many cases, images can be easily corrected for background by subtracting a background-only image from the image of the desired signal.

19.4.2.5 Noise

Noise is the random deviation between a pixel's actual value and its ideal value had the noise not been present (Figure 19.10B). It can arise from several sources: a low level of fluorescence signal from the specimen resulting in only a small number of photons recorded in the pixels; random variation in the response of the photomultiplier tube; and random background signal added by the detector electronics. Noise can be reduced, but not eliminated, by acquiring the signal at each pixel for a longer exposure time and/or using more intense excitation light. However, these approaches have drawbacks: because fluorescence molecules are gradually destroyed by the

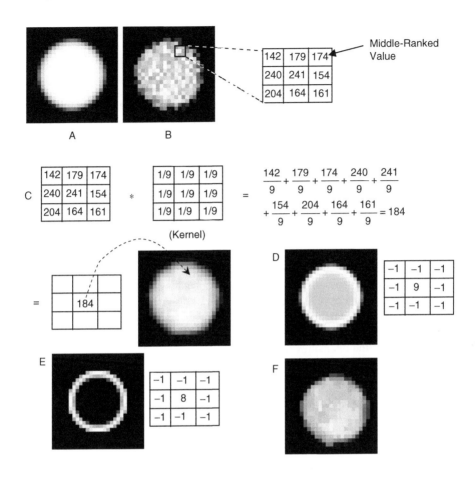

FIGURE 19.10 Neighborhood operations. (A) Image of a circle with a blurred edge. (B) Image A with noise added. The enlargement shows pixel intensities in a 3 × 3 array of the image and indicates the middle-ranked intensity (i.e., the 5th brightest intensity out of the 9). (C) Details of the filtering/convolution (*) process applied to the 3 × 3 neighborhood surrounding one of the pixels in image B using a "uniform" kernel, and the result after application of the filtering process to all pixels in image B in order to "smooth" out the noise. (D) Edge enhancement of image B and the kernel used. (E) Edge detection of image B and the kernel used. (The image was contrast stretched so that all pixels with zero or negative intensities are black.) (F) Image A after median filtering to reduce noise.

excitation light, live samples can be damaged by excessive excitation light and rapid changes in the fluorescence signal can be missed when using long exposure times.

19.4.2.6 Signal-to-Noise Ratio

Signal-to-noise ratio (snr) is a useful parameter for characterizing the quality of the signal in an image. Several definitions of snr exist, but a useful one for fluorescence microscope images that takes into account both noise and background[17] is

$$snr = \frac{i-b}{\sqrt{\sigma_i^2 + \sigma_b^2}} \; ,$$

where i is the intensity of a pixel or the mean intensity over a region of pixels where the sample fluorescence is believed to be uniform, b is the background intensity of the same pixel/region or a different pixel/region believed to have the same background, σ_i is the standard deviation of the intensity and σ_b is the standard deviation of the background. For a single pixel, σ must be measured from a series of measurements at the pixel, but for a region σ can be measured from the spatial variation of pixel intensities over the region.

19.4.3 IMAGE PROCESSING

Image processing is tasks implemented by computer algorithms that convert an input image to a different output image. Image processing algorithms fall into three categories: point operations that change pixel intensities in the image independent of the other pixel intensities; neighborhood operations that change a pixel's intensity as a function of its own intensity as well as the intensities of the neighboring pixels; and global operations that take into account the intensities of all pixels in the image when changing the intensities of each pixel.

19.4.3.1 Point Operations

The most common use of point operations is to enhance the visibility of images by reassigning the gray-level range (black to white) to a subset of the range of pixel intensities; a process known as contrast enhancement or gray-level transformation (Figure 19.8C and Figure 19.8D).

There are several other common uses for point operations:

1. Subtracting a background image from the acquired image.
2. Correcting an image for nonuniform illumination by division by an image of a uniform sample.
3. Geometrical operations that include rotating or translating (shifting) an image. Translating an image may be necessary following acquisition of multiple images of a specimen labeled with different dyes, because slight misalignment of the optical filters can result in the image of one of the dyes being shifted relative to the other.

19.4.3.2 Neighborhood Operations

Neighborhood operations, also known as "filtering," have several major applications, which include edge enhancement and smoothing to reduce noise. They fall into two categories: linear filtering and morphological filtering.

In linear filtering the intensity of each pixel is replaced with a weighted average of its own intensity and the intensities of its neighbors. The weights are fixed and collectively are known as a kernel, and the process of applying the weighted averaging is known as "convolution." Figure 19.10A, Figure 19.10B, and Figure 19.10C illustrate

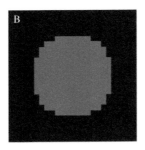

FIGURE 19.11 (A) Binary image of the circle shown in Figure 19.10A after thresholding at intensity 100. (B) Image A after binary erosion. (C) Image A after binary dilation.

and explain the use of linear filtering for reducing noise, but it can be at the expense of reduced visibility of small details. Figure 19.10D and Figure 19.10E illustrate the use of linear filtering for edge enhancement and edge detection respectively.

There are two basic types of morphological filters: erosion and dilation. The simplest version of erosion overlays a 3×3 kernel over each pixel in the image. At each pixel, its intensity in the output image is set to the minimum intensity of all 9 pixels in the input image that are under the kernel. Dilation, which is the reverse of erosion, sets the pixel intensity in the output image to the maximum intensity of the 9 pixels in the input image. These filters are often applied to binary images where they affect pixels at the edges of objects, while not affecting pixels distal from edges. Erosion shrinks objects, while dilation enlarges objects (Figure 19.11). (Larger kernels may be used, and Figure 19.13c shows the result of applying dilation with a circular kernel of 15 pixel diameter to the image in Figure 19.13b.)

The "median" filter is a type of morphological filter, where each pixel intensity in the output image is set to the middle-ranked intensity of the pixels in the input image that are under the kernel. It is a useful alternative filter for reducing noise while preserving small details (Figure 19.10F).

Only the most basic image processing tasks are described above, yet they form the foundations of more advanced image processing and image analysis operations. Thus far, only kernels of size 3×3, the smallest practical size, have been discussed. Kernel sizes can be any size up to the size of the image. At this maximum size, operations are called global operations. However, global operations are not widely used in image analysis, because generally the intensities of pixels separated by large distances are unrelated. On the other hand, global operations, such as Fourier transforms,[11] are frequently used to speed up the execution of linear filtering and other image processing operations.

In most cases, two-dimensional (2D) image processing operations readily extend to 3D operations for application to 3D images. For example, a 3×3 kernel extends to a $3 \times 3 \times 3$ kernel.

19.4.3.3 Deconvolution of Fluorescence Images

Deconvolution[8,11,12] is an image-processing procedure used for restoring optical microscope images by removal of the background haze from out-of-focus light, as well as improving the spatial resolution. Primarily, it is applied to 3D images acquired using conventional microscopy where out-of-focus light is significant. Deconvolution works by attempting to reverse the inherent convolution of the true distribution of fluorescence molecules in the sample and the PSF of the microscope that distorts the acquired image. However, also during acquisition there is inevitably addition of unknown noise to the image, causing direct reversal of the convolution to result in a highly erroneous image. This situation can be partially mitigated by imposing constraints on the resulting image; for example, requiring pixel intensities to be positive, or by limiting the intensity differences between neighboring pixels. Figure 19.12 shows examples of applications of different deconvolution algorithms.

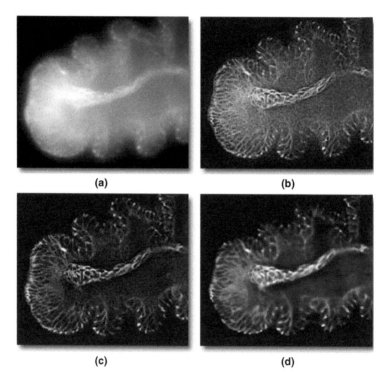

(a)   (b)

(c)   (d)

FIGURE 19.12 The results of applying three different deconvolution algorithms to the same data set. The original three-dimensional data are 192 optical sections of a fruit fly embryo leg acquired in 0.4-μm z-axis steps with a wide-field fluorescence microscope (1.25 NA oil objective). The images represent a single optical section selected from the three-dimensional stack. (a) The original (raw) image. (b) The results of deblurring by a nearest neighbor algorithm with processing parameters set for 95 % haze removal. (c) The same image slice is illustrated after deconvolution by an inverse (Wiener) filter and (d) by iterative blind deconvolution incorporating an adaptive point spread function method. (Blind deconvolution does not require exact information about the shape of the PSF.)

19.4.4 QUANTITATIVE IMAGE ANALYSIS

Image analysis is tasks implemented by computer algorithms that extract information from images. The usual image analysis procedure for images of cells is first to segment (identify) the individual structural components (objects) such as the cells themselves, structures within cells (nuclei and other organelles), or extracellular components (e.g., blood vessels, ducts). Following segmentation, several features of the objects are measured; for example, the amount of a labeled protein in each object. Features are often used to classify the objects into different types. There are many books[11, 18] online tutorials,[19] and downloadable image analysis packages[20, 21] to which the reader should refer to for further information about image analysis.

19.4.4.1 Image Segmentation

Here, we will describe two basic image-segmentation procedures that are well suited for images of fluorescence-labeled cells. The reader should refer to the cited literature for information about more advanced methods. The images we will use for illustration are confocal images of cells where the nuclei have been labeled with fluorescence dyes. One dye is specific to DNA (Yoyo-1[6]) (Figure 19.13a and Figure 19.14a) and another is a fluorescence in situ hybridization (FISH) probe specific to a specific DNA sequences (Figure 19.14c). Both procedures segment the cell nuclei using the Yoyo label.

In the first procedure, nuclei are segmented using a matched detector filter to identify the centers of nuclei, followed by the watershed algorithm to determine the edges of nuclei. Figure 19.13a is an acquired image of fluorescence labeled cell nuclei. A matched detector,[11] which is linear filtering with a kernel that represents a model object, is applied to the image. In this case, the kernel was a circular disk with a diameter equal to the average size of the nuclei. The matched detector filter is optimal for detecting known objects in the image and results in a smoothed image with a peak at approximately the center of each nucleus (Figure 19.13b). The next step is detection of the peaks, which serves as seeds for the watershed algorithm. Peaks are detected in two steps. First, gray-scale dilation is applied to the smoothed image using a circular kernel of the same diameter as the kernel used in the matched filter (Figure 19.13c). Using this sized kernel results in the peaks being spaced at distances at least the size the nuclei, thus reducing the chance of multiple peaks per nucleus. Pixel intensities in the dilated image are greater than pixels in the smoothed image, except at the peaks where they are equal. Therefore, subtraction of these two images results in zero intensity at the peaks and in areas of uniform background, and negative values elsewhere. Thus this image is thresholded at intensity level 0 to generate "seeds" for the watershed algorithm. Seeds (red) are shown overlaid on the original image in Figure 19.13d. The next step is to find the "intensity valleys" that serve as borders between nuclei, using the seeded watershed algorithm. The algorithm, which is illustrated in 1D in Figure 19.13e, is applied to the negative of the image, with seed pixels assigned much lower intensities than other pixels. Therefore, the algorithm finds ridges between nuclei. The algorithm treats each seed pixel as a source of water and as the water level rises, surrounding pixels are flooded. When a pixel is flooded, it is assigned to the same object as the seed supplying the water, hence the term "watershed." Locations

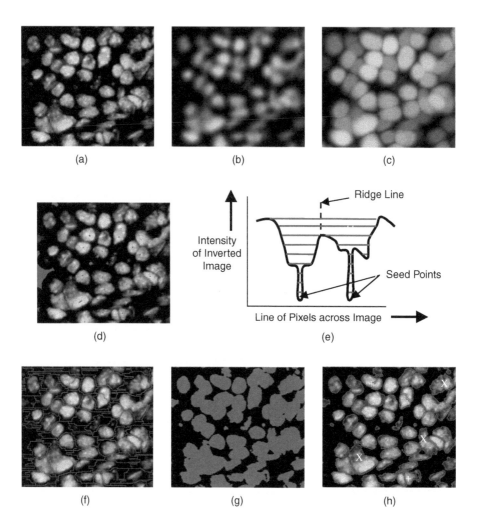

(a)   (b)   (c)

(d)   (e)

(f)   (g)   (h)

FIGURE 19.13 (Color insert follows page 204) Segmentation of cell nuclei using matched filtering and the watershed algorithm. (a) Image of fluorescence-labeled cell nuclei. (b) The image in A after application of a matched filter. (c) The image in B after application of the dilation filter. (d) The seed image (red) generated by subtracting image C from image B and thresholding at intensity 0, overlaid on the image in A. (e) One-dimensional schematic illustration of the watershed algorithm. "Water" gradually fills up the valleys starting at the seed points. All pixels in the same valley as the seed are assigned to the same object (green and red). Locations of adjacent pixels from different objects form ridge lines, which water cannot cross. (f) Watershed ridges overlaid on the original image. (g) Original image after automatic thresholding to determine object and background regions of the image. (h) Final borders overlaid on the original image. "X" and "+" indicate errors. "X" marks undivided clusters of nuclei and "+" marks incorrectly split nuclei.

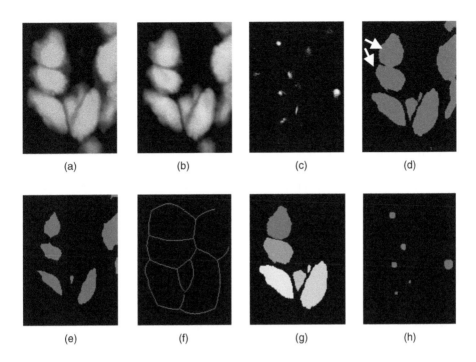

(a)   (b)   (c)   (d)

(e)   (f)   (g)   (h)

FIGURE 19.14 (Color insert follows page 204) (a) Image of cell nuclei (green) and FISH signals (blue). (b) Grey image of the nuclei. (c) Grey image of the FISH signals. (d) Image B following setting of a threshold intensity to separate bright nuclei (red) with intensities > threshold intensity from background (black) with intensities < threshold intensity. Arrows indicate two touching nuclei, which the computer would consider as one object. (e) Image D after erosion to shrink the nuclei so that touching nuclei separate. (f) "Skeletonization" of the background in image E in order to determine the lines midway between the objects, which serve as divisions between touching objects. (g) The individually detected nuclei. (h) The detected FISH signals.

where two pixels from different watersheds meet serve as barriers and define the borders of the objects. Figure 19.13f shows the watershed ridges overlaid on the original image. However, the watershed ridges extend into background regions of the image. Therefore, the original image is automatically thresholded using the "isodata threshold" algorithm,[19] to define object and background regions (Figure 19.13g). Isodata threshold can provide multiple thresholds, enabling the segmentation of dim objects in the presence of bright ones. (A variety of other automatic threshold algorithms exist.[22]) Figure 19.13h shows the borders around the image from thresholding and the watershed algorithm combined. This matched filter approach is appropriate for detecting approximately circular nuclei of similar size, but it tends to incorrectly divide elongated nuclei (oversegmentation, "+" in Figure 19.13h) and small nuclei may not get seeds leading to undivided clusters of nuclei (undersegmentation, "X" in Figure 19.13h). Usually these errors are corrected for by application of subsequent algorithms that merge oversegmented nuclei and split undersegmented nuclei.[23] The seeded watershed algorithm may also be used to segment whole cells where the cell cytoplasm has been labeled with one fluorescent dye and the nuclei has been labeled with a different dye. In this situation, the segmented nuclei serve as seeds for segmentation of the whole cells.[24]

An alternative segmentation method that has improved performance for less regularly shaped nuclei is as follows. First, the original image of the nuclei (Figure 19.14b) is thresholded (Figure 19.14d), but after this step some nuclei remain joined (arrows in Figure 19.14d). These nuclei are separated by successive binary erosions, which shrink the nuclei and cause clusters of nuclei to split apart at "neck" regions (Figure 19.14e). Next, the background was "skeletonized"[11] to find lines midway between the eroded objects (Figure 19.14f), and these lines served as boundaries between touching nuclei (Figure 19.14g). The FISH signals in Figure 19.14a were segmented using the "matched" filter method described above (Figure 19.14h).

19.4.4.2 Feature Measurement

Following segmentation, it is usual to classify the segmented objects into various types based on measurements of multiple quantitative features for each object. A large number of features can be measured for each object, and they fall into three categories: morphological, which measure the shape of the object; textural, which measure the variations in the pattern of the staining within the object; and contextual, which measure the spatial relationships of objects to each other. For the images shown in Figure 19.13 and Figure 19.14, relevant features would include the size of the nuclei, the regularity of the shape of the nuclei (usually measured by the shape factor = $perimeter^2/area$), and for Figure 19.14 the number of FISH signals per nucleus and the spatial organization of the FISH signals within their respective nuclei (e.g., do FISH signals tend to be near the edge of the nucleus?)[25]

19.4.4.3 Object Classification

It is beyond the scope of this chapter to describe object classification in detail. Instead, a simple example is given to illustrate the concepts and utility of classification for analyzing images.

Object classification is straightforward when it can be successfully performed using only one feature. For instance, classification of objects as "small" or "large" only requires the "size" feature. However generally, the desired classification does not directly correspond to one feature. For example, classification of cells as "normal" and "cancer" requires the combined use of multiple features to classify each cell. One of the features commonly used is size because often cancer cell nuclei are larger than normal nuclei. However, there exists inherent variation in the size of nuclei causing some normal nuclei to be larger than some cancer cell nuclei, thus leading to imperfect classification if size is the only feature used (Figure 19.15a). To improve the classification, other features are incorporated into the classification of the cells. Since cancer cell nuclei are more irregular in shape than normal nuclei, a second feature to include is shape in combination with size. As an example, Figure 19.15b shows 5 "normal" nuclei (+) and 5 "cancer" nuclei (*) that overlap in terms of their size and shape distributions. However, when both size and shape are combined using linear discriminant analysis[26] the two categories are correctly

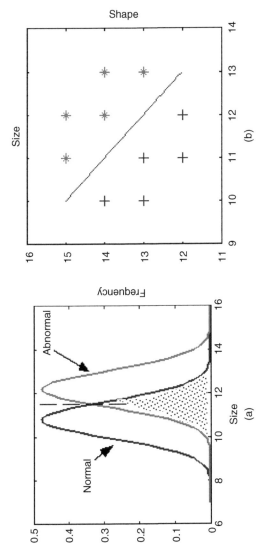

FIGURE 19.15 (a) Frequency distribution of the sizes of two populations of objects, "normal" and "abnormal." The size distributions of the two populations overlap (shaded area); therefore, it is not possible to correctly categorize all objects as "normal" or "abnormal" based on size alone. (b) Two populations of objects (+s and *s) that have overlapping size and shape distributions. However, all objects can be correctly categorized when size and shape are used in combination. The diagonal line, which was calculated by linear discriminant analysis, segregates the scatter plot into "+" objects and "*" objects.

separated (diagonal line). For further information about object classification, the reader should refer to Fukunaga[26] for statistical pattern recognition or Ripley[27] and Zheng, et al.[28] for artificial neural networks.

19.4.5 Visualization of 3D Images

Visualization of images is a vital part of image analysis, since the human eye–brain combination excels (and outperforms computer algorithms) at object recognition. While the presentation of 2D images on a computer screen is straightforward, display of 3D images of fluorescence-labeled specimens on a computer screen is not straightforward. This is because these 3D images of fluorescence-labeled samples are uniquely different from 3D scenes normally experienced in everyday life. In everyday life, the 3D objects we view are opaque, meaning we see only the outer surface of the facing side of objects. Such scenes are essentially 2D and can be readily displayed on a 2D screen. However, fluorescence-labeled specimens are transparent, which enables the microscope to detect signals from every 3D position in the specimen, resulting in images that are truly 3D. In this situation, not all information in the 3D image can be simultaneously displayed on a 2D screen. Therefore, it is necessary to select a subset of the information in a 3D image for display. Also, it is necessary to correct images for any anisotropy; otherwise their display will appear distorted. The display of higher-dimensional images, such as 4D images that have three spatial dimensions plus time, is more problematic.

There are four basic methods, which can be combined, for displaying 3D images: a gallery, orthogonal views, volume rendering, and surface rendering. They are described below.

19.4.5.1 Gallery and Orthogonal Views

The simplest way to display a 3D image is to display each 2D slice separately (Figure 19.16a). Such displays have the advantage of showing the intensity information at every voxel in the 3D image, but the spatial relationships between different slices can be difficult to follow. This problem can be partially solved by displaying each slice in rapid succession, a procedure known as "animation."

Orthogonal views display three 2D slices with each slice perpendicular to the other two (Figure 19.16b).

19.4.5.2 Volume Rendering

Volume rendering aims to project information at different depths in a 3D image into a 2D projection view. For fluorescence images, the most common flavor of projection is maximum intensity projection. Here, each pixel in the projection is traced back through the 3D image, and the maximum intensity encountered along each trace line is the intensity given to the pixel in the projection image (Figure 19.16c). However, a limitation of this method is that all depth information is lost in a given projection. In order to get around this problem, multiple projections are created by incrementally rotating the 3D image about an axis orthogonal to the projection direction (Figure 19.16d).

19.4.5.3 Surface Rendering

Surface rendering shows only the outer surfaces of the objects recorded in the 3D image, but produces the most visually pleasing displays (Figure 19.16e). However, it has the drawback that the objects must first be correctly segmented in order to define the surfaces. Direct display of the surfaces is disappointing (Figure 19.16e, left); thus it is necessary to add texture. This is typically done by reflecting virtual lights from the surfaces and displaying intensity of light that is reflected back to the viewer based on the orientation of the surface (Figure 19.16e, middle). In order to display (segmented) objects inside outer objects, the outer objects can be rendered using a "wire-frame" (Figure 19.16e, right).

19.5 MICROSCOPE TECHNIQUES TO STUDY PROTEIN DYNAMICS AND INTERACTIONS

The previous sections describe techniques for fluorescence labeling proteins in cells and image acquisition, visualization, and quantitative analysis of the cells. There are several microscopy and image analysis techniques for analyzing the interactions of proteins with each other and protein movements in cells. Such analysis provides important information for understanding the complex protein chemistry that drives cell behavior.

In this section we describe two techniques for studying protein movement within cells: time-lapse imaging and fluorescence recovery after photobleaching (FRAP), and two techniques for measuring the interactions of proteins with each other within cells: spatial colocalization and fluorescence resonance energy transfer (FRET).

19.5.1 TIME-LAPSE IMAGING

Time-lapse imaging (TLI) is the acquisition of a set of 2D or 3D images of a live sample over time. A major application of TLI is tracking the locomotion and shape changes of motile cells,[29] including analysis of motion in the presence of a chemoattractant[30] and measuring the traction forces exerted by cells on a substrate.[31,32] Several commercial vendors sell software for tracking cells and molecules within cells.[33]

19.5.2 FLUORESCENCE RECOVERY AFTER PHOTOBLEACHING

Fluorescence recovery after photobleach (FRAP) is one of a set of related techniques for measuring protein mobility and binding constants while the cell is at equilibrium. FRAP was originally described by Axelrod, et al.[34] and has been recently reviewed by Lippincott-Schwartz et al.[35] FRAP hinges on the idea that the fluorescent moiety (usually GFP or a variant) attached to the protein of interest can be destroyed (bleached) by intense light while leaving the activity of the protein and other cell functions unaltered. Generally, FRAP experiments are conducted by bleaching a small user-defined region of the cell, followed by acquisition of a series of time-lapse images of the entire cell using a low intensity of excitation light that does not cause significant additional bleaching. If the protein of interest is moving, then bleached

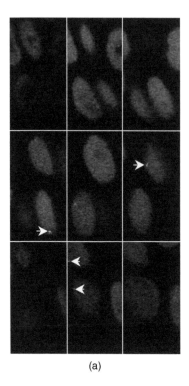

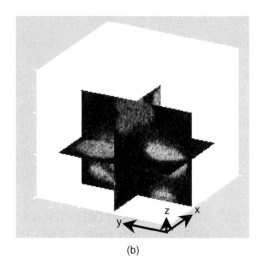

(b)

(a)

FIGURE 19.16 (Color insert follows page 204) (a) "Gallery" of 2D slices from a 3D image. The red signal is from a fluorescent DNA dye that labels the entire volume of each nucleus. The yellow dots are FISH signals at the centromere of chromosome 1 (arrows). (b) Orthogonal views through a 3D image. (c) Generation of projection images in volume rendering. The intensity at coordinate (x',y') in the projection image is a function of the set of intensities along the line (x',y', z_1) to (x', y', z_n) in the 3D image, where n is the number of slices in the 3D image. (For maximum intensity projection, the function selects the maximum intensity.) The same transformation is applied at all coordinates in the x-y plane. (d) Example of maximum intensity projections after rotating the 3D image in 30° increments from 0 to 150°. In these projections, all FISH signals are shown in each projection, except those FISH signals that are directly behind another. (e) Three segmented nuclei displayed using surface rendering. The display looks flat without using texture (left rendering versus middle rendering). Wire-frame rendering (right rendering) enables the display of objects (segmented FISH signals) internal to outer objects (nuclei). (Lockett, S.J. Three-dimensional image visualization and analysis. *Current Protocols in Cytometry.* Copyright 2004 Wiley.)

protein molecules in the bleached region will exchange with unbleached molecules outside the region. The rate of this exchange is quantified by measuring the increase of fluorescence intensity in the bleached subregion.

FRAP is used to measure a variety of types of protein movement, which include free diffusion, directed motion of a protein through a cell, movement of a protein that is confined to a small volume, and translocation rate of a protein from one cell compartment to another (Figure 19.17). With further analysis of FRAP data, the

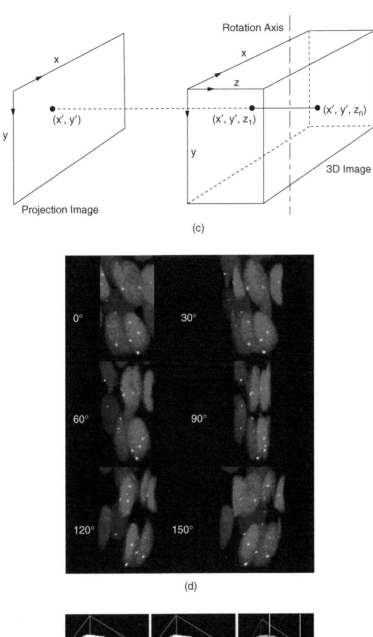

(c)

(d)

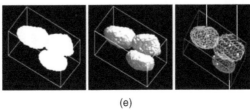

(e)

FIGURE 19.16 (Continued).

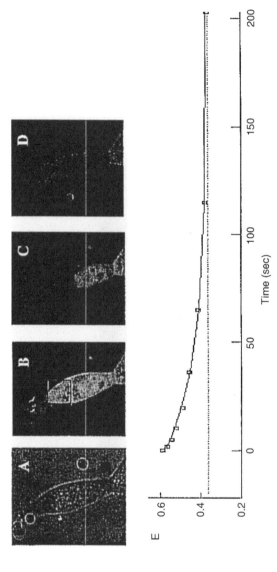

FIGURE 19.17 (Color insert follows page 204) Fluorescence recovery after photobleaching (FRAP) measures nuclear-cytoplasmic transport rate. (A) Differential interference contrast image of the cell. (B) Image of the cell expressing GFP showing a higher concentration of fluorescence in the nucleus vs. the cytoplasm. After recording this image, a rectangular region of the cytoplasm was bleached. (C) Image immediately after bleaching. GFP diffuses very rapidly in the cytoplasm, therefore, GFP outside the red rectangle is exchanging with GFP inside the rectangle during the brief bleaching period. This results in GFP in the entire cytoplasm being bleached. However, translocation of GFP between the nucleus and cytoplasm is much slower, resulting in no alteration in the nuclear fluorescence immediately after bleaching. (D) Image 290 seconds after bleaching, showing less fluorescence in the nucleus and increased fluorescence in the cytoplasm compared to C, because unbleached GFP in the nucleus has exchanged with bleach GFP in the cytoplasm. Note that the adjacent, attached cell was unaffected during the study, suggesting a very slow exchange of GFP between the cells. (E) Ratio of fluorescence from the nucleus vs. the entire cell as a function of time after bleaching measured at nine time points (squares). The red curve is the best-fit exponential decay from which the exchange rate of GFP between the cytoplasm and nucleus can be determined.

association and dissociation binding constants of proteins bound to a cell structure or another macromolecule can be calculated, because protein binding reduces the recovery rate.[36–38] Furthermore, FRAP measurements serve as a major source of experimental data for modeling cell biological processes.[39]

Rotational diffusion coefficients of fluorescence-labeled molecules can also be measured in cells by fluorescence polarization anisotropy measurement. In this technique, the sample is excited with linearly polarized light and the intensity of the emitted light is measured in the polarization directions parallel and perpendicular to the excitation light. The rotational diffusion coefficient is derived from the ratio of the intensities in the parallel and perpendicular directions.[40,41]

19.5.3 COLOCALIZATION ANALYSIS

Colocalization analysis is a method to detect and quantify the spatial association of two different proteins in a cell by labeling the two protein species with green and red fluorescent dyes and then visually observing yellow signal in the cell where the different proteins are colocalized. Colocalization can be from either interaction of the protein or from the two protein cohabiting subvisual compartments in the cell. Visual interpretation, however, is not quantitative and can be very misleading (Figure 19.18); thus, computational methods to quantify the colocalization of two proteins have been developed. Manders, et al.[42] developed a method to quantify the colocalization of two proteins after segmentation of the protein-positive regions of the image. More recently, Costes, et al.[43,44] have developed methods to determine the probability that two proteins are colocalized; and if this probability is considered high enough (e.g., >95%), then the proportion of each protein colocalized with the other protein is quantified. The method of Costes et al. has been recently commercialized by Bitplane.[33] Advantages of colocalization analysis over other methods to analyze protein interactions in the cell are that fixed samples may be analyzed and proteins associated at distances greater than those detected by FRET (see below) can be detected and quantified.

19.5.4 FLUORESCENCE RESONANCE ENERGY TRANSFER

Fluorescence resonance energy transfer (FRET) is direct, nonradiative transfer of energy from a donor fluorescence molecule to an acceptor fluorescence molecule. It occurs when the two molecules involved are closer than 10 nm and the emission spectrum of the donor significantly overlaps the excitation spectrum of the acceptor (Figure 19.19A). The major application of FRET is detection of two proteins directly binding to each other.[45,46] This is done by attaching a donor and acceptor to the different proteins, and the existence of a FRET signal strongly indicates that the proteins are directly bound to each other. FRET can be detected by several methods. The traditional method is to excite the sample with light at the excitation wavelength of the donor and detect light emission at the wavelength of the acceptor.[47] However, the detected light consists of extraneous signals that must be subtracted in order to know the true FRET signal. These extraneous signals consist of emissions from donor and acceptor following direct excitation and sample autofluorescence. Since these extraneous signals can be much larger than the weak FRET signals, very accurate corrections are necessary in order to obtain correct results.

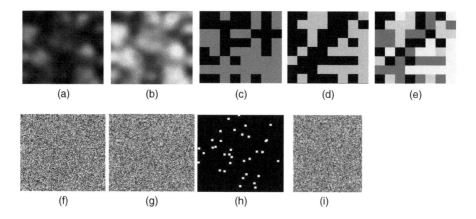

FIGURE 19.18 (Color insert follows page 204 Visual interpretation of colocalization is very misleading. (a) and (b) Contrast enhancement leads to misleading conclusions. (a) Simulation of red and green fluorescence signals, where the low amount of yellow (red overlaid with green) implies very little colocalization of the two colors. (b) The same image as A after contrast enhancement now shows a significant amount of yellow implying significant colocalization. (c), (d), and (e) Apparent colocalization is not real. (c) and (d) Two 8 × 8 pixel binary images where approximately half the pixels were randomly assigned signals. (e) Overlay of images (c) and (d) shows approximately one quarter of the pixels contain both red and green (yellow), which is as expected to occur by chance. Quantitative colocalization analysis, Costes, et al.,[44] however, report a probability of 25% that the colocalization is not random; i.e., the presence of real colocalization is not significant. (f), (g), (h), and (i) Real colocalization is not visually apparent. (f) and (g) Two random images, where the probability of the colocalization between the two images is not random is 22%. (h) Mask image showing the regions of image (f) that replaced the same regions in image (g) in order to introduce real (nonrandom) colocalization between the images. Image (i) is the result, which, when compared to image (g), does not show any visual evidence of colocalization. However, quantitative analysis reports a highly significant probability of over 99% that real colocalization exists between the images.

An alternative method, which does not require corrections to detect FRET, is the acceptor photobleaching method.[48] This method exploits the fact that FRET reduces the direct emission from the donor; therefore, eliminating the acceptor will eliminate FRET and thus increase emission from the donor. In this method, the donor is first imaged using excitation light at the appropriate wavelength and detecting light emitted at its emission wavelengths. Next, FRET is eliminated by photobleaching the acceptor with light at its excitation wavelength. This is followed by imaging the donor a second time. The difference between the donor's emission after photobleaching compared to before is attributed to FRET. This method has the advantage that correction for extraneous signals is unnecessary; however, it is essential to check that the donor was not photobleached as well as the acceptor.

When FRET is used to detect protein binding, the donor and acceptor are coupled to two proteins of interest, either via antibodies or directly tagging the

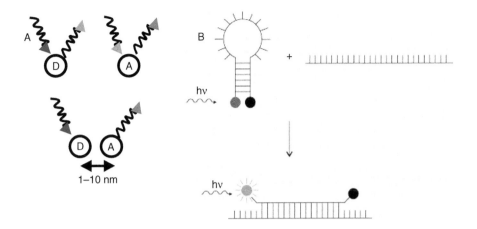

FIGURE 19.19 Fluorescence resonance energy transfer (FRET). (A) Schematic of FRET. When D and A are less than 10 nm apart, D will transfer its excitation energy directly to A. (B) Schematic representation of molecular beacons. In the hairpin loop structure, the quencher (black circle) forms a nonfluorescent complex with the fluorophore. Upon hybridization of the molecular beacon to a complementary sequence, the fluorophore and quencher are separated, restoring the fluorescence.

proteins with GFP and its variants (e.g., donor/acceptor = BFP/GFP, or CFP/YFP). (See Sekar and Periasamy[49] for a review FRET of detecting protein interaction in live cells.) A drawback of FRET for detecting protein binding is that the donor and acceptor can easily be more than 10 nm apart while the proteins are bound, which leads to a false-negative result. For the same reason, proteins that are part of a complex but not directly binding to each other are detected as unbound. Quantification of protein interactions from FRET signals is described by Hoppe et al.[50]

Another important application of FRET is for molecular beacons. Molecular beacons[6] (Figure 19.19B) are molecules that have donor and acceptor moieties in very close proximity so that when the donor is excited, the acceptor quenches light emission from the donor. Beacons are designed to change conformation by specific target molecules (e.g., enzymes, specific DNA sequences) and, upon activation, the donor and acceptor separate, removing the quenching of the donor. In the case of enzymes, the beacon is designed to be cleaved by the enzyme.[51]

19.6 EMERGING MICROSCOPE TECHNIQUES FOR PROTEIN ANALYSIS IN CELLS

This section outlines several emerging microscope techniques for protein analysis in cells.[52]

19.6.1 EMERGING LABELING METHODS: NANOPARTICLES

Nanoparticles refer to a number of types of particles that are approximately ≥ 5 nm in diameter and in most cases fluoresce. Due to their small size they readily enter live or fixed cells, and thus serve as useful labeling reagents.

Quantum dots,[53,54] which contain a fluorescent semiconducting crystal at their core, are a well-known type of nanoparticle, because their fluorescence properties have advantages over conventional fluorescent dyes. These advantages are high quantum yield, very low photobleaching, and a peak emission wavelength that is related to their size. Furthermore, their emission band is relatively narrow (~ 20 nm), enabling good spectral separation of particles of different sizes; yet, conversely, they have wide absorption spectrums that are independent of size, enabling the simultaneous excitation of beads of different sizes. Quantum dots can be coated and used to label specific proteins in cells by linking to immunoglobin or coating with strepavidin[55] or incorporated into phospholipid micelles.[56] Water-soluble quantum dots[57] can be used in vivo, and in a recent application they were shown to be readily endocytosed by motile cells and thus could be used to track the cells. Their applications are expected to grow rapidly as their surface chemistry becomes fully understood.

Another type of nanoparticles contain rare-earth doped yttria (Y_2O_3) coated in silica to produce nanoparticles about 25 nm in diameter.[58] Similar to quantum dots, they do not photobleach, but have the additional advantage of excited state lifetimes in the milliseconds. These very long lifetimes lead to increased sensitivity, because their phosphorescence can be readily detected long after autofluorescence in the cell has decayed. In addition, they exhibit step-wise multiphoton absorption at 974 nm, enabling imaging deep in tissue (see Section 19.6.2.2 below).

19.6.2 EMERGING ACQUISITION METHODS

19.6.2.1 Spectral Imaging

Generally in microscopy, each field of view in the specimen is acquired into a gray-level image, where the image intensities record the fluorescence signal from a predefined wavelength range determined by interference filters. For samples labeled with multiple fluorescent dyes, multiple gray-level images are taken of each field of view using interference filters that select different wavelength ranges. However, if the emission spectra overlap, and it is not possible to individually excite each dye, then each gray-scale image contains signals from multiple dyes. The GFP dyes are a good example of this phenomenon (Figure 19.2). Spectral imaging[59] followed by linear unmixing provides a solution to this problem by yielding a set of gray-level images where each image corresponds to the signal from only one dye. In spectral imaging, the emission spectrum at each pixel is acquired. A common method to acquire the spectrum is to disperse the emitted light with a diffraction grating or prism and to collect the dispersed light using a linear array of detectors. Usually the detected spectrum ranges from 400 to 700 nm and the intensity is measured at 10 nm increments using 32 detectors. Thus each pixel is composed of (up to) 32 intensity values. Linear unmixing is implemented by a mathematical algorithm, which is applied to each pixel independently. For each pixel the algorithm calculates the proportion of each dye contributing to its measured spectrum. The predominant application of spectral imaging is the quantification of dyes with overlapping emission spectra. Other applications include separation of dye emission from autofluorescence and detection of acceptor emission in the presence of donor emission in FRET experiments.

19.6.2.2 Multiphoton Microscopy

Multiphoton (MP) microscopy refers to a family of nonlinear optical acquisition modes where multiple photons are used to generate each emitted photon from the sample. The challenge with MP microscopy is in obtaining a high probability that two or more photons will arrive at the same fluorochrome virtually at the same time ($<10^{-18}$ sec interval). This is achieved by temporally compressing the infrared photons into extremely short (100 to 200 femtosecond width), extremely high-intensity pulses, and using a high numerical aperture objective lens to spatially compress the pulses into a small volume. An inherent advantage of MP microscopy is that 3D images are directly obtained, because the emitted photons can only arise from the small volume where the excitation photons have sufficient concentration.

Here, we briefly describe three of these modes, all of which may be used on live samples: two-photon fluorescence microscopy, second harmonic generation (SHG) imaging, and coherent anti-Stokes Raman spectroscopy (CARS) microscopy. Taken together, fluorescence, SHG, and CARS provide multiple channels of information about the cell, enabling more complex cellular functions to be dissected.

Two-photon fluorescence (2P) microscopy[60] is the most common form of MP microscopy. In 2P microscopy, a fluorescent molecule is excited using two photons arriving at the molecule at the same time. Generally, each photon has twice the wavelength (half the energy) normally used for single photon excitation. For example, EGFP, which is normally excited by blue light at a wavelength of 488 nm (Figure 19.2a) is instead excited by two infrared (IR) photons at a wavelength of 976 nm. The advantages of 2P microscopy are twofold: (1) IR photons are generally less damaging to live samples, and (2) thicker (300 to 500 μm) samples can be imaged, because IR photons scatter less in tissue. Furthermore, any scattering of the emitted photons is not a significant problem, because at any given time point only one small volume in the sample is exposed to light of sufficient intensity for 2P fluorescence to occur. Therefore, any photons emitted at that time must have originated from the same volume regardless of the path taken by the emitted photons out of the specimen. 2P microscopy is commonly used for imaging of live samples, including direct imaging into live animals.[61]

Second harmonic generation (SHG)[62] images ordered macromolecular structures. In SHG, an intense laser field induces a nonlinear polarization in a molecule or assembly of molecules, resulting in the production of a coherent wave that is emitted at exactly twice the incident frequency. Furthermore, the magnitude of the SHG wave is resonance enhanced when the energy of the second harmonic signal overlaps with an electronic absorption band. The major advantage of SHG is that such structures are detected without labeling, and in addition the techniques works in live cells. SHG has been used to detect noncentrosymmetric structures in cells and tissues such as collagen in the extracellular matrix[63,64] (Figure 19.20) and actin fibers, thus providing 3D structural information about the sample without labeling.

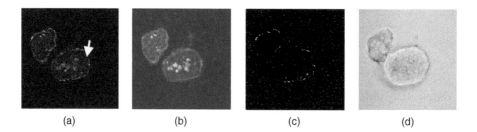

FIGURE 19.20 Second harmonic generation (SHG) imaging at 880 nm excitation of a spheroid of MCF10 cells. (a) SHG signal around the edges of the spheroid corresponding to the location of basement membrane proteins. Note the increased signal at the top and bottom edges of the spheroids due to the polarization of the laser light. The image also shows autofluorescence. (b) Image of the autofluorescence alone. (c) Harmonic signal alone after subtracting the autofluorescence using eigenanalysis.[11] (d) Bright-field image.

SHG has also been used to image lipid bilayers forming plasma membranes and intracellular organelle membranes[62]; however, the symmetry of the bilayer must be broken by labeling only one leaflet of the bilayer.

Coherent anti-Stokes Raman spectroscopy (CARS) microscopy is in its infancy, but promises to be a highly sensitive method to detect chemical bonds (e.g., –CH, –NH, and –SH bonds) via their vibrational (Raman) spectra with the spatial resolution of a confocal microscope.[65] CARS can be used to detect specific molecules by replacing hydrogen atoms in –CH bonds with deuterium atoms. This has the effect of shifting the Raman spectra to a frequency where naturally occurring bonds do not vibrate (Figure 19.21). This approach for detecting specific molecules, either small or large, has the following advantages over fluorescence labeling:

1. Infrared light is used, which can be tolerated by live cells
2. Deuterating –CH bonds to –CD in a molecule has far less effect on the molecules chemistry compared to attaching a fluorescent dye
3. –CD bonds do not suffer photodamage, which is an inherent problem when using fluorescent labels

19.6.3 EMERGING IMAGE ANALYSIS METHODS

19.6.3.1 Advanced Methods to Detect Cell Nuclei and Whole Cells in Tissue

The basic image analysis (IA) techniques described above generally perform satisfactorily when the cell nuclei or whole cells being detected are separated or slightly touching, and when some interactive correction is acceptable. However, more advanced methods are required when automation or analysis of large numbers of cells is desirable. Here, we outline some of the approaches researchers are taking to achieve improved segmentation of cells and nuclei.

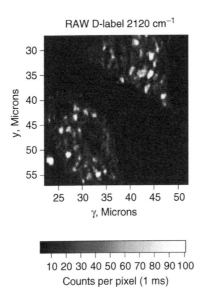

RAW D-label 2120 cm^{-1}

y, Microns

γ, Microns

10 20 30 40 50 60 70 80 90 100
Counts per pixel (1 ms)

FIGURE 19.21 Images of living RAW cells that have been exposed to deuterium-labeled lipids. (From Yoder, E.J. and Kleinfeld, D. *Microsc. Res. Tech.*, 56, 304–305, 2002. With permission.)

An advanced segmentation method adopts the strategy of exploiting multiple sources of information available from the image, in conjunction with a priori information about the objects, in order to achieve good segmentation. Either the method applies several algorithms in succession to first achieve an approximate segmentation followed by improvements by the subsequent algorithms. Alternatively, the method uses one algorithm that is able to unify multiple sources of image and object information to achieve segmentation.

The method of applying multiple algorithms in succession has been used with considerable success by Wählby, et al.[66] One of their approaches was to use a watershed algorithm to achieve an initial segmentation, then merge incorrectly divided objects and finally split clustered objects. In another approach,[67] cell nuclei in tissue sections were segmented by first identifying seeds inside nuclei, followed by an edge-based watershed algorithm that more accurately delineated the nuclear borders than the standard watershed algorithm. This method correctly segmented approximately 95% of the nuclei. In a subsequent study, Wählby, et al.[68] combined the above methods so that intensity, edge, and shape information were utilized for nuclear segmentation in tissue sections. This method was also implemented for 3D image segmentation.

Another approach to cell or nuclear segmentation is to segment the image several times using different algorithms or different variants of the same algorithm[69,70] and then automatically select the best segmentation result for each object from the set of segmentation results. Lockett, et al.[71] achieved approximately 95% correct segmentation for cell nuclei in tissue.

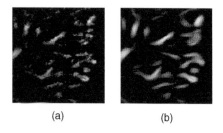

(a)   (b)

FIGURE 19.22 Noise reduction by anisotropic diffusion, which does not blur edges. (a) Input, noisy image. (b) Smoothed image.

Several segmentation algorithms that use multiple sources of information in unity have been developed. A significant subset of these algorithms is based on the image-processing technique of anisotropic diffusion,[72] which is a method to reduce noise in images without blurring edges by preferentially smoothing an image in directions perpendicular to the steepest edges (Figure 19.22). Following smoothing, it is much easier to delineate the edges of objects with standard algorithms. This approach has been used in several optical microscopy applications that include segmentation and dynamic tracking of nuclear processes in living cells[73] and tracking cell motility[74] and segmentation of whole cells in intact tissue where the cells have been labeled with a fluorescent cell surface marker.[75] The segmentation of whole cells in tissue is a particularly difficult problem because inherently all the cells are touching other cells.

Other segmentation methods utilize edge direction information measured from the image to detect object edges pointing toward the center of circular objects,[76,77] and another algorithm has extended the watershed algorithm to find smooth edges around objects.[78]

The combined development of automatic image analysis software together with quantitative image acquisition methods and high-throughput image-based cell screening is yielding large image databases. Consequently, bioinformatics tools for storage, analysis, mining, and modeling of information in large image databases are under development.[79]

On the hardware side, it will soon be possible to segment and analyze images as rapidly as they are acquired using multiprocessor computers or reconfigurable floating point gate array (FPGA) chips.[80]

19.6.3.2 Objective Analysis of the Subcellular Distribution of Proteins

Proteins expressed in a cell have characteristic spatial distributions within the cell that depend in part on their function and activity. Furthermore, changes in cell behavior (e.g., proliferation, differentiation, response to the external environment) can lead to both changes in the expression level and the subcellular distribution of some of the proteins in the cell. Thus, understanding the proteome of the cell must include analysis of the spatial distribution of proteins in cells. Recent work by Roques et al.[81] is achieving this goal through the combination of several technologies. The genetic technology of central dogma (CD) tagging,[82] which is a method to label an arbitrary (including an unknown) protein in cells with GFP, is used for labeling. This is followed by fluorescence microscopy, cell segmentation, quantitative measurement of features that characterize spatial intensity distributions, and classification of the protein patterns in each cell based on statistical classification of the features. This approach is high throughput, reproducible, and can resolve image sets indistinguishable by visual inspection.

19.6.4 EMERGING MICROSCOPE TECHNIQUES

Several new microscope techniques are becoming available for analyzing the dynamics of proteins and their interactions in cells and for high-throughput screening of large numbers of cells.

19.6.4.1 Fluorescence Correlation Spectroscopy and Fluorescence Intensity Distribution Analysis

Fluorescence correlation spectroscopy (FCS)[83,84] directly measures the fluctuating signal from a small number of fluorescence molecules (ideally one) as they pass through a confocal volume. By performing autocorrelation analysis on the measured fluctuations, two important measurements can be determined about the molecules in the confocal volume: their absolute concentration and their diffusion coefficient. The latter parameter may indicate binding of the observed molecules to each other or different, unlabeled molecules.[85] FCS, however, has two drawbacks: First, only very low concentrations of the molecules can be analyzed. Second, only one confocal volume in a cell can be analyzed at a time. Wiseman, et al.[86] have solved the latter with the introduction of image correlation spectroscopy (ICS). In ICS, measurements are made at every position in the image by rapid acquisition with a CCD camera, but the method can only measure relatively slow fluctuations compared to FCS.

Fluorescence intensity distribution analysis (FIDA)[87,88] uses the same experimental configuration as FCS, but it quantifies the intensity of each single molecule passing through the confocal volume. Molecular interactions and changes in the environment can be determined from analysis of the mean and shape of the intensity distribution.

19.6.4.2 Fluorescence Lifetime Imaging

Fluorescence lifetime imaging (FLIM)[89–91] measures the length of time (typically a few nanoseconds) that a fluorescence molecule is in its excited state. The methods for measuring lifetimes are reviewed by Wang, et al.,[92] but a common method is to measure the time from excitation of single molecules with a short pulse (less than a nanosecond) of laser light to the emission time. A drawback of this method is that only one photon per pulse should be excited; on the other hand, the same instrumentation can be used for fluorescence correlation spectroscopy.[93]

The lifetime of an excited fluorescence molecule is affected by its immediate chemical environment, and in particular FLIM is a sensitive method to detect FRET. This is because the close proximity of an acceptor fluorophore provides an additional nonradiative energy transfer pathway for the excited molecule to lose its energy, resulting in a reduced lifetime. In addition, FLIM can detect environmental factors such as pH, oxygen, and CA^{2+} concentrations, discriminate against autofluorescence, and detect fluorescence quenching.

19.6.4.3 Microscope-Based High-Throughput Screening (HTS) of Cells

Major advances in genomics, proteomics, and, particularly, the generation of large numbers of candidate pharmaceuticals, have created the need for rapid target validation of compounds in live cells.[94,95] This has become possible in recent years due to the convergence of several technical developments that include live-cell fluorescence probes, which report on a wide variety of cellular actions, and automatic image acquisition and analysis. Consequently, there are now several commercial instruments for rapid cell analysis, which screen samples in three different kinds of format:

1. Live cells of the same or different types are arrayed into 96 or 384 wells.[96–100] This format is commonly used in the early stage of drug development, when large numbers of compounds must be screened for response in many cell lines. In these applications, each well is automatically imaged over time following the addition of candidate drugs. Wells are analyzed for various endpoints, such as growth, differentiation, metabolic activity, or apoptosis that are indicated by fluorescence dyes.
2. Individual cells are dispersed over a standard microscope slide and a laser scanning cytometer[99–101] images each cell individually. This instrument is analogous to a flow cytometer, although the scan rate is considerably slower, but it has the advantage that cell morphology and the spatial distribution of fluorescence signals can be analyzed. Furthermore, one can later return to specific cells for re-examination.
3. Tissue arrays are a third format,[102–104] which consist of a 1000 small, cylindrical tissue biopsies (600 μm in diameter and over 1 mm in length) regularly arranged in a paraffin block (see Kononen, et al.[105] for further details). Following preparation, the array can be replicated hundreds of times by sectioning to standard 4 to 6 μm thickness. The key advantage of tissue array sections is that all 1000 samples can be identically labeled for a specific protein and all of them analyzed simultaneously, and hundreds of proteins can be screened using the same sample set. However, such high throughput is only achieved if images of the arrays are acquired and analyzed automatically. The image analysis needs to compartmentalize each sample into normal cells vs. tumor cells (in the case of tumor biopsies) and then quantify the level of protein/gene per cell in each compartment.

19.7 CONCLUSION

Cells are highly complex molecular machines. At any one time, a cell is processing many internal and external chemical and physical signals in a combined way in order to determine its course of action. Actions include growth, differentiation, metabolic activity, and apoptosis. In order to understand these extremely complex machines, proteomics needs to be taken inside the live cell. Using a wide variety of imaging techniques based on fluorescence microscopy and quantitative image analysis, it is possible to measure the interactions and dynamics of multiple proteins in cells from which molecular pathways can be deduced.

REFERENCES

1. Fields, S. Proteomics in Genomeland. *Science,* 291, 1221–1224, 2001.
2. Meyer, T. and Teruel, M.N. Fluorescence imaging of signaling networks. *Trends in Cell Biol.,* 13, 101–106, 2003.
3. Current Protocols: http://www.mrw2.interscience.wiley.com/cponline Accessed on 02/17/05.
4. Patterson, G.H. and Lippincott-Schwartz, J. A photoactivatable GFP for selective photolabeling of proteins and cells. *Science,* 297, 1873–1877, 2002.
5. Green Fluorescent Protein and DsRed. http://www.clontech.com/gfp Accessed on 02/17/05.
6. Fluorescence Probes. http://www.probes.com Accessed on 02/17/05.
7. Source of optical filters. http://www.chroma.com Accessed on 02/17/05.
8. General Microscopy. http://micro.magnet.fsu.edu/ Accessed on 02/17/05.
9. General Microscopy. http://www.cyto.purdue.edu Accessed on 02/17/05.
10. Inoué, S. and Spring, K.R. *Video Microscopy: The Fundamentals.* Plemun Press, New York, 1997.
11. Castleman, K.R. *Digital Image Processing.* Prentice Hall, Engelwood Cliffs, NJ, 1996.
12. van Kempen. G.M.P. Image Restoration in Fluorescence Microscopy, PhD Thesis. Delft Technical University, The Netherlands, 1999.
13. Pawley, J.B., Ed. *Handbook of Biological Confocal Microscopy.* Plenum Press, New York, 1990.
14. Sako, Y. and Uyemura, T. Total internal reflection fluorescence microscopy for single-molecule imaging in living cells. *Cell Struct. Funct.,* 27, 357–365, 2002.
15. Stout, A. and Axelrod, D. Evanescent field excitation of fluorescence by epi-illumination microscopy. *Appl. Optic.,* 28, 5237–5242, 1989.
16. Gao, H., Oberringer, M., Englisch, A., Hanselmann, R.G., and Hartmann, U. The scanning near-field optical microscope as a tool for proteomics. *Ultramicroscopy,* 86, 145–150, 2001.
17. Lockett, S.J., Jacobson, K., and Herman, B. Quantitative precision of an automated, fluorescence-based image cytometer. *Anal. Quant. Cytol. Histol.,* 14, 187–202, 1992.
18. Russ, J.C. *The Image Processing Handbook.* CRC Press, Boca Raton, FL, 1998.
19. Image Analysis Tutorial. http://www.qi.thw.tvdelft.nl/courses/FIP Accessed on 02/18/05.
20. NUM Insight Segmentation and Registration toolkit http://www.itk.org Accessed on 02/18/05.
21. The Visualization toolkit http://www.vtk.org Accessed on 02/18/05.
22. MacAulay, C. and Palcic, B. A comparison of some quick and simple threshold selection methods for stained cells. *Anal. Quant. Cytol. Histol.,* 11, 53–58, 1989.
23. Wählby, C., Lindblad, J., Vondrus, M., Bengtsson, E., and Björkesten, L. Algorithms for cytoplasm segmentation of fluorescence labeled cells. *Anal. Cell Pathol.,* 24, 101–111, 2002.
24. Lindblad, J., Wählby, C., Bengtsson, E., and Zaltman, A. Image analysis for automatic segmentation of cytoplasms and classification of Rac1 activation. *Cytometry,* 57A, 22–33, 2004.

25. Knowles, D., Ortiz de Solorzano, C., Jones, A., and Lockett, S.J. Analysis of the 3D spatial organization of cells and subcellular structures in tissue. *Proc. SPIE,* 3921, 66–73, 2000.

26. Fukunaga, K. *Introduction to Statistical Pattern Recognition,* 2nd ed., Academic Press, New York, 1990.

27. Ripley, B.D. *Pattern Recognition and Neural Networks.* Cambridge University Press, Cambridge 1996.

28. Zheng, Q., Milthorpe, B.K., and Jones, A.S. Direct neural network application for automated cell recognition. *Cytometry,* 57A, 1–9, 2004.

29. Zimmer, C., Labruyere, E., Meas-Yedid, V., Guillen, N., and Olivo-Marin, J.C. Segmentation and tracking of migrating cells in videomicroscopy with parametric active contours: A tool for cell-based drug testing. *IEEE Trans. Med. Imag.,* 21, 1212–1221, 2002.

30. Kimmel, A.R. and Parent, C.A. The signal to move: *D. discoideum* go orienteering. *Science,* 300, 1525–1527, 2003.

31. Vanni, S., Lagerholm, C., Otey, C., Taylor, L., and Lanni, F. Internet-based image analysis quanties contractile behavior of individual fibroblasts inside model tissue. *Biophys. J.,* 84, 2715–2727, 2003.

32. Roy, P., Rajfur, Z., Pomorski, P., and Jacobson, K. Microscope-based techniques to study cell adhesion and migration. *Nat. Cell Biol.,* 4, E91–E96, 2002.

33. Microscope Image Analysis www.bitplane.com Accessed on 02/20/05.

34. Axelrod, D., Koppel, D.E., Schlessinge, J., Elson, E., and Webb, W.W. Mobility measurement by analysis of fluorescence photobleaching recovery kinetics. *Biophys. J.,* 16, 1055–1069, 1976.

35. Lippincott-Schwartz, J. and Patterson, G.H. Development and use of fluorescent protein markers in living cells. *Science,* 300, 87–91, 2003.

36. McNally, J.G., Muller, W.G., Walker, D., Wolford, R., and Hager, G.L. The glucocorticoid receptor: Rapid exchange with regulatory sites in living cells. *Science,* 287, 1262–1265, 2000.

37. Cheutin, T., McNairn, A.J., Jenuwein, T., Gilbert, D.M., Singh, P.B., and Misteli, T. Maintenance of stable heterochromatin domains by dynamic HP1 binding. *Science,* 299, 721–725, 2003.

38. Carrero, G., McDonald, D., Crawford, E., de Vries, G., and Hendzel, M.J. Using FRAP and mathematical modeling to determine the in vivo kinetics of nuclear proteins. *Methods,* 29, 14–28, 2003.

39. National Resource for Cell Analysis and Modeling http://www.nrcam.uchc.edu Accessed on 02/20/05.

40. Diaspro, A., Chirico, G., Federici, F., Cannone, F., Beretta, S., and Robello, M. Two-photon microscopy and spectroscopy based on a compact confocal scanning head. *J. Biomed. Optic.,* 6, 300–310, 2001.

41. Steiner, R.F. Fluorescence anisotropy: Theory and applications, in *Topics in Fluorescence Spectroscopy: Principles,* Vol. 2, Lakowicz, J.R., Ed. Plenum, New York, 1991.

42. Manders, E.M.M, Verbeek, F.J., and Aten, J.A. Measurement of colocalization of objects in dual-color confocal images. *J. Microsc.,* 169, 375–382, 1993.

43. Costes, S., Cho, E., Catalfamo, M., Karpova, T., McNally, J., Henkart, P., and Lockett, S. Automatic 3D detection and quantification of co-localization. *Microsc. Microanal.,* 8, 1040CD–1041CD, 2002.

44. Costes, S.V., Daelemans, D., Cho, E.H., Dobbin, Z., Pavlakis, G., and Lockett, S. Automatic and quantitative measurement of protein–protein co-localization in live cells. Biophys J. 86, 6:3993–4003, 2004.

45. Gordon, G.W., Berry, G., Liang, X.H., Levine, B., and Herman, B. Quantitative fluorescence resonance energy transfer measurements using fluorescence microscopy. *Biophys. J.,* 74, 2702–2713, 1988.

46. Nagy, P., Bene, L., Balázs, M., Hyun, W., Lockett, S.J., Waldmann, F.M., Feuerstein, B.G., Damjanovich, S., and Szöllösi, J. EGF induces different redistribution of ErbB2 on breast tumor cells with low and high metastatic potential. Flow and image cytometric energy transfer measurements. *Cytometry,* 32, 120–131, 1998.

47. Nagy, P., Vámosi, G., Bodnár, A., Lockett, S.J., and Szöllösi, J. Intensity-based energy transfer measurements in digital imaging microscopy. *Eur. J. Biophys.,* 27, 377–389, 1998.

48. Kenwworthy, A.K. Imaging protein-protein interactions using fluorescence resonance energy transfer microscopy. *Methods,* 24, 289–296, 2001.

49. Sekur, R.B. and Periasamy, A. Fluorescence resonance energy transfer (FRET) microscopy imaging of live cell protein localizations. *J. Cell Biol.,* 160, 629–633, 2003.

50. Hoppe, A., Christensen, K., and Swanson, J.A. Fluorescence resonance energy transfer-based stoichiometry in living cells. *Biophys. J.,* 83, 3652–3664, 2002.

51. Takemoto, K., Nagai, T., Miyawaki, A., and Miura, M. Spatio-temporal activation of caspase revealed by indicator that is insensitive to environmental effects. *J. Cell Biol.,* 160, 235–243, 2003.

52. Special Issue on Biological Imaging *Science,* 300, 5616, 2003.

53. Alivisatos, A.P. Semiconductor clusters, nanocrystals, and quantum dots. *Science,* 271, 933–937, 1996.

54. Quantum Dot Corp http://www.qdots.com Accessed in 02/20/05.

55. Wu, X., Liu, H., Liu, J., Haley, K.N., Treadway, J.A., Larson, P.J., Ge, N., Peale, F., and Bruchez, M.P. Immunofluorescent labeling of cancer marker Her2 and other cellular targets with semiconductor quantum dots. *Nature Biotech.,* 21, 41–46, 2003.

56. Dubertret, B., Skourides, P., Norris, D.J., Noireaux, V., Brivanlou, A.H., and Libchaber, A. In vivo imaging of quantum dots encapsulated in phospholipid micelles. *Science,* 298, 1759–1762, 2002.

57. Larson, D.R., Zipfel, W.R., Williams, R.M., Clark, S.W., Bruchez, M.P., Wise, F.W., and Webb, W.W. Water-soluble quantum dots for multiphoton fluorescence imaging in vivo. *Science,* 300, 1434–1436, 2003.

58. Bergey, E. and Prasad, P. Small spheres, big potential. *OEMagazine,* 3, 26–29, 2003.

59. Zimmermann, T., Reitdorf, J., and Pepperkok, R. Spectral imaging and its applications in live cell microscopy. *FEBS Letters,* 546, 87–92, 2003.

60. Denk, W., Strickler, J.H., and Webb, W.W. 2-Photon laser scanning fluorescence microscopy. *Science,* 248, 73–76, 1990.

61. Yoder, E.J. and Kleinfeld, D. Cortical imaging through the intact mouse skull using two-photon excitation laser scanning microscopy. *Microsc. Res. Tech.,* 56, 304–305, 2002.

62. Campagnola, P.J., Wei, M.-D., Lewis, A., and Loew, L.M. High-resolution nonlinear optical imaging of live cells by second harmonic generation. *Biophys. J.,* 77, 3341–3349, 1999.

63. Zoumi, A., Yeh, A., and Tromberg, B.J. Imaging cells and extracellular matrix in vivo by using second-harmonic generation and two-photon excited fluorescence. *Proc. Natl. Acad. Sci. USA,* 99, 11014–11019, 2002.

64. Wang, W., Wyckoff, J.B., Frohlich, V.C., Oleynikov, Y., Hüttelmaier, S., Zavadil, J., Cermak, L., Bottinger, E.P., Singer, R.H., White, J.G., Segall, J.E., and Condeelis, J.S. Single cell behavior in metastatic primary mammary tumors correlated with gene expression patterns revealed by molecular profiling. *Canc. Res.,* 62, 6278–6288, 2002.

65. Holtom, G.R., Thrall, B.D., Chin, B.-Y., Wiley, S.H., and Colson, S.D. Achieving molecular selectivity in imaging using multiphoton Raman spectroscopy techniques. *Traffic*, 2, 781–788, 2001.
66. Wählby, C. Algorithms for Applied Digital Image Cytometry, PhD Thesis. Uppsala University, Sweden, 2003.
67. Wählby, C. and Bengtsson, E. Segmentation of Cell Nuclei in Tissue by Combining Watersheds with Gradient Information, in Proc. 13th Scandinavian Conference on Image Analysis (SCIA), vol. 2749 of Lecture Notes in Computer Science. Göteborg, Sweden, 408–414, July 2003.
68. Wählby, C., Sintorn, I.-M., Erlandsson, F., Borgefors, G., and Bengtsson, E. Combining intensity, edge, and shape information for 2D and 3D segmentation of cell nuclei in tissue sections. J. Micros. 215, 1: 67–76, 2004.
69. Peng, S., Urbanc, B., Cruz, L., Hyman, B.T., and Stanley, H.E. Neuron recognition by parallel Potts segmentation. *Proc. Natl. Acad. Sci. USA*, 100, 3847–3852, 2003.
70. Nattkemper, T.W., Twellmann, T., Ritter, H., and Schubert, W. Human vs. machine: Evaluation of fluorescence micrographs. *Comput. Biol. Med.*, 33, 31–43, 2003.
71. Lockett, S.J. and Herman, B. The automatic detection of clustered, fluorescent-stained nuclei by digital image-based cytometry. *Cytometry*, 17, 1–12, 1994.
72. Weickert, J. Anisotropic Diffusion in Image Processing. Teubner, Stuttgart, 1998.
73. Tvarusko, W., Bentele, M., Misteli, T., Rudolf, R., Kaether, C., Spector, D.L., Gerdes, H.H., and Eils, R. Time-resolved analysis and visualization of dynamic processes in living cells. *Proc. Natl. Acad. Sci. USA*, 96, 7950–7955, 1999.
74. Uttenweiler, D., Weber, C., Jähne, B., Fink, R.H.A., and Scharr, H. Spatiotemporal anisotropic diffusion filtering to improve signal-to-noise ratios and object restoration in fluorescence microscopic image sequences. *J. Biomed. Optic.*, 8, 40–47, 2003.
75. Ortiz de Solorzano, C., Malladi, R., Lelièvre, S.A., and Lockett, S.J. Segmentation of nuclei and cells using membrane related protein markers. *J. Microsc.*, 201, 404–415, 2001.
76. Ortiz de Solorzano, C., Rodriguez, E.G., Jones, A., Sudar, D., Pinkel, D., Gray, J.W., and Lockett, S.J. Automatic nuclear segmentation for 3D thick tissue confocal microscopy. *J. Microsc.*, 193, 212–226, 1999.
77. Yang, Q. and Parvin, B. Harmonic cut and regularized centroid transform for localization of subcellular structures. IEEE Trans. *Biomed. Eng.*, 50, 469–475, 2003.
78. Nguyen, H.T., Worring, M., and van den Boomgaard, R. Watersnakes: Energy-driven watershed segmentation. *IEEE Trans. Patt. Anal. Mach. Intell.*, 25, 330–342, 2003.
79. Swedlow, J.R., Goldberg, I., Brauner, E., and Sorger, P.K. Informatics and quantitative analysis in biological imaging. *Science*, 300, 100–102, 2003.
80. Soldek, J. and Mantiuk, R. A reconfigurable processor based on FPGAs for pattern recognition, processing, analysis and synthesis of images. *Patt. Recog. Lett.*, 20, 667–674, 1999.
81. Roques, E.J. and Murphy, R.F. Objective evaluation of differences in protein subcellular distribution. *Traffic*, 3, 61–65, 2002.
82. Jarvik, J.W., Fisher, G.W., Shi, C., Hennen, L., Hauser, C., Adler, S., and Berget, P.B. In vivo functional proteomics: Mammalian genome annotation using CD tagging. *Biotechniques*, 33, 852–854, 856, 858–860, 2002.
83. Foldes-Papp, A., Demel, U., Domej, W., and Tilz, G.P. A new dimension for the development of fluorescence-based assays in solution: From physical principles of FCS detection to biological applications. *Exp. Biol. Med.*, 227, 291–300, 2002.
84. Meseth, U., Wohland, T., Rigler, R., and Vogel, H. Resolution of fluorescence correlation measurements. *Biophys. J.*, 76, 1619–1631, 1999.

85. Larson, D.R., Ma, Y.M., Vogt, V.M., and Webb, W.W. Direct measurement of Gag–Gag interaction during retrovirus assembly with FRET and fluorescence correlation spectroscopy. *J. of Cell Biol.*, 162, 1233–1244, 2003.

86. Wiseman, P.W., Capani, F., Squier, J.A., and Martone, M.E. Counting dendritic spines in brain tissue slices by image correlation spectroscopy analysis. *J. Microsc.*, 205, 177–186, 2002.

87. Chen, Y., Wei, L.-N., and Müller, J.D. Probing protein oligomerization in living cells with fluorescence fluctuation spectroscopy. *Proc. Natl. Acad. Sci. USA*, 100, 15492–15497, 2003.

88. Chen, Y., Müller, J.D., Ruan, Q., and Gratton, E. Molecular brightness characterization of EGFP in vivo by fluorescence fluctuation spectroscopy. *Biophys. J.*, 82, 133–144, 2002.

89. Lakowicz, J.R. and Berndt, K.W. Lifetime selective imaging using an RF phase-sensitive camera. *Rev. Sci. Instrum.*, 62, 1727–1734, 1991.

90. Lakowicz, J.R. *Principles of Fluorescence Spectroscopy.* Plenum Press, New York, 1999.

91. Perisamy, A., Wodnicki, P., Wang, X.F., Kwon, S., Gordon, G.W., and Herman, B. Time resolved fluorescence life-time imaging microscopy using picosecond pulsed tunable dye laser system. *Rev. Sci. Instrum.*, 67, 3722–3731, 1996.

92. Wang, A.F., Periasamy, A., and Herman, B. Fluorescence lifetime imaging microscopy (FLIM): Instrumentation and applications. *Crit. Rev. Anal. Chem.*, 23, 369–395, 1992.

93. Becker and Hickl GmbB http://www.becker-hickl.com Accessed on 02/20/05.

94. Taylor, D.L., Woo, E.S., and Giuliano, K.A. Real-time molecular and cellular analysis: The new frontier of drug discovery. *Curr. Opin. Biotech.*, 12, 75–81, 2001.

95. Eggeling, C., Brand, L., Ullmann, D., and Jäger, S. Highly sensitive fluorescence detection technology currently available for HTS. *Drug Discov. Today*, 8, 632–641, 2003.

96. Cellomics, Inc. http://www.cellomics.com Accessed on 02/20/05.

97. BD Biosciences Biological Imaging http://www.atto.com Accessed on 02/20/05.

98. Molecular Devices Corp. http://www.moleculardevices.com Accessed on 02/20/05.

99. Amersham Biosciences http://www4.amershambiosciences.com Accessed on 02/20/05.

100. Compucyte Corp. http://www.compucyte.com Accessed on 02/20/05.

101. Applied Imaging Corp. http://www.appliedimagingcorp.com Accessed on 02/20/05.

102. BD Biosciences http://www.bdbiosciences.com Accessed on 02/20/05.

103. Liotta, L. and Petricoin, E. Molecular profiling of human cancer. *Nature Rev.*, 1, 48–56, 2000.

104. Lakhani, S.R. and Ashworth, A. Microarray and histopathological analysis of tumours: The future and the past? *Nature Rev.*, 1, 151–175, 2001.

105. Kononen, J., Bubendorf, L., Kallioniemi, A., Barlund, M., Schraml, P., Leighton, S., Torhorst, J., Mihatsch, M.J., Sauter, G., and Kallioniemi, O.P. Tissue microarrays for high throughput molecular profiling of tumor specimens. *Nature Med.*, 4, 844–847, 1998.

Index

T - #0328 - 071024 - C16 - 234/156/21 - PB - 9780367392550 - Gloss Lamination